Preface

Robotics is a natural source of many difficult and beautiful geometrical problems. The interest in this field has grown very rapidly in recent years and considerable attention has been paid to robotics in the mathematical and computer science literature. We felt the need to bring together mathematicians, computer scientists and practitioners of robotics, to compare viewpoints and methods, and to try to identify new robotics problems with a strong geometrical flavour.

The French "Geometry and Robotics" workshop was held at the LAAS-CNRS in Toulouse (France), May 26-28, 1988. The scientific program consisted of 22 contributions of which 20 are contained in the present volume.

The contributions covered several important areas. Algebraic geometry has yielded methods that are very general but appear to be extremely difficult to implement. Results in this area are presented from both a theoretical and a practical point of view. Several papers are concerned with computational complexity issues. These issues are of major concern in robotics and the work in this area is now recognized as an autonomous discipline, christened computational geometry. Projective geometry and algebraic topology find natural applications in computer vision. More surprisingly, Grassmann geometry is also useful in analyzing the mechanics of some kind of manipulators, the so-called "parallel manipulators".

On a more practical side, impressive results have been obtained for planning motions. Several papers present approximate and/or heuristic approaches that have been found to be of interest in solving large scale problems. Other questions that arise in practice, most notably the problem of dealing with uncertainties, are also presented.

The workshop has been jointly organized by INRIA and LAAS-CNRS. It was founded by the SPI department of the CNRS, INRIA and LAAS-CNRS.

We would like to thank all the participants who have contributed to the success of this workshop and to this volume.

Valbonne, March 1989 Jean-Daniel Boissonnat (INRIA)
Jean-Paul Laumond (LAAS/CNRS)

Lecture Notes in Computer Science

Edited by G. Goos and J. Hartmanis

391

J.-D. Boissonnat   J.-P. Laumond   (Eds.)

Geometry and Robotics

Workshop, Toulouse, France, May 26–28, 1988
Proceedings

Springer-Verlag
Berlin Heidelberg New York London Paris Tokyo Hong Kong

Editorial Board
D. Barstow W. Brauer P. Brinch Hansen D. Gries D. Luckham
C. Moler A. Pnueli G. Seegmüller J. Stoer N. Wirth

Editors

Jean-Daniel Boissonnat
INRIA, 2004, route des Lucioles
F-06565 Valbonne Cedex, France

Jean-Paul Laumond
LAAS-CNRS, 7, avenue du Colonel Roche
F-31077 Toulouse Cedex, France

CR Subject Classification (1987): F.2.2, I.1.2, I.2.9–10, I.3.5

ISBN 3-540-51683-2 Springer-Verlag Berlin Heidelberg New York
ISBN 0-387-51683-2 Springer-Verlag New York Berlin Heidelberg

This work is subject to copyright. All rights are reserved, whether the whole or part of the material is concerned, specifically the rights of translation, reprinting, re-use of illustrations, recitation, broadcasting, reproduction on microfilms or in other ways, and storage in data banks. Duplication of this publication or parts thereof is only permitted under the provisions of the German Copyright Law of September 9, 1965, in its version of June 24, 1985, and a copyright fee must always be paid. Violations fall under the prosecution act of the German Copyright Law.

© Springer-Verlag Berlin Heidelberg 1989
Printed in Germany

Printing and binding: Druckhaus Beltz, Hemsbach/Bergstr.
2145/3140-543210 – Printed on acid-free paper

Table of contents

Effective Semialgebraic Geometry
M. Coste 1

Curves and Computer Algebra
D. Duval, M.-F. Roy 28

Placement of Polygons
J.-J. Risler 43

The algorithm by Schwartz, Sharir and Collins on the piano mover's problem
J. Marchand 49

A practical exact motion planning algorithm for polygonal objects amidst polygonal obstacles
F. Avnaim, J.-D. Boissonnat, B. Faverjon 67

Motion planning for manipulators in complex environments
B. Faverjon, P. Tournassoud 87

Planning collision free trajectories by a configuration space approach
T. Siméon 116

Trajectory planning and motion control for mobile robots
J.-P. Laumond, T. Siméon, R. Chatila, G. Giralt 133

Motion planning for a mobile robot with a kinematic constraint
P. Tournassoud 150

Motion from point matches : multiplicity of solutions
O. Faugeras, S. Maybank 172

Singular configurations of parallel manipulators and Grassmann geometry
J.-P. Merlet 194

Applications of geometric homology
H. Crapo 213

Some examples of algorithm analysis in computational geometry by means of mathematical morphological techniques
M. Schmitt 225

An optimal algorithm for the boundary of a cell in a union of rays
P. Alevizos, J.-D. Boissonnat, F. Preparata 247

Triangulation in 2d and 3d space
M. Yvinec 275

Hamiltonian cycles in Delaunay complexes
H. Crapo, J.-P. Laumond 292

Models of robot manipulators
B. Gorla, M. Renaud 306

Modelling positioning uncertainties
I. Mazon 336

Contact manipulation and geometric reasoning
A. Giraud, D. Sidobre 361

Geometric reasoning in motion planning
C. Laugier 377

EFFECTIVE SEMIALGEBRAIC GEOMETRY

Michel Coste
Institut Mathématique de Rennes

Summary : The first part of the paper introduces the decision problem and the quantifier elimination problem for elementary algebra. Tarski's algorithm, which is based on criteria for the existence of solutions of systems of polynomial equations and inequalities generalizing Sturm's theorem, is presented. Then the text is devoted to the cylindrical algebraic decomposition algorithm of Collins, which relies on the basic methods of the theory of semialgebraic sets initiated by Lojasiewicz. Special attention is paid to the topological aspect of the cylindrical algebraic decomposition, which provides a general method (surely too general to be efficient in practice) for robot motion planning. The techniques and results presented here have no pretention to originality. The references given at the end of the paper are just a sample of the literature on the subject, far from being exhaustive.

1) THREE PROBLEMS

We shall be guided in this paper by the consideration of the three following problems :

(i) Is the ellipse with equation $16(x-\frac{1}{2})^2 + \frac{49}{36}y^2 = 1$ in the interior of the unit circle ?

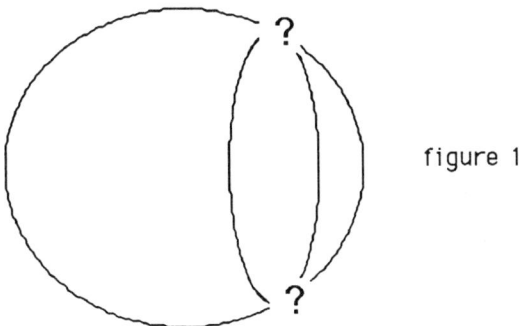

figure 1

(ii)   Give a necessary and sufficient condition, in the form of a boolean combination of polynomial equations and inequalities on the numbers a, b (0<a, 0<b), x_0, y_0, for the ellipse with equation :

$$\frac{(x-x_0)^2}{a^2}+\frac{(y-y_0)^2}{b^2}=1$$

to be contained in the interior of the unit circle.

(iii) Can an ellipse with semiaxes 1 and $\frac{1}{3}$ pass a right angle corner in a corridor with width 1 ?

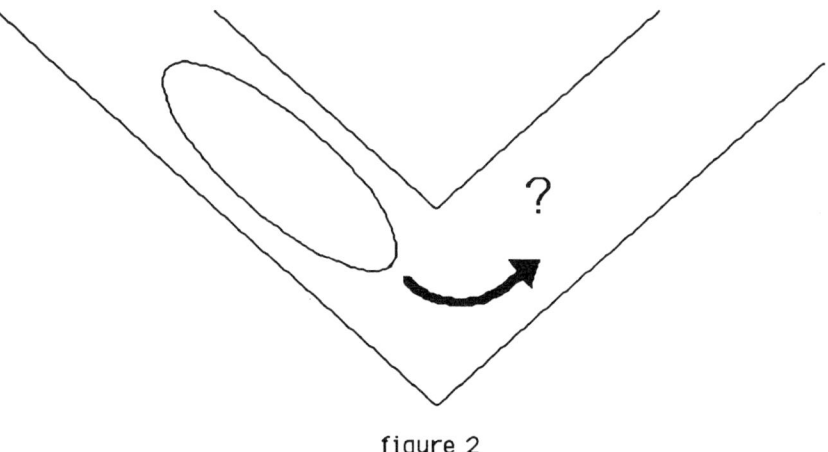

figure 2

The problem (ii) has a name : this is the " Kahan's ellipse problem", which has already been considered in some papers ([AM], [LAZ], [PAU]). The problem (iii) is a simple problem of robotics.

2) DECISION PROBLEM, QUANTIFIER ELIMINATION

2.1 Decision problem for elementary algebra

Let us discuss problem (i). This problem may be reformulated as follows : "Is it the case that all x, y satisfying the equation of the ellipse also satisfy $x^2+y^2<1$?". With symbolic notation, one has to decide whether the formula :

(1) $\qquad \forall x\, \forall y\, \left(\, 16(x-\frac{1}{2})^2+\frac{49}{36}\,y^2=1\, \Rightarrow\, x^2+y^2<1\, \right)$

is true. This formula is an example of what we shall call "formulas of elementary algebra" : these are the formulas built up from polynomial equations and inequalities using the logical connectives AND, OR, NOT, ⇒ and the quantifiers ∃ and ∀ on real variables (let us remark that the connective ⇒ is superfluous, since "$A \Rightarrow B$" is equivalent to "NOT A OR B"). Formula (1) is said to be closed, since all its variables x, y are quantified. Problem (i) is an instance of the *decision problem for elementary algebra* : given a closed formula Φ of elementary algebra, decide whether Φ is true or not. Of course, we are looking for a decision algorithm and not for a soothsayer.

2.2 Elimination of quantifiers for elementary algebra.

Let us consider now problem (ii). Taking into account the first step of the analysis of problem (i), one sees that the fact that the ellipse is inside the interior of the unit circle may be expressed by the formula :

(2) $\qquad \forall x \; \forall y \; \left\{ \dfrac{(x-x_0)^2}{a^2} + \dfrac{(y-y_0)^2}{b^2} = 1 \Rightarrow x^2 + y^2 < 1 \right\}$

The problem is to find a combination (using the connectives AND, OR and NOT) of polynomial equations and inequalities in the variables x_0, y_0, a and b which is equivalent to formula (2). In others terms, we are looking for a quantifier-free formula of elementary algebra which is equivalent to formula (2). So problem (ii) is an instance of the *quantifier elimination problem for elementary algebra*. An other instance of this problem is well known : a second degree polynomial ax^2+bx+c with real coeficients has a real root if and only if its discriminant b^2-4ac is positive or null. That is to say : the formula with quantifier "$a \neq 0$ AND $\exists x \; ax^2+bx+c=0$" is equivalent to the quantifier-free formula "$a \neq 0$ AND $b^2-4ac \geq 0$".

It is not at all obvious that it is possible to eliminate the quantifiers for any formula of elementary algebra (in particular, for formula (2)). This is a non trivial result.

THEOREM 2.2.1 (Tarski-Seidenberg [TAR], [SEI]) : *For any formula of elementary algebra, there is an equivalent quantifier-free formula.*

We must say that we are actually interested in the problem of effective elimination of quantifiers : we are looking for an algorithm which, given any formula of elementary algebra, produces an equivalent quantifier-free formula. We present in this text two such algorithms.

An algorithm for quantifier elimination gives a decision algorithm : if the quantifiers of a closed formula of elementary algebra are eliminated, one obtains a combination of equations and inequalities involving only constants, and we can decide its truth. One has to mention here that if we start with a formula in which all the constants are rational, then the elimination of quantifiers in this formula will produce a new formula in which all constants are again rational : so there will be an exact answer to the problem of comparison of two constants.

2.3 Motivation : theoretical or practical ?

We have mentioned above the name of Tarski. He was the first to describe a decision method for elementary algebra, via an explicit algorithm for the elimination of quantifiers. Tarski found his results around 1930, but the publication, delayed by the war, took not place before 1948. Tarski's motivation was originally from logic : the decision problem for different theories was a central problem, and even the central problem in the metamathematics following Hilbert. Let us recall the impact of the results of Gdel-Church-Rosser, obtained also in the thirties, and which say among others that there *cannot* be a decision method for elementary arithmetic (the arithmetic of natural numbers with addition and multiplication). Tarski signals a consequence of this negative result : if the elementary algebra is extended by adding the sine function, then there is no more decision method, since the sine function, with its periodicity, makes possible the characterization of natural numbers.

Besides the theoretical motivation, perfectly legitimate, Tarski's paper contains a reference to the feasibility of a *true* decision machine. I do not know whether Tarski actually believed it, but anyhow his text was first published by the Rand Corporation. And in the long run there is nowadays a decision machine which works, even if its algorithm is very different from Tarski's, and if it has very limited performances.

3) STURM AND COMPANY

3.1 From quantifier elimination to criteria for the existence of solutions of systems of polynomial equations and inequalities

Let us come back to the problem of elimination of quantifiers for elementary algebra. Any formula of elementary algebra may be transformed in an equivalent formula of the type $Q_1 x_1 ... Q_n x_n \, \Phi$ where $Q_1,...,Q_n$ are existential or universal quantifiers, and Φ a quantifier-free formula (we say that the formula is now in prenex form). This is a very general syntactic transformation, which is not specific to elementary algebra. Once the formula is in prenex form, it is clear that it is sufficient, for eliminating all quantifiers, to know how to eliminate them one by one : first one eliminates the quantifier $Q_n x_n$ out of the formula $Q_n x_n \, \Phi$ to obtain a quantifier-free formula Ψ_{n-1} , then one eliminates the quantifier $Q_{n-1} x_{n-1}$ out of the formula $Q_{n-1} x_{n-1} \, \Psi_{n-1}$ and so on, and at the end one arrives to a formula Ψ_0 without quantifier. Moreover, one may suppose that the quantifier to be eliminated is an existential quantifier, since "$\forall x ...$" means "NOT $\exists x$ NOT...". So the problem reduces to eliminating the quantifier in a formula $\exists x \, \Phi$, where Φ is quantifier-free. Let us see how Φ looks : we have said that Φ is built up from polynomial equations and inequalities using the connectives AND, OR and NOT. Up to a syntactic transformation which does not change the meaning of the formula, it is possible to write Φ in the form "Σ_1 OR ... OR Σ_k", each Σ_i being a conjunction (or : a system) of polynomial equations and inequalities. At the end we have to eliminate the quantifier in the formula which says : "there exists a real solution to the system Σ of polynomial equations and inequalities", that is to find a necessary and sufficient condition for the existence of a real solution of the system, which is in the form of a boolean combination of polynomial equations and inequalities on the coefficients of the polynomials of the system Σ. This is the old and classical problem of elimination, except that one insists on the fact that the solution must be real, and that inequalities come in.

One may even be more precise concerning the form of the systems to be considered. In the real case, a finite system of equations $P_1=...=P_l=0$ reduces to the sole equation $P_1^2+...+P_l^2=0$. On the other hand a non strict inequality may be disjoined in a strict inequality or an equation. At the end the quantifier elimination problem will be solved if one knows how to produce criteria for the existence of a real solution of a system containing at most one polynomial equation, and a finite list (maybe empty) of strict polynomial inequalities.

3.2 <u>Sturm's theorem and its variants</u>

We arrive now to Sturm : his theorem enables us to count the number of real roots of a polynomial in x, according to the signs of certain polynomial expressions of the coefficients. The problem of elimination, for a system consisting of only one equation, is solved !

Let us recall the statement of Sturm's theorem. First we introduce some notations. Let P and Q be two polynomials in the variable x, with coefficients in a field. *We always suppose in this section that P is squarefree (without multiple root) and that P and Q are relatively prime.*

a) We denote by ST(P,Q) (the Sturm's sequence) the list of polynomials $(P_0, P_1, ..., P_m)$ such that $P_0=P$, $P_1=Q$, P_{i+2} is the opposite of the rest of the euclidian division of P_i by P_{i+1}, and P_m is a non null constant of the field of coefficients of P and Q. This is actually the algorithm for the calculation of the g.c.d. of P and Q (up to the signs), and this algorithm terminates with a constant since the polynomial are relatively prime.

b) Let us suppose now that P and Q have real coefficients. Let ST(P,Q) = $(P_0, P_1, ..., P_m)$, and choose $a \in \mathbb{R}$ which is not a root of P. We denote by $v(P,Q;a)$ the number of changes of sign in the list $(P_0(a), P_1(a), ..., P_m(a))$. This definition extends straitghforwardly when $a=+\infty$ or $a=-\infty$; in these cases one has to consider the signs of the leading coefficients.

c) Still under the assumption that P and Q have real coefficients, and if the real numbers a, b (satisfying $a<b$) are not roots of P, we denote by $c(P;a,b)$ the number of real roots ζ of P lying between a and b. We denote by $c(P)$ the total number of real roots of P.

THEOREM 3.2.1 (Sturm's theorem): *Suppose that P is squarefree, Then :*
$$c(P;a,b) = v(P,P';a) - v(P,P';b),$$

and in particular :
$$c(P) = v(P,P';-\infty) - v(P,P';+\infty).$$

Let us also recall a generalization of this theorem. We denote by $c(P,Q>0)$ (resp. $c(P,Q<0)$) the number of real roots of P where Q is positive (resp. negative).

PROPOSITION 3.2.2 : *Suppose that the polynomials P and Q are relatively prime, and that P is squarefree. Then :*
$$c(P,Q>0) - c(P,Q<0) = v(P,P'Q;-\infty) - v(P,P'Q;+\infty).$$

Since by Sturm's theorem we know that $c(P) = c(P,Q>0) + c(P,Q<0) = v(P,Q;-\infty) - v(P,Q;+\infty)$, we are able to calculate $c(P,Q>0)$ and $c(P,Q<0)$.

3.3 An example

Let us see how to solve problem (i) using the tools introduced in the preceding section – this is certainly not the most direct way. Problem (i) has a positive answer if and only if the system :

$$\begin{cases} 16\left(x - \frac{1}{2}\right)^2 + \frac{49}{36}y^2 - 1 = 0 \\ x^2 + y^2 - 1 \geq 0 \end{cases}$$

has no real root. Let P be the first polynomial (equation of the ellipse), and Q the second one (equation of the unit circle). Consider the system as a system in y, with x as parameter. The equation $P=0$ has two distinct real roots if and only if $16(x-\frac{1}{2})^2 < 0$, i.e. if and only if $\frac{1}{4} < x < \frac{3}{4}$. This can also be obtained "mechanically" by calculating $\mathrm{ST}(P, \frac{\partial P}{\partial y})$ to obtain the number of real roots of P. Let us calculate now $\mathrm{ST}(P, \frac{\partial P}{\partial y} Q)$. We get :

$\mathrm{ST}(P, \frac{\partial P}{\partial Y} Q) = (P_0 = P,$

$P_1 = \frac{\partial P}{\partial y} Q = \frac{49}{18} y (x^2 + y^2 - 1),$

$P_2 = -P,$

$P_3 = \frac{1}{18} y (527x^2 - 576x + 157),$

$P_4 = 16(x - \frac{1}{2})^2 - 1 \ ).$

So if $16(x-\frac{1}{2})^2 - 1 < 0$ and $527x^2 - 576x + 157 < 0$, then $v(P, \frac{\partial P}{\partial y} Q; -\infty) - v(P, \frac{\partial P}{\partial y} Q; +\infty) = 2$, and so Q is positive at the two roots of P. To know whether there exists x satisfying the two inequalities above, one can count the roots of the polynomial $R = 527x^2 - 576x + 157$ between $\frac{1}{4}$ and $\frac{3}{4}$. One finds $v(R,R';\frac{1}{4}) - v(R,R';\frac{3}{4}) = 2$, which shows that R has negative values for some x between $\frac{1}{4}$ and $\frac{3}{4}$. At the end, problem (i) has a negative answer : the ellipse protrudes beyond the circle.

3.4 The general case

We have seen up to now how to treat (under certain hypothesis) the case of a system consisting of one equation, or of one equation and one strict inequality. One may add strict inequalities. For a system in the form :

$$\begin{cases} P = 0 \\ Q > 0 \\ R > 0 \end{cases}$$

with the hypothesis that P is squarefree, and that Q and R are relatively prime to P, the calculation of the number of changes of sign at $-\infty$ and $+\infty$ for $\text{ST}(P,P')$, $\text{ST}(P,P'Q)$, $\text{ST}(P,P'R)$, $\text{ST}(P,P'QR)$ will give us $c(P,Q>0,R>0)$ (with the obvious notation), hence a criterium for the existence of a solution to the system. One can go further ; it then appears that the number of Sturm sequences to consider is 2 to the number of inequalities, but the remark that the total number of real roots of P is certainly bounded by its degree allows to break this exponential growth [BKR].

Consider now the case of a system

$$\begin{cases} Q_1 > 0 \\ Q_2 > 0 \\ \ldots\ldots\ldots \\ Q_l > 0 \end{cases}$$

which contains only inequalities. An inspection of the signs of the leading coefficients of the polynomials $Q_1,\ldots,Q_l$ shows whether the inequalities are simultaneously satisfied on an unbounded interval. If they are simultaneously satisfied on a bounded interval, then the bounds of this interval are roots of the product $Q_1 \cdots Q_l$ and on this interval the derivative $(Q_1 \cdots Q_l)'$ is going to have a root ; so one can introduce an equation, and we are reduced to something which we know how to handle.

We have up to now considered only the "good" case (P squarefree, P and Q relatively prime,...). Besides, we have not taken into account that, in general, the coefficients of the polynomial in a system depend upon other variables than the variable being eliminated, and so that we can encounter exceptions with respect to the calculation of the Sturm sequence in the general case ; these exceptional cases meet the preceding ones. To be more precise, in the example of section 3.3 above, the calculation of the Sturm sequence $\text{ST}(P, \frac{\partial P}{\partial Y} Q)$ is incorrect when $527x^2 - 576x + 157 = 0$ or when $16(x-\frac{1}{2})^2 - 1 = 0$. This was pointless, since it was sufficient to explore one branch of the tree of possibilities to get an answer to our decision problem. If we had wanted to write down a necessary and sufficient condition on x for the existence of a solution of the considered system (a quantifier elimination problem), we would have to do the calculations in all the particular cases. We shall not enter into the details.

We have just described the main lines of Tarski's algorithm for the elimination of quantifiers. It is not the only one. Some others are quoted in the references. Here we shall present the algorithm proposed by Collins [COL], [ACM a]; its main ideas are at the base of semialgebraic geometry, and it is the only one to be implemented up to now (in SAC 2).

4) SEMIALGEBAIC SETS AND CYLINDRICAL ALGEBRAIC DECOMPOSITION

4.1 Definition and first properties

We have considered in the preceding section systems of polynomial equations and inequalities. It is known that algebraic geometry, in its most naive definition, is the study of sets of solutions of systems of polynomial equations : the algebraic sets. Here are now the objects of semialgebraic geometry :

DEFINITION 4.1.1 : *A semialgebraic subset of* $\mathbb{R}^n$ *is the set of points of* $\mathbb{R}^n$ *which satisfy a combination (obtained using the connectives* AND, OR, *and* NOT*) of polynomial equations and inequalities.*

Let us give now some examples of semialgebraic sets.
 a) Any algebraic set is semialgebraic.
 b) The semialgebraic subsets of $\mathbb{R}$ are finite unions of points and intervals. It is easily seen remarking that the finite set of polynomials in one variable appearing in the definition of the semialgebraic set has a finite number of roots, and that on each interval between these roots the polynomials have constant signs. This description works also for the semialgebraic subsets of any line.
 c) The solids of C.S.G. (Constructive Solid Geometry) are semialgebraic subsets : the primitives (spheres, cylinders, parallelepipeds, cones,...) may surely be described by means of polynomial inequalities, and the operations on these primitives (union, intersection, difference) may be translated using the connectives AND, OR and NOT.
 The last example emphasizes the stability properties of the class of semialgebraic sets : stability under finite union and intersection, under complementation. Here is

another stability property, which is very important and does not immediately follow from the definition :

THEOREM 4.1.2 : *Let $S\subset\mathbb{R}^{n+1}$ be a semialgebraic subset. Then the image of S under the projection $\Pi : \mathbb{R}^{n+1}=\mathbb{R}^n\times\mathbb{R} \longrightarrow \mathbb{R}^n$ is a semialgebraic subset of $\mathbb{R}^n$.*

This theorem is nothing but the geometric version of the theorem of Tarski-Seidenberg. Indeed, let us suppose that S is described by the combination $\Phi(x_1,...,x_{n+1})$ of polynomial equations and inequalities in the variables $x_1,...,x_{n+1}$. Then $\Pi(S)$ is described by the formula $\exists x_{n+1}\Phi(x_1,...,x_{n+1})$, which is equivalent to a quantifier-free formula, that is to say to a combination of polynomial equations and inequalities on the variables $x_1,...,x_n$.

The argument above actually gives the following result, which is very useful.

COROLLARY 4.1.3 : *Let $\Theta(x_1,...,x_n)$ be a formula of elementary algebra. Then*
$$\{ (x_1,...,x_n)\in\mathbb{R}^n;\, \Theta(x_1,...,x_n)\}$$
is a semialgebraic subset of $\mathbb{R}^n$.

Let us apply this corollary to our problem (iii). We are going to show that the set of allowable positions of the ellipse in the corridor is a semialgebraic set. The corridor with a corner may be described as follows :

$$(x\geq 0 \text{ AND } (y-x)^2-\frac{1}{2}<0) \text{ OR } (x\leq 0 \text{ AND } (y+x)^2-\frac{1}{2}<0).$$

The position of the ellipse may be marked with the coordinates (x_0,y_0) of its center and the slope t of its big axis. In this way the vertical position of the ellipse is not obtained, but there is no trouble since the ellipse cannot stand vertical in the corridor. The equation of the ellipse is then :

$$[(x-x_0) + t(y-y_0)]^2 + 9[(y-y_0) - t(x-x_0)]^2 - (1+t^2) = 0 \,.$$

The set of allowable positions of the ellipse inside the corridor is then the set L of points (x_0,y_0,t) of $\mathbb{R}^3$ such that :

(3) $\quad \forall x \forall y \, \Big\{ [(x-x_0) + t(y-y_0)]^2 + 9[(y-y_0) - t(x-x_0)]^2 - (1+t^2) = 0$
$\Rightarrow (x\geq 0 \text{ AND } (y-x)^2-\frac{1}{2}<0) \text{ OR } (x\leq 0 \text{ AND } (y+x)^2-\frac{1}{2}<0) \Big\}$

and so it is a semialgebraic set. Problem (iii) is then to know whether there exists, inside L, an arc joining the point $(-1,1,-1)$ (ellipse in the left part of the corridor) to the point $(1,1,1)$ (ellipse in the right part of the corridor). This is a problem concerning the topology of L, and we shall come back later to this problem.

Let us continue for the moment our catalog of semialgebraic sets. Let us consider a B-spline function $N_i{}^k(x)$; for instance in the case of integer nodes we have $y = N_0{}^3(x)$ if and only if :

$0 \le x \le 1$ AND $y = \dfrac{x^2}{2}$

OR $\quad 1 \le x \le 2$ AND $y = \dfrac{(-2x^2 + 6x - 3)}{2}$

OR $\quad 2 \le x \le 3$ AND $y = \dfrac{(3-x)^2}{2}$

OR $\quad (x \le 0$ OR $3 \le x)$ AND $y = 0$.

This description shows that the graph of $N_0{}^3$ is semialgebraic. This is true for any B-spline.

DEFINITION 4.1.4: *A function is said to be semialgebraic when its graph is semialgebraic.*

If $\{P_i \ ; \ i \in I\}$ is a finite set of points in the plane or in the space, then the function $S_k(t) = \sum_{i \in I} P_i N_i{}^k(t)$, which parametrizes the corresponding spline curve of order k, is semialgebraic, and the spline curve itself is a semialgebraic subset (the image of a semialgebraic function is a semialgebraic set). Also belong to the semialgebraic realm Bézier curves and surfaces, etc.

Let us give an example of an object which is not semialgebraic : the pencil P of lines $y = nx$, for all integers n. If we cut P by the line $x = 1$, we get the set of points $(1,n)$ for $n \in \mathbb{N}$, which is not semialgebraic (remember what we said about semialgebraic subsets of a line). Yet P is described by the formula $\exists n \in \mathbb{N} \ y = nx$; but this formula is not a formula of elementary algebra, since the quantification is on a variable which lives in $\mathbb{N}$, and not on a real variable.

4.2 Taking semialgebraic sets to pieces : the cylindrical algebraic decomposition (c.a.d.)

A *cylindrical algebraic decomposition* (c.a.d. in abbreviated form) of $\mathbb{R}^n$ will consist in a partition of $\mathbb{R}^k$ (for any $k, 1 \le k \le n$) into a finite family $\mathscr{C}_k$ of semialgebraic subsets (which will be called cells) of $\mathbb{R}^k$, satisfying the following properties :

a) For any k, each cell $C \in \mathscr{C}_k$ is homeomorphic to a box $]0,1[^d$ (with the convention that $]0,1[^0$ is a point).

b) For any k, $1 \le k < n$, and for any $C \in \mathscr{C}_k$, the cylinder $C \times \mathbb{R} \subset \mathbb{R}^{k+1}$ is a disjoint union of cells of $\mathscr{C}_{k+1}$ which are :
- either the graph of continuous semialgebraic function from C to $\mathbb{R}$,
- or a slice of the cylinder bounded from below and from above by the graph of a function as above, eventually unbounded from below or/and from above.

Remark that condition *a*) for $k=1$ simply says that $\mathscr{C}_1$ is a partition of $\mathbb{R}$ into a finite number of open intervals and points, and then that condition *b*) implies condition *a*) for $k\geq 1$, by induction on k. Indeed if C is homeomorphic to $]0,1[^d$, a graph in the cylinder $C\times\mathbb{R}$ will be homeomorphic to $]0,1[^d$, and a slice to $]0,1[^{d+1}$. Also, the semialgebraicity of the cells follows from the fact that the functions which cut the cylinders are semialgebraic.

We have now to say what we want to do with c.a.d.. We shall use the following terminology : let $P_1,...,P_r$ be polynomials in $x_1,...,x_n$, and let C be a subset of $\mathbb{R}^n$; we say that C is $(P_1,...,P_r)$-*invariant* if each polynomial P_i has a constant sign (>0, <0, or =0) on C. Given a finite family $P_1,...,P_r$ of polynomials in n variables $x_1,...,x_n$, we want to produce a c.a.d. of $\mathbb{R}^n$ such that :

c) each cell $C\in\mathscr{C}_n$ is $(P_1,...,P_r)$-invariant.

A c.a.d. of $\mathbb{R}^n$ satisfying this property will be called *adapted to* $(P_1,...,P_r)$.

Let us see the advantages of such a c.a.d.. First, condition *c*) shows that any semialgebraic subset of $\mathbb{R}^n$ which is described by a combination of equations $P_i=0$ and inequalities $P_j>0$ or $P_j<0$ involving polynomials among $P_1,...,P_r$ is a union of certain cells of $\mathscr{C}_k$. This enables us to take any semialgebraic set to "simple" pieces (from the topological point of view, these are simply boxes $]0,1[^d$). Moreover the cylindrical arrangement (property *b*) shows that any semialgebraic subset of $\mathbb{R}^k$ described by a formula $Q_{k+1}x_{k+1}...Q_nx_n\ \Phi$, where $Q_{k+1},...,Q_n$ are universal or existential quantifiers and Φ a combination of equations $P_i=0$ and inequalities $P_j>0$ or $P_j<0$, is a union of certain cells of $\mathscr{C}_k$. So the use of such a c.a.d. for decision problems or for elimination of quantifiers is obvious.

4.3 What makes c.a.d. work

The description of c.a.d. above emphasizes the role of the semialgebraic functions which cut the cylinders. Since we want the slices of a cylinder contained in $\mathbb{R}^n$ to be $(P_1,...,P_r)$-invariant, these functions have to describe the real roots of the polynomials P_i, as functions of $x_1,...,x_{n-1}$. Here is a first result in this direction.

PROPOSITION 4.3.1 : *Let $P(x_1,...x_n)$ be a polynomial in $\mathbb{R}[x_1,...x_n]$. Let $k\in\mathbb{N}$, and let $C\subset\mathbb{R}^{n-1}$ be a connected semialgebraic subset such that, for every point $(a_1,...,a_{n-1})\in C$, the number of distinct real or complex roots of $P(a_1,...,a_{n-1},x_n)$ is equal to k. Then there exist $l\leq k$ continuous semialgebraic functions $\xi_1<...<\xi_l : C\longrightarrow\mathbb{R}$ such that, for every point $(a_1,...,a_{n-1})\in C$, the set of real roots of $P(a_1,...,a_{n-1},x_n)$ is exactly $\{\xi_1(a_1,...,a_{n-1}),...,\xi_l(a_1,...,a_{n-1})\}$.*

A proof of this proposition uses the fact that the integral

$$I(a\,;\gamma) = \frac{1}{2i\pi} \int_\gamma \frac{\frac{\partial P}{\partial x_n}(a,z)}{P(a,z)} dz \quad \text{where } a = (a_1,...,a_{n-1}),$$

taken on a circle γ in the plane of the complex variable on which $P(a,z)$ has no zero, counts the number of real or complex roots of $P(a,x_n)$ in the interior of γ, with their multiplicities. Choose now a point a in C, and k small circles $\gamma_1,...,\gamma_k$ in $\mathbb{C}$ which surround the k distinct roots $z_1,...,z_k$ of $P(a,x_n)$, so that $I(a;\gamma_i)$ is the multiplicity m_i of the root z_i. If z_i is not real, choose γ_i which does not cut the real axis, and if z_i is real, choose the center of γ_i on the real axis. If we replace a by a point $a' \in C$ sufficiently near to a, the value of $I(a';\gamma_i)$ remains by continuity equal to the same integer and so there are, in the interior of γ_i, m_i roots of $P(a',x_n)$ counted with their multiplicities. Since on the whole there are still k distinct roots, each γ_i contains exactly one root, with multiplicity m_i, of $P(a',x_n)$, and so we get all the roots of $P(a',x_n)$. If γ_i surrounded a non real root of $P(a,x_n)$, the root of $P(a',x_n)$ which it contains is surely non real. If γ_i surrounded a real root of $P(a,x_n)$, it cannot contain a non real root of $P(a',x_n)$, since then it would also contain its conjugate. All this shows that the number of real roots of $P(a',x_n)$ is the same that the number of real roots of $P(a,x_n)$. Using the connexity of C, we get that the number of real roots of $P(a,x_n)$ is constant when a runs through C ; let l be this number. For $1 \leq i \leq l$, we denote by $\xi_i : C \longrightarrow \mathbb{R}$ the function of $x=(x_1,...,x_{n-1})$ which gives the ith real root (for the natural order) of $P(x,x_n)$. The argument using small circles also shows the continuity of the functions ξ_i. If C is described by the formula $\Theta(x)$, we have :

(4) $y = \xi_i(x) \Leftrightarrow \Theta(x)$ AND $\exists y_1...\exists y_l \ (y_1<...<y_l$ AND $P(x,y_1)=0$ AND...AND $P(x,y_l)=0$ AND $y=y_i)$,

which shows that the functions ξ_i are semialgebraic.

It remains now to express the fact that the number of distinct real or complex roots of a polynomial is equal to k.

PROPOSITION 4.3.2 : *Let $P(x) = a_0 x^d + a_1 x^{d-1} +...+a_d$ be a polynomial with real or complex coefficients, with $a_0 \neq 0$, and let $k \leq d$. Then there exists a combination of polynomial equations and inequations (...$\neq 0$) in $a_0,...,a_d$ with integer coefficients, which is satisfied if and only if P has exactly k distinct real or complex roots.*

The polynomial P has k distinct roots if and only if the g.c.d. of P and its derivative P' has degree $d-k$. Let us see how to express that two polynomials P as above and $Q(x) = b_0 x^e + b_1 x^{e-1} +...+b_e$ have a g.c.d. with degree l. This may be done using the method of subresultants, which is detailed in [LOO]. Consider the Sylvester matrix (square $(d+e) \times (d+e)$ matrix) whose lines are the coefficients of $x^{e-1}P,...,P,x^{d-1}Q,...,Q$:

$$\begin{pmatrix} a_0 & a_1 & \cdot & \cdot & \cdot & a_d & 0 & \cdot & \cdot \\ 0 & a_0 & a_1 & \cdot & \cdot & \cdot & a_d & 0 & \cdot \\ \cdot & \cdot & \cdot & \cdot & \cdot & \cdot & \cdot & \cdot & \cdot \\ \cdot & \cdot & 0 & a_0 & a_1 & \cdot & \cdot & \cdot & a_d \\ b_0 & b_1 & \cdot & \cdot & b_e & 0 & \cdot & \cdot & \cdot \\ 0 & b_0 & b_1 & \cdot & \cdot & b_e & 0 & \cdot & \cdot \\ \cdot & \cdot & \cdot & \cdot & \cdot & \cdot & \cdot & \cdot & \cdot \\ \cdot & \cdot & \cdot & 0 & b_0 & b_1 & \cdot & \cdot & b_e \end{pmatrix}$$

For $j \leq \inf(d,e)$, denote by $\mathrm{PSC}_j(P,Q)$ the minor obtained by deleting the j last lines of coefficients of P, the j last lines of coefficients of Q, and the $2j$ last columns (PSC is for "principal subresultant coefficient"). If $j=0$, then $\mathrm{PSC}_0(P,Q)$ is the determinant of the Sylvester matrix, that is to say the resultant of P and Q. We have the following result :

PROPOSITION 4.3.3 : *The g.c.d. of polynomials P and Q has degree l if and only if $\mathrm{PSC}_0(P,Q)=...=\mathrm{PSC}_{l-1}(P,Q)=0$ and $\mathrm{PSC}_l(P,Q) \neq 0$.*

It is equivalent to say that P and Q have a g.c.d. with degree $\geq l$ and that there exist non null polynomials U and V, with degrees respectively $\leq e-l$ and $\leq d-l$, such that $UP+VQ=0$. This last equality may be translated into a homogeneous linear system $(*)$ of $d+e+1-l$ equations in $d+e+2-2l$ unknowns (the coefficients of U and V). The $d+e+2-2l$ first equations of the system $(*)$ form a square system $(**)$, which expresses that the degree of $UP+VQ$ is strictly less than $l-1$; the determinant of this system is nothing but $\mathrm{PSC}_l(P,Q)$. Suppose that we already know that P and Q have a g.c.d. with degree $\geq l-1$ if and only if $\mathrm{PSC}_0(P,Q)=...=\mathrm{PSC}_{l-1}(P,Q)=0$. If the g.c.d. of P and Q has degree $\geq l$, then the system $(**)$ has a non trivial solution, and so its determinant $\mathrm{PSC}_l(P,Q)$ is null. In the other direction, if $\mathrm{PSC}_0(P,Q)=...=\mathrm{PSC}_l(P,Q)=0$, then $(**)$ has a non trivial solution and so we have $UP+VQ=R$, with U and V non null, $\deg(U) \leq e-l$, $\deg(V) \leq d-l$, and $\deg(R)<l-1$; since, according to the induction hypothesis, the g.c.d. of P and Q has degree $\geq l-1$, and must divide R, we have $R=0$ and so P and Q have a g.c.d. with degree $\geq l$. The proposition is proved.

Up to now, we did not worry about the fact that the polynomials we have to consider are polynomials in x_n whose coefficients are polynomials in $x_1,...,x_{n-1}$. So for instance if the leading coefficient of P vanishes, one has to resume the calculations of PSC's with the polynomial P truncated, i.e. without its leading coefficient.

At the end, we are lead to perform the following construction, starting from a finite family $(P_1,...,P_r)$ of polynomials in $\mathbb{R}[x_1,...,x_n]$ considered as polynomials in x_n with coefficients in $\mathbb{R}[x_1,...,x_{n-1}]$: if Q and R are polynomials from the initial list, or which may

be obtained from polynomials in this list by the vanishing of some number of head coefficients, take :
- the leading coefficient of Q,
- all $\mathrm{PSC}_j(Q,Q')$ if $\deg(Q) \geq 2$,
- all $\mathrm{PSC}_j(Q,R)$ if $\deg(Q) \geq 1$ and $\deg(R) \geq 1$.

We get in this way a new list of polynomials in $x_1,...,x_{n-1}$ which we shall denote by $\mathrm{PROJ}(P_1,...,P_r)$. We can now sum up what we have seen in this section.

THEOREM 4.3.4 : *Let $C \subset \mathbb{R}^{n-1}$ be a semialgebraic connected subset which is $\mathrm{PROJ}(P_1,...,P_r)$-invariant. Then the total number of real roots of $P_1,...,P_r$ (once those among these polynomials which vanish on C have been taken out) is constant on C, and these real roots are given by continuous semialgebraic functions $\xi_1 < ... < \xi_l : C \longrightarrow \mathbb{R}$.*

4.4 The algorithm for c.a.d.

The algorithm accepts as input a finite list of polynomials $P_1,...,P_r \in \mathbb{Q}[x_1,...,x_n]$ and gives as output the list of cells of a c.a.d. adapted to $P_1,...,P_r$ (with moreover information on the cylindrical arrangement) and, for each cell, a "testing point" in the cell with rational or real algebraic coordinates. Let us indicate now an easy way to number the cells of a c.a.d. which gives all the information about their cylindrical arrangement. A cell $C \in \mathcal{C}_k$ ($C \subset \mathbb{R}^k$) is encoded by a list $(\alpha_1,...,\alpha_k)$ of k positive integers. If $k=1$, α_1 is the number of the cell (open interval or point of $\mathbb{R}$), counting from left to right on the line ; α_1 is even for a point, odd for an open interval. Suppose the codification made up to k included. If $C \in \mathcal{C}_{k+1}$ is a cell in the cylinder $D \times \mathbb{R}$, $D \in \mathcal{C}_k$, the code of C is $(\alpha_1,...,\alpha_k,\alpha_{k+1})$ where $(\alpha_1,...,\alpha_k)$ is the code of D, and α_{k+1} the number of C in the cylinder $D \times \mathbb{R}$ counting "from bottom to top"; α_{k+1} is even for a graph, odd for a slice.

The algorithm works as follows :
- given a list of polynomials in one variable x_1, count, and isolate in intervals with rational bounds, all the real roots of these polynomials (this is done using Sturm). The cells of $\mathcal{C}_1$ are the roots and the intervals between these roots. The roots are characterized by the polynomial which they nullify, and the interval with rational bounds which isolates them. These bounds of intervals may also serve as testing points for the intervals between the roots (but one may have more convenient choices).
- given a list $(P_1,...,P_r)$ of polynomials in $x_1,...,x_n$ with $n>1$, produce $\mathrm{PROJ}(P_1,...,P_r)$ and call the algorithm, with input this list of polynomials in $x_1,...,x_{n-1}$. One gets $\mathcal{C}_{n-1}$ which is a partition of $\mathbb{R}^{n-1}$ into $\mathrm{PROJ}(P_1,...,P_r)$-invariant cells, together with a testing point a_C for each $C \in \mathcal{C}_{n-1}$. To such a cell C one may apply theorem 4.3.4, in order to cut the cylinder $C \times \mathbb{R}$ into $(P_1,...,P_r)$-invariant cells. To know how many cells there are in the cylinder, and to get a testing point for each of them, one has to look for the real roots of $P_1(a_C,x_n),...,P_r(a_C,x_n)$. That is again Sturm's theorem, but one has to take into account the fact that the coefficients may be real algebraic numbers. Here comes the problem of

coding and performing operations on real algebraic numbers. In the c.a.d. algorithm as it is implemented, the coordinates of the testing point a_C are expressed in function of a primitive element of the extension of $\mathbb{Q}$ they generate, this primitive element being given by its minimal polynomial and an interval with rational bounds isolating it.

4.5 An example

Let us see how the algorithm works on the example of problem (i). Our starting list of polynomials is ($P = 16(x-\frac{1}{2})^2 + \frac{49}{36} y^2 - 1$, $Q = x^2 + y^2 - 1$). Its PROJ is ($R = 527x^2 - 576x + 157$, $S = x^2 - 1$, $T = 16x^2 - 16x + 3$); one cheats a little, since the resultant of P and Q is actually $\frac{(527x^2 - 576x + 157)^2}{1296}$, but one can add in the algorithm a procedure which "cleans up" the lists of polynomials obtained, in order that the polynomials are squarefree, relatively prime, etc.. The roots of these polynomials are (in order) $-1, \frac{1}{4}$, the two roots ζ_1 and ζ_2 of R, $\frac{3}{4}$, 1. Using dichotomy, ζ_1 and ζ_2 may be isolated in the intervals $]\frac{1}{2}, \frac{9}{16}[$ and $]\frac{9}{16}, \frac{5}{8}[$. This gives, for $\mathcal{C}_1$ the list of cells (1), (2),...,(13) with testing points $-2, -1, 0, \frac{1}{4}, \frac{1}{2}, \zeta_1, \frac{9}{16}, \zeta_2, \frac{5}{8}, \frac{3}{4}, \frac{7}{8}, 1, 2$. Let us describe now the cylinder above the cell (7).

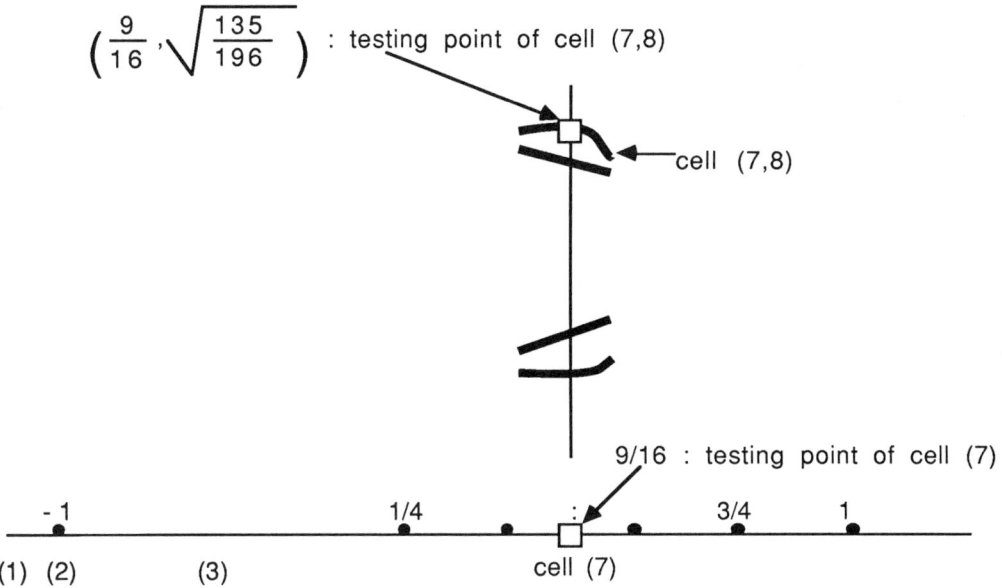

figure 3

We are looking for the roots of $P(\frac{9}{16},y) = \frac{196y^2-135}{144}$ and of $Q(\frac{9}{16},y) = \frac{256y^2-175}{256}$. We arrive, for the cylinder, to the decomposition into cells (7,1), (7,2),...,(7,9) together with the testing points which have for first coordinate $\frac{9}{16}$ and for second coordinates $-1, -\rho, -\frac{53}{64}, -\tau, 0, \tau, \frac{53}{64}, \rho, 1$ with $\tau = \sqrt{\frac{175}{256}}$ and $\rho = \sqrt{\frac{135}{196}}$. Evaluation at the testing point of cell (7,8) gives $P(\frac{9}{16},\rho) = 0$ and $Q(\frac{9}{16},\rho) = \frac{65}{12544} > 0$, which shows that the ellipse protrudes beyond the circle.

4.6 Other properties of semialgebraic sets

The c.a.d. gives a clear idea of what should be the dimension of a semialgebraic set. If a semialgebraic set A has been decomposed as a finite union of cells C_i homeomorphic to boxes $]0,1[^{d_i}$, the dimension of A must be the sup of the d_i's. One may check that the dimension so defined does not depend on the c.a.d.. The coding of cells indicated in section 4.4 gives very easily their dimensions : the dimension of a cell with code $(\alpha_1,...,\alpha_k)$ is the sum of the parities of the integers $\alpha_1,...,\alpha_k$. It is shown by induction on k, remarking that according to α_{k+1} being even or odd, the cell $(\alpha_1,...,\alpha_{k+1})$ is a graph or a slice, and so has dimension respectively equal to the dimension of the cell $(\alpha_1,...,\alpha_k)$, or greater by 1.

Taking a semialgebraic set into a finite union of cells also shows that the connected components of this set must be unions of some of these cells.

PROPOSITION 4.6.1 : *A semialgebraic set has a finite number of connected components, which are semialgebraic.*

The theory of semialgebraic sets was initiated by [LOJ], where one can find the results of section 4.3 (subresultants excepted). For more information on semialgebraic geometry, see [BCR].

5) WHAT IS THE USE OF C.A.D. ?

5.1 : c.a.d. and elimination of quantifiers

The c.a.d. algorithm described in section 4.3 gives a decision algorithm for elementary algebra. We have seen it working on the example of problem (i). But, as it stands, it is not an algorithm for the elimination of quantifiers. For this, one should also have a description of cells by quantifier-free formulas. Coming back to example 4.4, the sign conditions on the polynomials R, S, T satisfied at the testing points do not suffice to distinguish cell (5) from cell (9) : at both places we have $R>0$, $S<0$, $T<0$.

We are looking for a description of the graphs and of the slices between the graphs in a cylinder. Without loss of generality we can consider the case where the cylinder is cut by the roots of only one polynomial. A description of these graphs (simply saying the number of the root) was given in the proof of 4.3.1, formula (4) ; but it is a description with quantifiers, and here we want to eliminate the quantifiers. The way to do that is to introduce the derivatives of the considered polynomial. We present here a method which is not the quickest one, but which relies on a result interesting by itself.

THEOREM 5.1.1 (Thom's lemma) : *Let $P \in \mathbb{R}[x]$ have degree $d \geq 1$, and let $P',...,P^{(d-1)}$ be its non constant derivatives. Let $A \subset \mathbb{R}$ be a semialgebraic subset in the form*

$$A = \bigcap_{i=0}^{d-1} \{x \in \mathbb{R} ; P^{(i)}(x) \varepsilon_i 0\} \quad \text{where} \quad \varepsilon_i \in \{<, >, =, \leq, \geq\}.$$

Then A is either empty, or a point, or an interval. In the case where A is not empty, its adherence is obtained by relaxing the strict inequalities appearing in its definition.

This result may be obtained by induction on the degree d. The induction step from $d-1$ to d is taken by applying the induction hypothesis to P' and by remarking that, on an interval where P' has a constant sign, P is monotonic.

Let us make precise how Thom's lemma gives a way to describe the cells without quantifier. We begin in dimension 1, supposing that $\mathbb{R}$ was cut into cells by means of all the roots of $P, P',...,P^{(d-1)}$. Then each cell is entirely described by the signs (>0, <0 ou $=0$) of $P, P',...,P^{(d-1)}$ at the testing point of the cell : there is a quantifier-free description, by means of a conjunction of d equations or strict inequalities. In the description of an interval appear only strict inequalities, and in the description of a point there is at least one equation. Then we easily get quantifier-free descriptions for the intervals between two roots of P, which are unions of cells of the preceding decomposition.

Suppose now that P is a polynomial in $x_1,...,x_n$, with degree d with respect to x_n, and consider what happens above a cell $C \subset \mathbb{R}^{n-1}$, supposed to be PROJ($P,P',...,P^{(d-1)}$)-invariant (here we derive with respect to x_n). Then the cylinder $C \times \mathbb{R}$ is cut by the continuous semialgebraic functions which give the roots of $P,P',...,P^{(d-1)}$ and so we are, with the graphs and the slices of the cylinder, exactly in the same position as above with points and intervals in dimension 1. Hence we get a quantifier-free description of the cells inside the cylinder, and if we know a quantifier-free description of C (induction hypothesis), we have a quantifier-free description of the cells in $\mathbb{R}^n$.

We shall develope here no example of application of c.a.d. to a quantifier elimination problem. Some may be found in [AM], and specially problem (ii) (Kahan's ellipse problem) in the particular case where $y_0=0$ (the ellipse has an axis which coincides with the x-axis).

5.2 c.a.d. and the topology of semialgebraic sets : the problem of adjacency relation between cells

Let us come back to problem (iii). We have reformulated the problem in the following way : are the points $(-1,1,-1)$ and $(1,1,1)$ in the same connected component of the semialgebraic set L of points (x_0,y_0,t) in $\mathbb{R}^3$ satisfying formula (3) :

$$\forall x \forall y \left\{ [(x-x_0)+t(y-y_0)]^2 + 9[(y-y_0)-t(x-x_0)]^2 - (1+t^2) = 0 \right.$$
$$\left. \Rightarrow (x \geq 0 \text{ AND } (y-x)^2 - \frac{1}{2} < 0) \text{ OR } (x \leq 0 \text{ AND } (y+x)^2 - \frac{1}{2} < 0) \right\}.$$

One can (at least theoretically) produce a c.a.d. adapted to the polynomials $[(x-x_0)+t(y-y_0)]^2 + 9[(y-y_0)-t(x-x_0)]^2 - (1+t^2)$, $(y-x)^2 - \frac{1}{2}$, $(y+x)^2 - \frac{1}{2}$, x, the variables being considered in order x_0, y_0, t, x, y. One gets then L as a union of certain cells of $\mathscr{C}_3$, and one can find the cells C and D which contain respectively $(-1,1,-1)$ and $(1,1,1)$. It remains to know whether there exists a chain $C_0=C$, $C_1,\ldots,C_l=D$ of cells contained in L such that C_{i-1} and C_i are adjacent for $i=1,\ldots,l$. We say that the cells C_{i-1} and C_i are *adjacent* when C_{i-1} meets the adherence of C_i, or vice-versa; it is then possible to join, by an arc contained in $C_i \cup C_{i-1}$, any point of C_{i-1} to any point of C_i. This leads to the following problem : determine the adjacency relations between cells of a c.a.d.. There is no problem in the case of cells in a same cylinder : two such cells are adjacent if and only if they have consecutive numbers in the cylinder. What is much more delicate is the case of cells in different cylinders.

5.3 Adjacency relation : the case of c.a.d. of the plane

Consider the example of a c.a.d. of the plane adapted to the polynomial $P = y^4 - 2y^2 + 4xy + 1 - x^2$. We find PROJ($P$) = (x, x^2+4), after cleaning up. Hence $\mathscr{C}_1 = \{(1),(2),(3)\}$ with testing points $-1, 0$ et 1. Then, we realize that each of the three cylinders of $\mathbb{R}^2$ is cut by two roots of P. A root of P above the cell (1) ($x<0$) cannot tend to infinity as x tends to 0, since there is the cassical majoration of roots by $3 + |4x| + |1-x^2|$; its limit is then one of the two roots $-1, 1$ of $P(0,y)$. The same argument applies to the roots above cell (3). Besides we know that the number of half-branches of the curve $P=0$ at the points $(0,-1)$ and $(0,1)$ is even (cf. Duval-Roy, this volume). But all this leaves four possibilities.

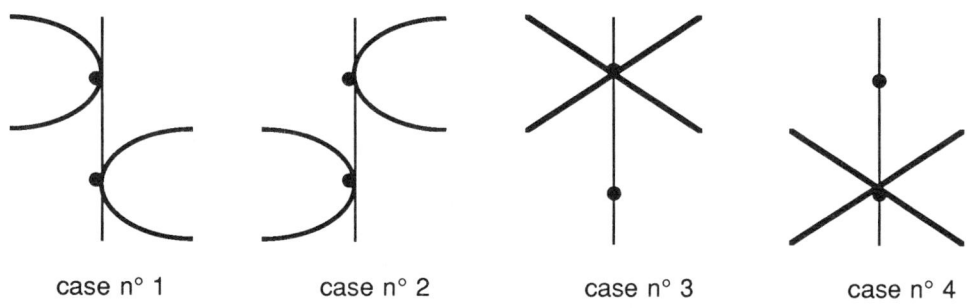

figure 4

To choose between these possibilities, [ACM b] proposes the following method : put the point (0,−1) in a box]−ε,ε[×]−2,0[with ε>0 sufficiently small so that neither $P(x,-2)$ nor $P(x,0)$ have a root in [−ε,ε]. Then, calculate the number of roots of $P(-\varepsilon,y)$ in]−2,0[: this will give the number of roots of P above cell (1) which tend to −1. Do the same thing for $P(\varepsilon,y)$.

The method above is consistent with the coding of real algebraic numbers which depends on the isolation of roots in an interval with rational bounds. Let us give here a variant of this calculation, which consists in considering ε as infinitely small positive. We calculate $\text{ST}(P, \frac{\partial P}{\partial y}) = ($

$P_0 = y^4 - 2y^2 + 4xy + 1 - x^2$,
$P_1 = 4y^3 - 4y + 4x$,
$P_2 = y^2 - 3xy + x^2 - 1$,
$P_3 = -8x^2y + 3x^3 - 4x$,
$P_4 = -\dfrac{x^4 + 8x^2 + 16}{64x^2}$).

If we make $x=0_+$ (infinitely small positive) then $v(P, \frac{\partial P}{\partial y}; -2) - v(P, \frac{\partial P}{\partial y}; 0) = 4-2 = 2$, whereas if we make $x=0_-$ (infinitely small negative) then $v(P, \frac{\partial P}{\partial y}; -2) - v(P, \frac{\partial P}{\partial y}; 0) = 2-2 = 0$. So −1 is the limit of two roots above cell (3) and of no root above cell (1) : we are in the case n°1. Let us see for instance how the calculation of the sign of $P_3(x,0) = 3x^3 - 4x$ for $x=0_-$ or $x=0_+$ reduces to calculations for $x=0$: we have $P_3(0,0) = 0$, which gives no answer, but $\frac{\partial P_3}{\partial x}(0,0) = -4 < 0$, which gives $P_3(0_-,0) > 0$ and $P_3(0_+,0) < 0$. This should convince the reader that it is possible to perform these "infinitesimal" calculations at the left or at the right of any real algebraic number.

5.4. Adjacency problems in dimension ≥ 3

Consider now c.a.d. in dimension 3. Let us for instance calculate a c.a.d. adapted to the polynomial $P = z^2x^2 + y^4 + x^4 - x^2$. We have, after cleaning up, $\text{PROJ}(P) = (x, y^4+x^4-x^2)$ and $\text{PROJ}(\text{PROJ}(P)) = (x, x-1, x+1)$. The c.a.d. obtained is shown on figure 5.

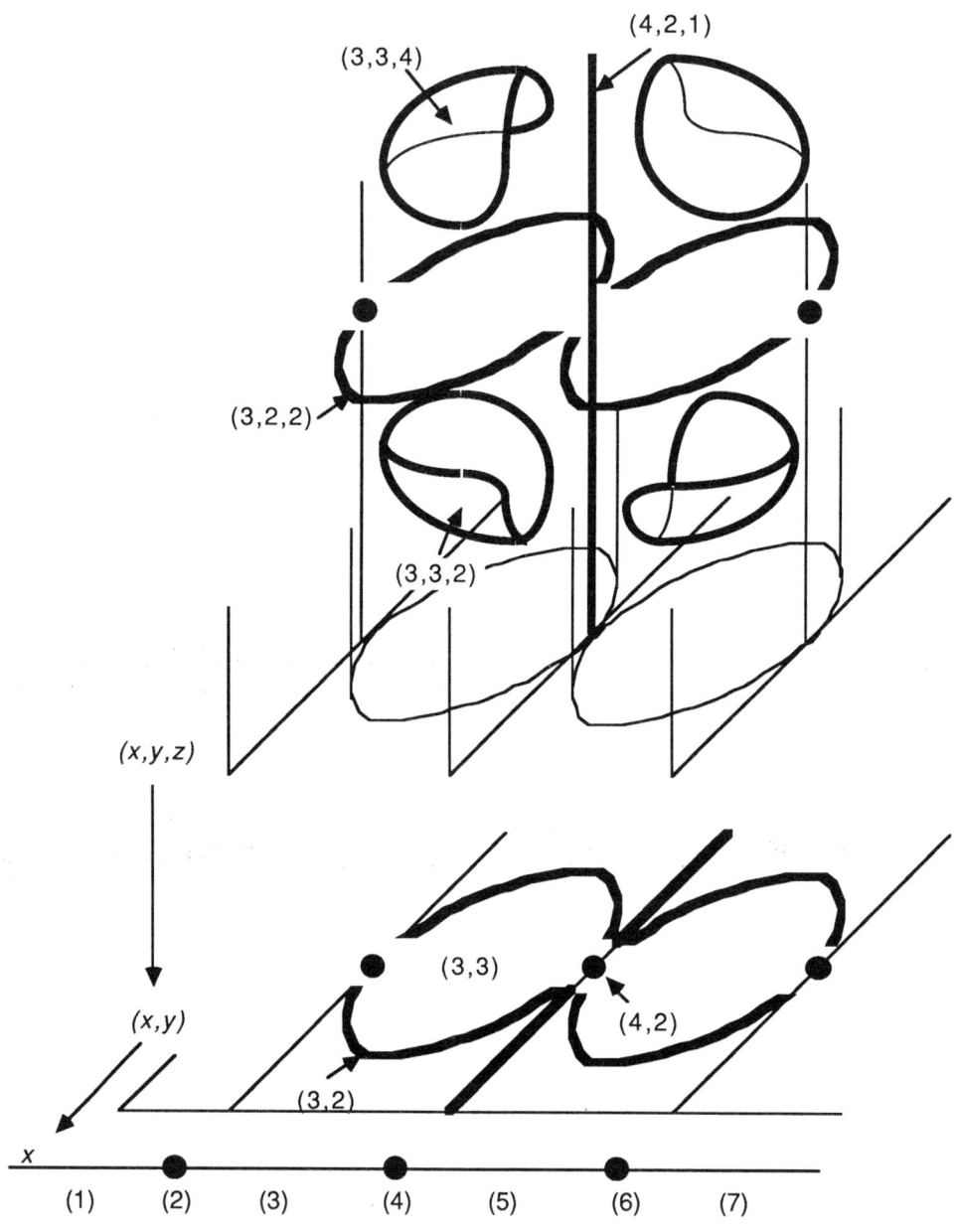

figure 5

Some adjacency relations in $\mathscr{C}_3$ are easily found. For instance examine what happens to a root of $P(a,z)$ when a, element of the cell (3,3) in $\mathbb{R}^2$ gets closer to cell (3,2) ; this root cannot go to infinity (thanks to the majoration by $\dfrac{1+|y^4+x^4-x^2|}{x^2}$ and so tends to the unique root of P above cell (3,2). So the cells (3,3,2) and (3,3,4) are adjacent to the cell (3,2,2).

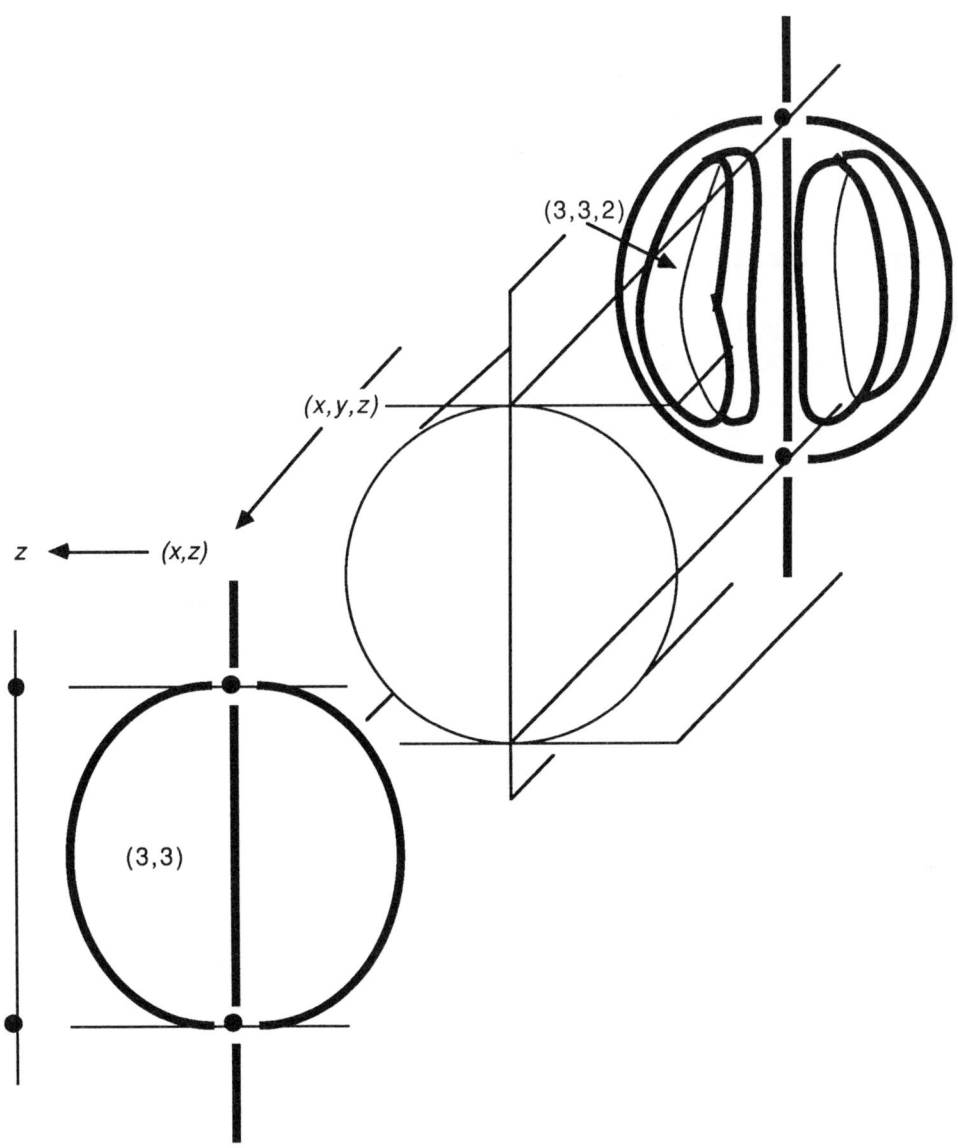

figure 6

On the other hand we can say nothing about how the cylinder above cell (3,3) and the cylinder above cell (4,2) (the origin in $\mathbb{R}^2$) glue together.

Let us change our point of view, and consider the variables in order z,x,y. We get now PROJ(P) = $(x^2+ z^2 - 1, x)$ and PROJ(PROJ(P)) = $(z-1, z+1)$. The new c.a.d. is shown on figure 6, and this time the adjacency relations are all easy to find. The difference between the two situations is that, in the first case, the polynomial P vanishes above cell (4,2), whereas in the second case P is identically null above no cell in the (z,x)-plane, for the good reason that it is monic with respect to y (the leading coefficient is a constant).

The configuration obtained with the help of the second c.a.d. tells us what happened in the first : the intersection of the adherence of graph (3,3,2) with the z-axis (the cylinder above cell (4,2)) is the segment $[-1,0]$ of this axis, whereas the axis is a cell by itself (the cell (4,2,1)).

5.5 The good case : monic polynomials

Working with polynomials which are monic with respect to the last variable ensures that the kind of situation we just described cannot happen.

PROPOSITION 5.5.1 : *Suppose that all polynomials $P_1,...,P_r \in \mathbb{R}[x_1,...,x_n]$ are monic with respect to x_n, and suppose moreover that the* PROJ$^i(P_1,...,P_r)$ (*i-th iteration of* PROJ, $1 \leq i \leq n-1$) *consist of polynomials which are all monic with respect to x_{n-i}. Then*

(i) *if two cells D and D' of $\mathscr{C}_k$ ($1 \leq k \leq n$) are adjacent, we have either $D \subset \text{adh}(D')$, or $D' \subset \text{adh}(D)$,*

(ii) *if D and D' are two cells of $\mathscr{C}_k$ ($1 \leq k < n$) with $D' \subset \text{adh}(D)$, and if C is a graph in the cylinder $D \times \mathbb{R}$, then the intersection of the adherence of C with the cylinder $D' \times \mathbb{R}$ is a graph of this last cylinder.*

When we are in the situation of the proposition above, we can (always theoretically) effectively determine all the adjacency relations between cells. Such an algorithm is described in [SS]. It is relatively easy to give an idea of this algorithm in the case of two cells C and C' of $\mathscr{C}_n$ with dimensions respectively n and $n-1$. The discussion of section 4.6 shows that only one of the integers in the code $(\alpha'_1,...,\alpha'_n)$ of C' is even, say α'_k, whereas all the integers in the code $(\alpha_1,...,\alpha_n)$ of C are odd. If C and C' are adjacent, then it is also the case of the cells $(\alpha'_1,...,\alpha'_k)$ (graph of dimension $k-1$) and $(\alpha_1,...,\alpha_k)$ (slice of dimension k) of $\mathscr{C}_k$. This happens if and only if $\alpha'_i = \alpha_i$ for $1 \leq i < k$ and $\alpha'_i = \alpha_i \pm 1$; say for instance that $\alpha'_i = \alpha_i -1$, which means that the cell $(\alpha'_1,...,\alpha'_k)$ is the graph which bounds the slice $(\alpha_1,...,\alpha_k)$ from below. Let now $a' = (a'_1,...,a'_n)$ be the testing point of C', and e_k the vector $(0,...,0,1,0,...,0) \in \mathbb{R}^n$ with the 1 in k^{th} position. We have to think that the vector e_k points in the direction from C' to C, and indeed C' is adjacent to C if and only if $a' + te_k$ is in C for small positive t (even, if preferred, infinitely small positive). This last condition may be tested for instance if we have a quantifier-free description of C, but there are other ways to do.

The calculation of adjacency relations is more delicate in the other cases. The following result (which differs from the method of [SS]) offers a solution, theoretically elegant. This is a generalization of Thom's lemma to the case of several variables, which may be found in [BCR].

THEOREM 5.5.2 : *Let* $\mathcal{F} = \{ P_{i,j} ; 1 \leq i \leq n, 1 \leq j \leq r_i \}$ *be a family of polynomials satisfying the following properties* :

(i) *For fixed* i, *the family* $\mathcal{F}_i = \{ P_{i,j} ; 1 \leq j \leq r_i \}$ *is a family of polynomials in* $\mathbb{R}[x_1,...,x_i]$, *all monic with respect to* x_i ; *for any* j, *if* $\dfrac{\partial P_{i,j}}{\partial x_i}$ *is not constant, it belongs to* $\mathcal{F}_i$.

(ii) *For any* $i<n$, $\mathrm{PROJ}(\mathcal{F}_{i+1}) \subset \mathcal{F}_i$.

Let, for $1 \leq k \leq n$, $\mathcal{C}_k$ *be the finite semialgebraic partition of* $\mathbb{R}^k$ *given by the non empty subsets in the form*:

$$C = \bigcap_{i=1}^{k} \bigcap_{j=1}^{r_i} \{(x_1,...,x_k) \in \mathbb{R}^k ; P_{i,j}(x_1,...,x_k) \varepsilon_{i,j} 0 \} \quad \text{where} \quad \varepsilon_{i,j} \in \{<, >, =\}.$$

Then the $\mathcal{C}_k$ *form a c.a.d. (in particular any* $C \in \mathcal{C}_k$ *is homeomorphic to a box), and moreover the adherence of a cell* $C \in \mathcal{C}_k$ *is obtained by relaxing the strict inequalities in its definition.*

Finding the adjacency relations between the cells of the c.a.d. of the theorem is very simple : it is enough to compare the signs of the polynomials $P_{i,j}$ ($i \leq k$) at the testing points a_C and $a_{C'}$ of two cells C and C' of $\mathcal{C}_k$. If we find that any sign >0 or <0 obtained at $a_{C'}$ is also obtained at a_C, then we have $C' \subset \mathrm{adh}(C)$. The inconvenient of the method using the preceding theorem is obviously the addition, at each step, of the derivatives, which of course increases the number of polynomials and cells.

Once the adjacency relations between cells have been completely calculated, we know everything about the topology of a semialgebraic set which is an union of cells. Without entering into details, let us note that if the considered semialgebraic set is compact, one can produce, starting with the list of its cells and the adjacency relations between these cells, a triangulation of the set. This triangulation is built up by induction on the dimension of the ambient space (cf. [BCR]). Classical combinatorial calculations allow then to recover the homology groups.

5.6 And in the bad case ?

All what we have done in the preceding section was under the hypothesis that, at each step, the polynomials are monic with respect to the last variable. We may always find changes of variables to be in this situation (which is, in a sense which can be made precise, the general situation). But such a change may have the side effect of complicating the polynomials you started with. Besides there are situations where you have no choice on the variables : typically, when you eliminate quantifiers or when you handle a decision problem, since then it is impossible to tilt the quantified variables. But, says the gentle reader, why bother about adjacency relations between cells when you just want to eliminate the quantifiers ?

However this approach is sensible, and even useful (cf. [ARN]). Suppose for instance that one knows how to group the cells of $\mathscr{C}_{n-1}$ into clusters whose union is a connected $\text{PROJ}(P_1,...,P_r)$-invariant set. Then the structure of the cylinders $C \times \mathbb{R}$ is the same for all the cells C of a same cluster, and so one has to perform calculations only above one testing point for each cluster. This is the reward for the time spent in determining adjacency relations.

In the situation where no change of variables is allowed, there exists an algorithm for the determination of adjacency relations which works up to dimension 3 included [ACM c]. Beyond 3, it is still possible to say something about the adjacency relations between cells of dimensions n and $n-1$ in $\mathbb{R}^n$. Here is a statement analoguous to the one of proposition 5.5.1. To encompass the case where the roots "go to infinity", we shall consider completed cylinders $C \times \overline{\mathbb{R}}$ where $\overline{\mathbb{R}} = \mathbb{R} \cup \{-\infty\} \cup \{+\infty\}$ with the obvious topology, and a graph of the completed cylinder will be either a graph of the cylinder $C \times \mathbb{R}$, or $C \times \{-\infty\}$, or $C \times \{+\infty\}$.

PROPOSITION 5.6.1 : *Suppose that for $0 \leq i \leq n-2$, and for any polynomial $Q \in \text{PROJ}^i(P_1,...,P_r)$, the coefficients of Q (considered as a polynomial in x_{n-i} with coefficients in $\mathbb{R}[x_1,...,x_{n-i-1}]$) have no common factor. Then :*

(i) *If two cells D and D' of $\mathscr{C}_k$ ($1 \leq k \leq n$), with dimensions k and $k-1$ respectively, are adjacent, then $D' \subset \text{adh}(D)$; moreover any point of D' has a basis of neighborhoods whose intersections with D are connected.*

(ii) *If D and D' are two cells of $\mathscr{C}_k$ ($1 \leq k < n$), with dimensions k and $k-1$ respectively, such that $D' \subset \text{adh}(D)$, and if C is a graph in the cylinder $D \times \mathbb{R}$, then the intersection of the adherence of C with the completed cylinder $D' \times \overline{\mathbb{R}}$ is a graph of this cylinder*

The hypothesis comes in in the step of the proof which consists in showing that (i) for k implies (ii) for k : if C is a graph of a root of a polynomial $P \in \text{PROJ}^{n-k}(P_1,...,P_r)$ on D, the polynomial P cannot be identically null on D', since otherwise, as D' has dimension $k-1$, the coefficients of P would have a common factor. Once the result is established, one proceeds as indicated after 5.5.1 to find the adjacency relations between cells of $\mathscr{C}_k$ of dimensions k et $k-1$.

One can always retreat to a situation where the hypothesis of 5.6.1 applies ; it is sufficient to include in PROJ a procedure which finds the g.c.d. of the coefficients of the polynomials, and eventually factors out this g.c.d..

Even if we do not get in this way a complete information about the topology, the knowledge of what happens between cells of dimensions n and $n-1$ in $\mathbb{R}^n$ is sufficient to solve some problems. For instance, let us come back to our ellipse mover's problem (problem (iii)). We want to know whether the points $(-1,1,-1)$ and $(1,1,1)$ are in the same connected component of L, which is obviously open in $\mathbb{R}^3$. If this is true, and if we have not the bad luck that one of the two points is in a cell of dimension ≤ 1, then they can be joined inside L by an arc which runs only through cells of dimensions 2 or 3 : the information on the adjacency relations between these cells is sufficient.

5.7 c.a.d. and ray casting

The visualisation of a scene (composed for instance of objects of C.S.G.) often appeals to ray casting, that is, from the geometrical point of view, a central projection. The idea of cylindrical arrangement for the projection $\mathbb{R}^3 \longrightarrow \mathbb{R}^2$ transposes easily to the central projection : one can pass from one situation to the other by a simple change of projective coordinates. The cylinders are transformed into cones with the center of projection as vertex.

6) FEASABILITY AND COMPLEXITY OF THE ALGORITHMS

6.1. The c.a.d. algorithm

We refer to [MAR] for a discussion of the possibilities and the limits of the implemented algorithm. See also [DAV b] which relates an unlucky experiment on a problem simpler than problem (iii), with the ellipse replaced by a bar of length 3. The calculations on real algebraic numbers are extremely expensive in the algorithm. One can try to lower the cost by modifying the coding and the manipulations on real algebraic numbers (an alternative method is proposed in [CR]), or by avoiding as much as possible such calculations : we have already signalled in 5.6 the technique of clustering of cells which reduces the number of testing points. An other idea for amelioration consists in reducing the number of polynomials to consider in PROJ ([MC], [RIS]).

On the other hand, the c.a.d. algorithm is a "general purpose" algorithm. It may be advantageous to have specialized algorithms for certain types of problems, which give just the information which is strictly necessary to the solution of the problem : we have seen for instance that for the ellipse in the corridor, the consideration of the "big" cells is sufficient.

6.2 Complexity

The estimation of the asymptotic complexity of the c.a.d. algorithm gives a bound which is doubly exponential in the number of variables, and, for a fixed number of variables, polynomial in the size of the list of polynomials $(P_1,...,P_r)$ of the input (this size takes into account the number of polynomials, the degrees, the size of the coefficients, cf. [COL], [DAV]). This doubly exponential feature explains why the general algorithm seems practically useless when the number of variables is not very small. An example of [DH] shows that the complexity of a quantifier elimination algorithm cannot be less (in the worst case) than doubly exponential in the number of quantifiers to eliminate, at least in the context of a dense representation of polynomials. So we cannot expect to avoid the double exponential. But note that [GRI] gives a decision algorithm which is doubly exponential in the number of *alternations* of quantifiers, which is better (and there is hope for having a quantifier elimination algorithm in the same line). On the other hand [FGM] considers the *parallel* complexity of quantifier elimination, and gives a bound which is simply exponential in the number of variables.

REFERENCES

[AB] D.S. Arnon, B. Buchberger, ed., *Algorithms in real algebraic geometry*, Academic Press (1988).

[ARN] D.S. Arnon, *A cluster-based cylindrical algebraic decomposition algorithm*, pp. 189-212 in [AB].

[ACM a] D.S. Arnon, G.E. Collins, S. McCallum, *Cylindrical algebraic decomposition I : the basic algorithm*, SIAM J. Comp **13** (1984), 865-877.

[ACM b] D.S. Arnon, G.E. Collins, S. McCallum, *Cylindrical algebraic decomposition II : an adjacency algorithm for the plane*, SIAM J. Comp **13** (1984), 878-889.

[ACM c] D.S. Arnon, G.E. Collins, S. McCallum, *An adjacency algorithm for cylindrical algebraic decompositions of three-dimensional space*, pp. 163-188 in [AB].

[AM] D. Arnon, M. Mignotte, *On mechanical quantifier elimination for elementary algebra and geometry*, pp 237-260 in [AB].

[BKF] M. Ben-Or, D. Kozen, J. Reif, *The complexity of elementary algebra and geometry*, J. of Comput. and System Sci. **32** (1986) 251-264.

[BCR] J. Bochnak, M. Coste, M-F. Roy, *Géométrie algébrique réelle*, Ergebnisse der Math. 12, Springer-Verlag (1987).

[COL] G.E. Collins, *Quantifier elimination for real closed fields by cylindrical algebraic decomposition*, Proc. 2nd GI Conf. Automata Theory & Formal Languages, Springer, Lecture Notes in Computer Science 33 (1985), 134-183.

[CR] M. Coste, M-F. Roy, *Thom's lemma, the coding of real algebraic numbers and the computation of the topology of semialgebraic sets*, pp. 121-130 in [AB].

[DAV a] J.H. Davenport, *Computer algebra for cylindric algebraic decomposition*, TRITA-NA-8511, NADA, KTH, Stockholm (1985).

[DAV b] J.H. Davenport, *A "piano movers" problem*, preprint (1985).

[DH] J.H. Davenport, J. Heintz, *Real quantifier elimination is doubly exponential*, pp. 29-36 in [AB].

[FGM] N. Fitchas, A. Galligo, J. Morgenstern, *Algorithmes rapides en séquentiel et en parallèle pour l'élimination de quantificateurs en géométrie élémentaire*, Séminaire Structures Algébriques Ordonnées, Univ. Paris VII (1987)

[LAZ] D. Lazard, *Quantifier elimination : optimal solution for two classical examples*, pp. 261-266 in [AB].

[LOJ] S. Lojasiewicz, *Ensembles semi-analytiques*, multigraphed, I.H.E.S. (1965).

[GRI] D.J. Grigoriev, *Complexity of deciding Tarski algebra*, pp. 65-108 in [AB].

[LOO] R.G.K. Loos, *Generalized polynomial remainder sequences*, Computer Algebra - Symbolic and algebraic computation, Springer-Verlag (1982)

[MAR] J. Marchand, *Elimination des quantificateurs et décomposition algébrique cylindrique,* in Séminaire "Calcul formel et outils algébriques pour la modélisation géométrique", C.N.R.S. (1988).

[MC] S. McCallum, *An improved projection operation for cylindric algebraic decomposition*, Computer Science Tech. Report 548, Univ. Wisconsin at Madison (1985), see also pp. 141-162 in [AB].

[PAU] A. Paugam, *Comparaison entre 3 algorithmes d'élimination des quantificateurs sur les corps réels clos*, Thèse, Univ. Rennes 1 (1986).

[RIS] J-J. Risler, *About the piano mover's problem*, preprint (1987).

[SS] J. Schwartz, M. Sharir, *On the "piano movers" problem II. General techniques for computing topological properties of real algebraic manifolds*, Advances in Applied Mathematics **4** (1983), 298-351.

[SEI] A. Seidenberg, *A new decision method for elementary algebra*, Annals of Math. **60** (1954) 365-371.

[TAR] A. Tarski, *A decision method for elementary algebra and geometry*, 2nd ed., Univ. Calif. Press, Berkeley (1951).

Curves and computer algebra

D. Duval (Institut Fourier, Grenoble)
M.-F. Roy (Institut Mathématique, Rennes)

In order to give a flavour of the general relationship between computer algebra and geometry we describe without technical details some simple and general ideas concerning the possibilities given by computer algebra in the exact geometric study of algebraic curves.

1) Real algebraic curves: definitions and examples

In this paragraph we give the definitions and first properties of real algebraic curves. For proofs see for example [B C R].

a) Plane real algebraic curves: definitions

A plane real algebraic curve is an algebraic set of $\mathbb{R}^2$ of dimension 1, which means:
- that it is the zero set $Z(P)$ of a non null polynomial $P(X,Y)$ with real coefficients,
- and that $Z(P)$ contains a subset homeomorphic to the open interval $]0,1[$.

For example, $Z(X^2+Y^2-1)$ is a plane real algebraic curve (it is the unit circle), but not $Z(X^2+Y^2+1)$ (which is empty) neither $Z(X^2+Y^2) = \{0\}$ (which is of dimension 0).

It is often supposed that the curve is irreducible, which means that it is defined by an irreducible polynomial P.

A simple criterion to determine whether an irreducible polynomial P in two variables defines a plane real algebraic curve is the following sign change criterion: an irreducible polynomial defines a real algebraic curve if and anly if its sign changes, that is if there exist points M_1 and M_2 of $\mathbb{R}^2$ with $P(M_1) > 0$ et $P(M_2) < 0$.

b) Singular points of a plane real algebraic curve

A point M of a plane curve $C = Z(P)$ is singular if the two partial derivatives P'_X and P'_Y vanish at M.

A plane real algebraic curve cannot have more than a finite number of singular points.

There are several kinds of singular points: real isolated points, cusps, or multiple points, also called branching points.

c) Examples of plane real algebraic curves

Let us give immediately some examples of real algebraic curves:
- lines, circles and more generally conics (ellipses, hyperbolas and parabolas),
- cubics, defined by an equation of the third degree ; cubic $Y^2+X^2-X^3$ (*figure a*) has a real isolated point at O, cubic $Y^2-X^2-X^3$ (*figure b*) has a double point at O, cubic Y^2-X^3 (*figure c*) has a cusp at O, cubic$^2-X^3 + X$ (*figure d*) has two conected components and is nevertheless algebraically irreducible.

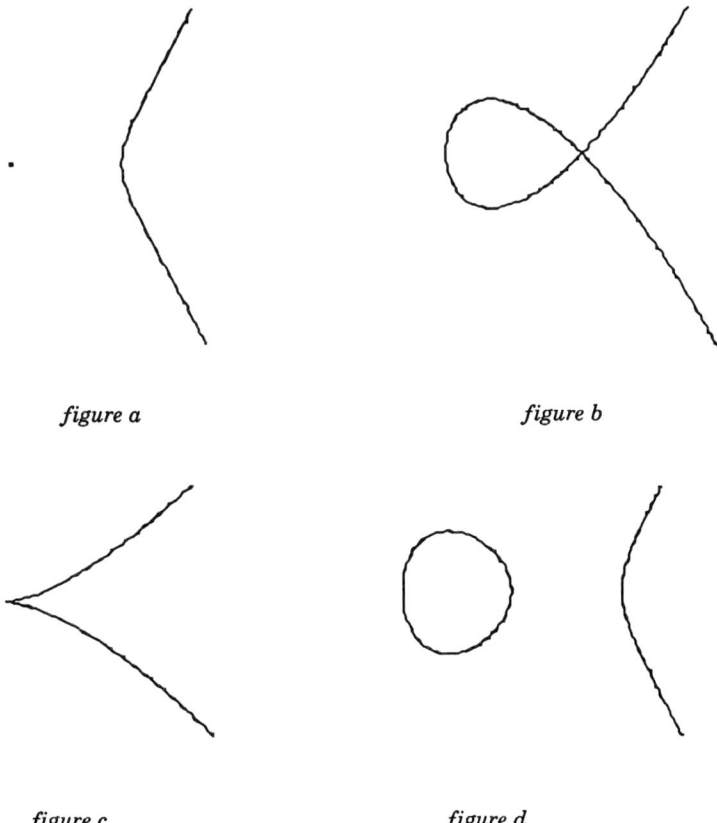

figure a   *figure b*

figure c   *figure d*

Some real algebraic curves can be rationally parametrized: when there exist rational fractions f and g such that the point with coordinates $(f(t),g(t))$ describes the curve. It is the case for conics, or cubics with a singular point. But this is not true in general. On the other hand, a plane real algebraic curve can always be parametrized locally by analytic functions. This will be developed in point 4.

d) Real algebraic curves of $\mathbb{R}^n$

A real algebraic curve of $\mathbb{R}^3$ or more generally of $\mathbb{R}^n$, is an algebraic subset of $\mathbb{R}^3$ (resp. $\mathbb{R}^n$) of dimension 1, which means:
- that it is the zero set $\mathcal{Z}(P_1,...,P_k)$ of polynomials $P_1(X_1,...,X_n),...,P_k(X_1,...,X_n)$ with real coefficients
- and that $\mathcal{Z}(P_1,...,P_k)$ contains a subset homeomorphic to $]0,1[$, and no subset homeomorphic to $]0,1[^2$.

A real algebraic plane of the space can for example be defined as the intersection of two surfaces.

It is the case of Viviani's window, intersection of the cylinder $\mathcal{Z}((X-1/2)^2+Y^2-1/4)$ and of the unit sphere.

A singular point of an algebraic real curve is a point where the rank of the jacobian matrix of the system of equations $P_1,...,P_k$ is smaller than $n-1$.

In the example of Viviani's window the normal vectors to the cylinder and the sphere coincide at point $(1,0,0)$, which means that the jacobian matrix is there of rank 1, and that point $(1,0,0)$ is singular (a double point in this case). If we modify slightly the radius of the sphere we obtain starting from this singular situation two qualitatively different situations: one connected component if the radius R of the sphere is slightly smaller than 1 ($R=0,99$ *figure f*), two connected components if it is slightly bigger than 1 ($R=1,01$ *figure g*).

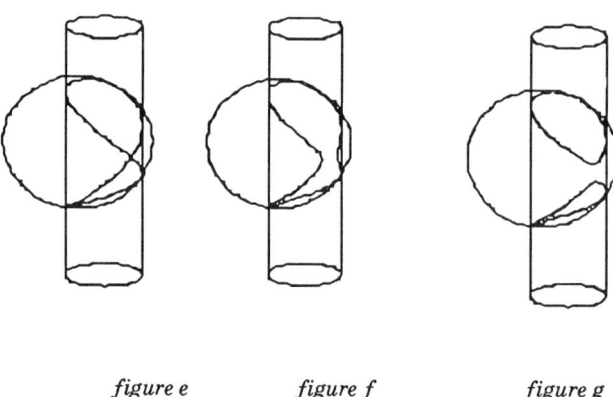

figure e   *figure f*   *figure g*

e) Topological properties of real algebraic curves

Real algebraic curves have particular topological properties : every point M of a real algebraic curve C has a neighbourhood U such that $U-\{M\} \cap C$ is homeomorphic to a finite union of an even number of open intervals. That is, a real algebraic curve has in every point an even number of half-branches. A closed interval for example cannot be homeomorphic to a real algebraic curve.

An irreducible real algebraic curve can have several connected components. A real algebraic curve has always a finite number of connected components. These components are no more algebraic sets but in general semi-algebraic sets (for this notion see [C], this volume).

2) Computer algebra and geometric modelling

Properties of symbolic computations are complementary of those of numerical computations. For example, it is very easy to compute with $\sqrt{2}$ so that $(\sqrt{2})^2$ is exactly equal to 2, but it is a bit more difficult to distinguish $\sqrt{2}$ from $-\sqrt{2}$. We shall come back to this question later but for the moment we want to explain on some examples how computer algebra can be used for geometric modelling in the plane.

a) Does a point belong to a curve?

If the inputs (point and curve) are exact, computer algebra answers *exactly*. For example, point $(\sqrt{2}, \sqrt{2}/2)$ belongs to ellipse $x^2+4y^2 = 4$ since
$(\sqrt{2})^2+4\ (\sqrt{2}/2)^2 = 2+4\cdot(2/4) = 4$.
But point $(\sqrt{2}+1/100, \sqrt{2}/2-1/200)$ does not since
$(\sqrt{2}+1/100)^2+4\ (\sqrt{2}/2-1/200)^2 = ...$
$\qquad ... = 2 + 2\sqrt{2}/100 + 1/10000 + 4(1/2 - \sqrt{2}/200 + 1/40000)$
$\qquad ... = 4 + 2/10000 \neq 4$

b) Decomposition of a curve in algebraically irreducible components.

Factorization of polynomials is one of the best developed subjects in computer algebra (see [D S T]).

It is necessary to distinguish two sorts of factorizations. *Factorisation on integers* or rational numbers, in general implemented on computer algebra systems, allows to know whether a polynomial P in two variables is irreducible over the rationals. If it is not, then the algebraic curve is not irreducible. But if it is irreducible over rationals, nothing can be concluded. Another existing factorization technique, for the moment rarely implemented, is the *absolute factorisation*, which allows to know if the polynomial is irreducible over the complex numbers.

For example, we know this way that curve C_2 (*figure i*) with equation $P_2(X,Y)=Y^4-2Y^3+Y^2-3X^2Y+2X^4$ is irreducible, though looking very similar to the union of two ellipses (curve C_1) (*figure h*).

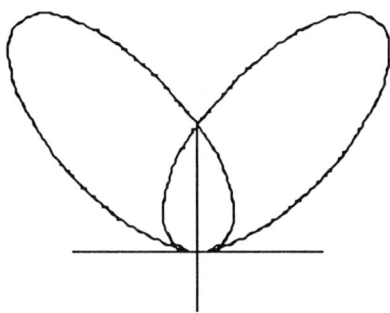

figure h

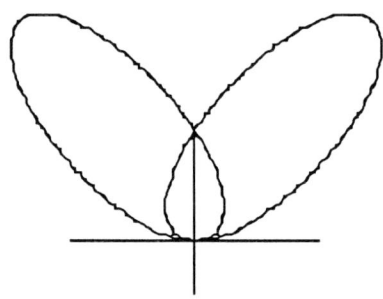

figure i

c) *Avoid to forget connected components.*

In the sequel we shall consider only plane real algebraic curves with equation a polynomial $P(X,Y)$ monic in Y (the leading coefficients of P as polynomial in Y is 1). It is always possible to realize this situation by a linear change of coordinates. This algebraic condition implies that curve $Z(P)$ contains no line parallel to the Y- axis.

Every connected component which is not a real isolated point must either cut the Y- axis or have a point with vertical tangent or have a singular point. A real isolated point is always singular.

The *dicriminant* of polynomial $P(X,Y)$ (considered as polynomial in Y) disc(P) is a polynomial in X with roots the abscisses of points where "something happens" : real isolated point, point with vertical tangent, cusp or branching point for example.

So the set of points (x,y) where x is equal 0 or is a real root of the discriminant of P, and where y is a real root of $P(x,Y)$, has at least one point on every connected component of the curve.

So that we do not forget small connected components or isolated points.

For example if $P(X,Y) = Y^2 - 2/5\ Y - X^3 - 2X^2 - 99/100\ X + 2/5$, then $\text{disc}(P) = -4X(X^2 + 2X + 99/100)$ has 3 real roots 0, 9/10, 11/10. We obtain on the curve 3 points (0, 1/5), (9/10, 1/5), (11/10, 1/5). In fact the curve has 2 connected components, a big one passing through (0,1/5) and a small one passing through the other points.

d) Does a point belong to a set delimited by a curve?

Let us take again the example of curve C_1 union of two ellipses of respective equations E_1 and E_2. It is possible to chose these equations so that point M is in the intersection of the interior of the two ellipses if and only if $E_1(M)$ and $E_2(M)$ are both negative.

If we consider now curve C_2, condition $P_2(M) > 0$ is not sufficent to characterize the hatched part (*figure j*).

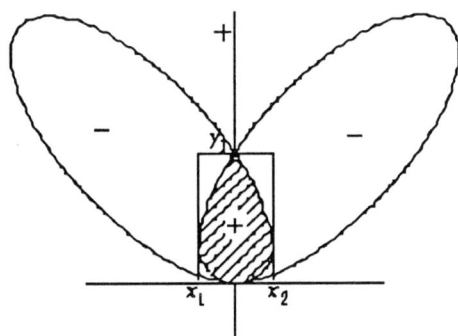

figure j

It can be characterized as : $P_2(M) > 0$, and $x_1 < x < x_2$, and y between the second and third root of $P_2(x,Y)$. The determination of x_1 et x_2 uses the *discriminant* of P_2. This is an application of the method of *cylindrical decomposition* ([Co] or [C]).

e) Description of the curve near a singular point.

Thanks to symbolic computations it is possible to use the powerful tools developed by mathematicians to study a curve in the neighbourhood of a singular point. Let us mention particularly *blowing-ups* and *Puiseux expansions*, that we shall study in paragraph 4.

f) Global study of curves: topollogy and analytic components

This point is developed in 6.

3) Computer algebra and real algebraic numbers

The interest of the methods we just considered lies in the *exactness* properties of symbolic computations: it is possible to test the equality of two numbers, determine multiplicities of the roots of a polynomial, or work at a singular point of a curve (and not near, at a non singular point!).

This supposes that the inputs themselves are exactly given. Let us suppose that all the inputs are integer (or rational, which is equivalent). Computer algebra systems know how to commpute with arbitrarily long integers. But for example the roots of the discriminant of P are not integers even if P is with integer coefficients. Coefficients of Puiseux expansions neither, as we will see in next paragraph. They are *algebraic numbers* like i, $\sqrt{2}$, or any root of X^5-X+1. Some are *real algebraic numbers* like $\sqrt{2}$, $(1+\sqrt{5})/2$, or the unique real root of X^5-X+1.

To explain how it is possible to make exact computations with such numbers let us take the example of $\sqrt{2}$. It is represented by a symbol, say sqrt(2) (but α would be as good), together with a computation rule: (sqrt(2))**2 = 2, which allows to simplify expressions with sqrt(2) at a power $n \geq 2$, or with sqrt(2) at denominator (since the rule implies 1 / sqrt(2) = sqrt(2) / 2). This is quite analogous to what we do usually to compute with complex numbers by introducing symbol i and rule $i**2 = -1$.

But then, in the same way that i can represent any root of X^2+1, symbole sqrt(2) can represent as well $\sqrt{2}$ as $-\sqrt{2}$. Sor some computations this ambiguty is not a problem and can even avoid unnecessary computations : if a computation has to be done both for $\sqrt{2}$ and $-\sqrt{2}$, it can be sufficent to do it for sqrt(2).

But if we are interested in real situations, where order is important the question is different. One may want to determine the sign of sqrt(2), for example. It is then necessary to distinguish between the two square roots of X^2-2. There are several possible solutions.

First we need to know the exact number of real roots of P, which is done by using the classical Sturm theorem (see for example [B C R]).

The more elementary solution consists in distinguishing the real roots of a given polynomial $F(X)$ by ordering them : so $\sqrt{2}$ is the biggest real root of X^2-2. But this method does not always allow to continue the computations.

It is also possible to define *isolating intervals* for the roots: then $\sqrt{2}$ is the unique root of X^2-2 in the interval]1,413, 1,415[, or even in the half line]0,+∞[. Such roots are very easily ordered. This method is implemented in Collins algorithm [Co]. This method needs polynomials with integer coefficients.

Another method, based on Thom's lemma [C R], characterizes the real roots of $F(X)$ by the signs taken at this root by the derivatives of F: for example, $\sqrt{2}$ is the unique root X^2-2 where the derivative $2X$ is positive. It is shown that this allows to distinguish effectiviely the roots, and to compare them.

This method is also valid for polynomials with real algebraic coefficients, which appear naturally in geometric situations.

It is remarkable that the notion of real number, generally considered as more natural is the more difficult to manipulate in computer algebra. Not only more difficult than rational numbers, but also more difficult than complex numbers!

For the moment an efficient system for computing with complex algebraic numbers is available : this is system D5, on Reduce and soon on Scratchpad II [D-D].

A system for manipulating real algebraic numbers characterized by the sign of derivatives is being implemented [G O R].

4) Algebraic tools for the study of real curves

Some tools have already been introduced and are useful in the general study of semi-algebraic sets: that is elimination theory, precisely properties of discriminant, Sturm theorem and Thom's lemma.

Elimination theory correspond geometrically to a projection. The discriminant of P is obtained by eliminating variable Y between polynomials P and P'_Y. Zeroes of the discriminant are points of the X–axis above which the polynomial $P(x,Y)$ has a multiple root. This means that "something happens" to the curve above these roots : real isolated point, point with a vertical tangent, cusp or multiple point (not necessarily real).

Just to the left and just to the right of a zero of the discriminant, the situation may be different (the number of real zeroes of $P(x,Y)$ may have changed). Sturm theorem is the tool to know the number of these real roots.

Cylindrical decomposition method is based on these two tools.

Finally Thom's lemma ([C R], [C]), useful as we have seen for coding real algebraic numbers, gives also some global information on topology or analytic structure.

Some tools are more specific to the local study of plane curves: blowing-up and Puiseux expansions.

a) Blowing up

Using succesive changes of variables (translations ou changes of the form $Z = Y/X$) one constructs a non singular curve above the singular one we started with.

Let us consider for example the cubic with double point C with equation $X^2-Y^2+X^3$ and let $Z=Y/X$. One obtains (by divising by X^2) curve C' of equation $1-Z^2+X$ which is no more singular. Above point O there are two points of C' corresponding to values $Z=1$ and $Z=-1$ (which are the limit of Y/X lorsque when a point M tends to O on C) (*figure k*).

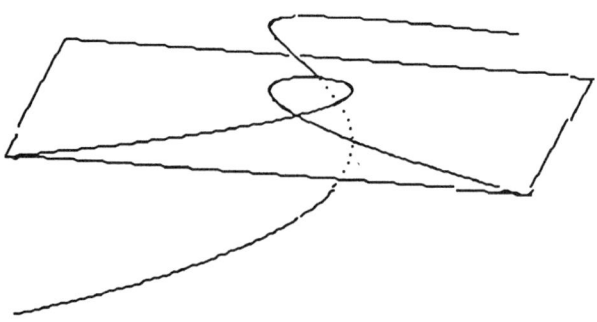

figure k

When blowing up real isolated point may disappear (take the cubic with isolated real point of equation $X^2+Y^2-X^3$).

b) Puiseux expansions

Let C be a plane curve defined by equation $P(X,Y)=0$. Near x_0 we try to express the different values of y such that $P(x,y)=0$ as a function of x.

By making if necessary a translation, let us suppose $x_0=0$. It is very often possible to express y as a formal sery in x (consider for example $Y^2-1-X=0$), but not always. It is any case possible to express y as a Puiseux sery in x, that is as a formal sery in $x^{1/q}$ for a well chosen integer $q>0$ (for example $q=2$ for $Y^2-X=0$). Moreover, these formal series are converging and hence define local (complex) analytic parametrisations of the curve above a neighbourhood of $x=0$: to a Puiseux expansion $y = \sum_{n\geq 0} \alpha_n x^{n/q}$ correspond parametrisation $(x=t^q, y = \sum_{n\geq 0} \alpha_n t^n)$.

Even if the equation we started with is with integer coefficients the coefficients α_n in Puiseux expansions are no more integers but are algebraic numbers. When one has systems for exact computation on algebraic numbers one can compute effectively Puiseux expansions, by a method essentially due to Newton ([W]).

One defect of Puiseux expansions is that the expansions of a real branch are not necessarily with real coefficients. For example, expansions of $Y^2+X=0$ are $y=\pm i\, x^{1/2}$.

A more recent notion is that of *rational Puiseux expansions* [D]. They are parametrisations of branches of the curve in the neighbourhood of $x=0$ of the form $(x=\lambda t^q, y= \sum_{n\geq 0} \alpha_n t^n)$ where λ is not necessarily equal to 1. These expansions enjoy the same properties than classical Puiseux expansions, they are easier to compute, and verify: a branch is real if and only if λ and all α_n are real. The verification that all α_n are real can be made in a finite number of steps because it is sufficent to verify that $\alpha_1, \alpha_2,...,\alpha_R$ are real (where R is an integer easy to compute) to insure that all α_n are real.

Combining rational Puiseux expansions with formal computation on inequalities allowing to verify whether the roots of a polynomial with real algebraic coefficients are real (for example methods based on Sylvester's lemma [C R]) we get a convenient tool for local study of real algebraic curves.

5) Local study of curves.

Let us give two ewamples of these technique:

a) The parabola $Y^2+X=0$ has a unique rational Puiseux expansion : $(x=-t^2, y=t)$, which correspond to the unique real branch of the curve.

b) We have already seen that curve C_1, union of two ellipses tangent to the X-axis and curve C_2 (cf *figures f* et *g*) of equation
$$P_2(X,Y)=Y^4-2Y^3+Y^2-3X^2Y+2X^4$$
are different since one is algebraicallly reducible, and the other not.

The study of the singular point (0,0) shows another difference : following branches of the curves (using parametrisations given by rational Puiseux expansions) through the point the two branches cross each other in case of C_1 (this is clear since one branch is coming from an ellipse and the second from the other), but do not cross in the case of C_2 because their rational Puiseux expansions are respectively
$(x=t, y=2t^2+ \ldots..)$
and $(x=t, y=t^2+ \ldots.)$.

6) Global properties of curves: topology and analytical components.

a) Topology

Let us consider the following problem: given an equation $P(X,Y) = 0$, describe the topological type of $Z(P)$.

We know that $Z(P)$ has at most a finite number of isolated points, and a finite number of branching points so that in each branching point there is a finite even number of half-branches.
The goal of the algorithm is to draw a planar graph composed of a finite union of points and open intervals such that there exist an homeomorphism of the plane mapping it to $Z(P)$.

Cylindrical algebraic decomposition provides a partition of the X-axis in a finite union of A_i that are (homeomorphic to) open intervals or points such that
above each A_i (homeomorphic to) an open interval the curve is composed of a finite number of pieces homeomorphic to open intervals
above each A_i reduced to a point the curve is composed of a finite number of points.

It may happen that two curves with the same cylindrical decomposition have two different topologies as we can see in the examples of curve C_3 with equation $P_3(X,Y)=X-(Y-1)^2)(X+(Y+1)^2)$, union of two parabolas (*figures l*) and of curve C_4 with equation $(X^2+(Y-1)^2)(Y-X)(Y+X)$, union of two lines and an isolated point (*figure m*):

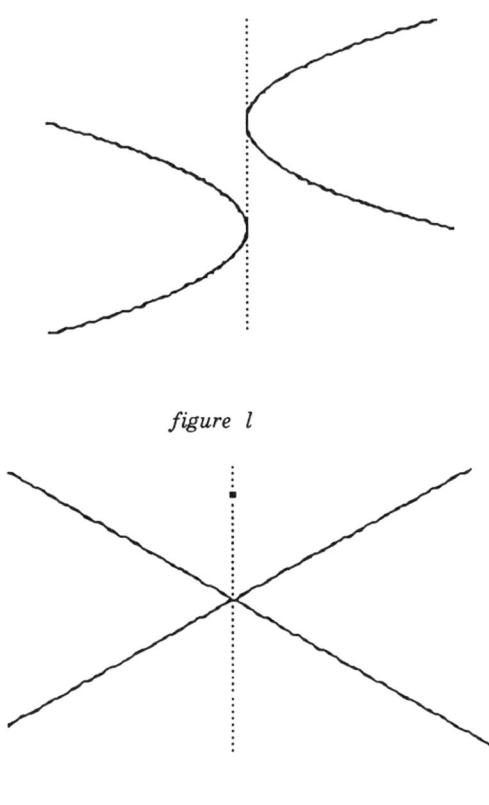

figure l

figure m

which have both the following cylindrical decomposition:

$A_1 \qquad A_2 \qquad A_3$

In order to get the complete topology it is sufficient to determine adjacency relations between pieces of cylindrical decomposition homeomorphic to open intervals and points.

In the case when the curve has at most a singular point above each point of the X-axis, it is enough to know the multiplicity of the root y of polynomial $P(x,Y)$ to determine adjacency relations ([G T], [R]).

In the other cases, multiplicity of roots is not enough and one can use again Thom's lemma [R].

Let us take the example of curve C_3: the two roots of $P_3(0,Y)$ (1 et -1 !) are roots of multiplicity two. They are not separated by $P_3'_Y(0,Y)$ which is zero for these two roots, neither by $P_3''_Y(0,Y)$ which is strictly positive for these two roots. They are separated by $P_3^{(3)}_Y(0,Y)=Y$ which is strictly positive for one and strictly negative for the other. If x is just at the right of 0, óne can compute (using Sylvester lemma) that Y is positive for the two half-branches above A_3. If x is just at the left of 0, one can compute (using Sylvester lemma) that Y is negative for the two half-branches above A_1. This is enough to determine adjacency relations and compute topology of C_3.

So again we have to make exact computations with real algebraic numbers (computation of multiplicity for example).

b) Analytic components

A real algebraic equation is sometimes analytically reducible, that is its equation is the product of two analytic equations, even if it is algebraically irreducible. It is the case when the curve has several connected components as the cubic defined by Y^2-X^3+X : each connected component has an analytic equation. There are other cases of curves algebraically irreducible and analytically reducible as curve defined by $Y^2-X^2-X^4$ which has two analytic components (*figure n*).

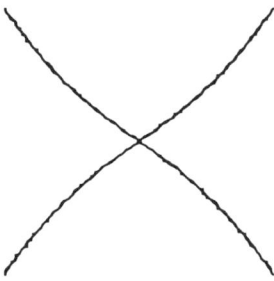

figure n

Using rational Puiseux expansions and formal computation with inequalities it is possible to follow analytic branches at branching points and hence to describe the global analytic structure of the curve [Cu P 3R].

In example C_1 and C_2 the algorithm in section Topology will give the following graph (the same for the two curves) *figure o*).

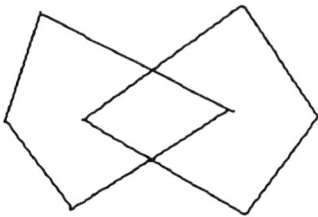

figure o

The algorithm in section Analytic components will give : for curve C_1 two analytic components (*figure p*) and for curve C_2 just one, described in the sense of arrows (*figure q*).

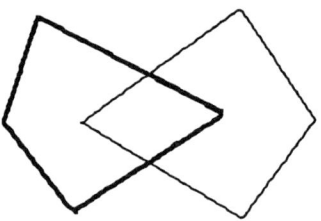

figure p

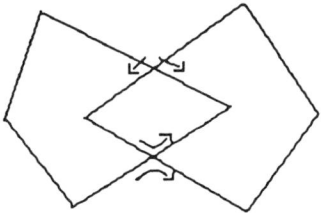

figure q

7) Some non real algebraic geometry

As we have noticed in point 4, real numbers are not very easy to use in symbolic computations.

On the other hand several geometric questions on other fields (complex numbers, finite fields) have been developed and implemented. A remarkable example is that of formal integration. In order to compute the integral of algebraic functions computation on complex algebraic curves are necessary. Some of the geometric problems raised are difficult and need sometimes computations on curves with points in finite fields [T].

Integration algorithms implemented in Reduce or Scratchpad II include very powerful algorithms of geometry on curves, even if this does not appear at all to the user.

8) Conclusions:
Relations between symbolic and numerical computations

We think that the preceeding developments illustrate the following general ideas:

- it is quite well known that using floating point arithmetic it is impossible to decide whether a number is zero or not: it is just equal to zero up to a given precision,

- it is nevertheless possible to make exact algebraic computations (for example decide if a number is zero or not) if the inputs are exactly known and if the length of the data structure is not limited in advance (which is possible using lists),

- getting an exact answer may be expensive ,

- singular situations where numerical methods have precision problems are perhaps the situations where computer algebra techniques are the most powerful and useful (multiplicity of roots, study of branching points).

Bibliography

[B C R] J. Bochnak, M. Coste, M.-F. Roy: *Géométrie algébrique réelle*. Ergebnisse der Mathematik vol. 12. Berlin Heidelberg New York: Springer (1986).

[Co] G. Collins: *Quantifier elimination for real closed fields by cylindrical algebraic decomposition*. Lecture Notes in Computer Science, vol. 33, 134-183.

[C] M. Coste: Effective semi-algebraic geometry (1988). This volume.

[C R] M. Coste, M.-F. Roy :*Thom's lemma, the coding of real algebraic numbers and the topology of semi-algebraic sets*. J. Symbolic Computation 5 (1988) 121-129.

[Cu P 3R] F. Cucker, L. M. Pardo, M. Raimondo, T. Recio, M.-F. Roy: *On the computation of the local and global analytic branches of real algebraic curves*. To appear in AAECC Minorca (1987).

[D S T] J. Davenport, Y. Siret, E. Tournier : *Calcul formel*. Paris: Masson (1986).

[D-D] C. Dicrescenzo, D. Duval: *Computation with algebraic numbers- the D5 system*. Preprint (1987)

[D] D. Duval: *Rational puiseux expansions.*Preprint (1987).

[G T] P. Gianni, C. Traverso: *Shape determination for real curves and surfaces*. Ann. Univ. Ferrara vol. XXIX 87-109 (1983).

[G O R] L. Gonzalez, M. Ollazabal, M.-F. Roy : *Real algebraic numbers by Thom's lemma: an implementation*. Preprint (1988).

[R] M.-F. Roy: *Computation of the topology of a real algebraic curve*. To appear in Sevilla's congress on Computational Topology and Geometry (1987).

[T] B. Trager: *Integration of algebraic functions.*Thèse M.I.T. (1984).

[W] R. Walker: *Algebraic curves*. Berlin Heidelberg NewYork: Springer (1978).

PLACEMENT OF POLYGONS

Jean-Jacques RISLER

Analyse Numérique , Université P. et M. Curie ,
4 place Jussieu , 75252   Paris CEDEX 05
and
DMI , Ecole Normale Supérieure ,
45 Rue d'Ulm , 75005 Paris.

§1   INTRODUCTION.

In the paper $[A-V]_1$, Francis Avnaim and Jean-Daniel Boissonnat give an algorithm which computes the set of translations for the disjoint placement of one or two polygonal regions inside a third one E. For the placement of three polygons , their algorithm works only when E is a parallelogram.

In this note , I will explain how , in the frame of "c.a.d." (cylindric algebraic decomposition) , or more precisely in the frame of the "mover's problem" one can give a general solution for the placement of any number of polygons inside a polygonal region E , with the same order of complexity as in $[A-B]_1$ for the case of one or two polygonal regions.

I will also give some indication for the case of placement of one polygon under translation and rotation , however without describing any precise algorithm.

This shows that the Mover's problem's point of view is interesting for the case of placement of polygonal regions under translation (this is the piecewise linear case of the mover's problem) , but also the piecewise linear case should give a simple model for the general mover's problem , because in the linear case the elimination procedure (i.e. the computation of resultants and discriminants) becomes trivial ,and some original approach , as utilisation of convexity for instance , is possible.

§2 CASE OF ONE POLYGON.

Let us fix some notations: let I be a polygonal region with n vertices, E a polygonal region with m vertices; if A is a polygonal region, $A \subset \mathbb{R}^n$, $\underline{A}$ will denote the exterior of A (so A and $\underline{A}$ are closed sets). Let $T \simeq \mathbb{R}^2$ be the set of translations of the plane, π: $T \times \mathbb{R}^2 \to T$ the natural projection, h: $T \times \mathbb{R}^2 \to \mathbb{R}^2$ defined by $h(t,x) = t(x) = t + x$; one have so the following diagram:

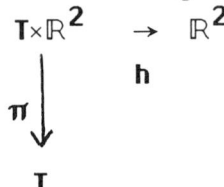

Lemma 1: Let U be the set of translations t such tat $t(I) \subset \underline{E}$; then, if $A = h^{-1}(\underline{E}) \cap (T \times I)$, one has $U = \underline{\pi(\underline{A})}$.

Proof: Immediate.

The method to find U is then to determine $h^{-1}(\underline{E})$ (immediate if one knows $\underline{E}$), $A = h^{-1}(\underline{E}) \cap (T \times I)$, $\pi(A)$, and finally $\underline{\pi(\underline{A})}$.

1) Let us describe an algorithm directly based on the c.a.d.'s method (as in Collins's work, or as in [Schw-Sh]$_2$).

Let us assume that E and I are described by the equations of their edges: $f_1, \ldots, f_m$ for E, $g_1, \ldots, g_n$ for I (this gives a presentation of E and I as union of convex regions). A is then defined as a semi-algebraic set of $T \times \mathbb{R}^2 \simeq \mathbb{R}^4$ (with coordinates (t,x)), with the help of the linear forms $f_1(t+x), \ldots, f_m(t+x), g_1(x), \ldots, g_n(x)$.

a) Let $x = (x_1, x_2)$. One eliminates x_1 beetween all the pairs of equations ($f_i = 0$, $g_j = 0$); this gives $N = (m+n)(m+n-1)/2$ linear equations in (t, x_2).

b) One eliminates x_2 beetween all the pairs of these N equations; this gives $P = N(N-1)/2$ linear equations in t.

c) These P equations give a partition of $T \simeq \mathbb{R}^2$ in $P(P+1)/2 + 1$ regions (at least in the generic case).

d) For each of these regions, one has to make a

test to know if it is (or not) in $\pi(A)$ (by construction , each of these regions is entirely in $\pi(A)$ or in $\underline{\pi(A)}$) .

For that , let R be one of these regions and $t_0 \in R$; $\pi^{-1}(t_0)$ is then a plane , partitioned in $(m+n)(m+n+1)/2$ regions by the linear equations $f_i(t_0,x) = 0$, $g_j(x) = 0$; for each of these regions , one tests if it is in A or not (by construction , each region is entirely in A or $\underline{A}$, so one test by region is enough).

e) One finds U as the union of the regions of T which are not in $\pi(A)$.

2) **Estimation of the complexity**.

a) The cost of this step is $O(m+n)^2$.
b) The cost of this step is $O(m+n)^4$.
c) The cost of this step (i.e. find a point in each region) is $O(m+n)^8$.
d) One has now $O(m+n)^8$ regions (in the plane T) ; for each of these regions , on has to make $O(m+n)^2$ tests : the cost of this step is $O(m+n)^{10}$.

The cost of all the others operations is surely of order less than this $O(m+n)^{10}$, so the complexity of this algorithm is $O(m+n)^{10}$, very far from the complexity $O(m^2n^2\log mn)$ of $[A-B]_1$ and of the next section.

3) **Amelioration of the algorithm**.

a) One decomposes I and $\underline{E}$ in union of convex parts; this can be done in time $O(m+n)\log mn$, $\underline{E}$ being the union of $O(m)$ convex parts , and I the union of $O(n)$ convex parts (cf. $[A-B]_1$).

b) Let I_1 be a convex part of I , E_1 a convex part of $\underline{E}$; $A_1 = h^{-1}(E_1) \cap (T \times I_1)$ is then a convex polyhedra with at most $n+m$ faces of dimension 3 and mn vertices.

Each one of these vertices is of the form (t,y_1) , where y_1 is a vertice of I_1 , and where $t = y_1 - e_1$, e_1 being a vertice of E_1.

The projection of these vertices on T is immediate to obtain , and $\pi(A_1)$ is obtained as the convex hull of these points, in time $O(mn\log mn)$.

c) One has so found $\pi(A)$ in time $O(m^2n^2\log mn)$, and $\pi(A)$ is obtained as the union of $O(mn)$ convex regions. One finds now $\pi(A)$ in time $O(mn\log mn)$ (cf. $[A-B]_1$) , and the total complexity of the algorithm is $O(m^2n^2\log mn)$, as in $[A-B]_1$.

§2 CASE OF SEVERAL POLYGONS.

The above method can be generalised and is (in theory) able to solve the placement problem under translation of any number of polygonal regions.

Let us , for simplicity , take three polygonal regions I_1 , I_2 , I_3 , to be placed inside the polygon E.

Let , as above , $h_i : T_i \times \mathbb{R}^2 \to \mathbb{R}^2$ $(1 \leq i \leq 3)$ be defined as $h_i(t,x) = t+x$, where $T_i \simeq \mathbb{R}^2$ is the space of translations applicated to I_i.

Set $A_i = h_i^{-1}(E \cap (T_i \times I_i))$, $B_i = \pi_i(A_i)$, π_i being the first projection : $T_i \times \mathbb{R}^2 \to T_i$. $B_1 \times B_2 \times B_3$ is then the set of translations (t_1,t_2,t_3) such that $t_i(I_i) \subset E$ $(1 \leq i \leq 3)$.

Let $C_1 \subset T_1 \times T_2 \times T_3$ be the set of translations (t_1,t_2,t_3) such that $t_2(I_2) \cap t_3(I_3) = \emptyset$; C_2 and C_3 are defined in a simular way (by circular permutation of the indices).

On has clearly:

Lemma 2 : The set U of wanted translations (i.e. translations which place I_1 , I_2 and I_3 in E in a disjoint way) is equal to $B_1 \times B_2 \times B_3 \setminus (C_1 \cup C_2 \cup C_3)$.

Let us now determine C_1 for instance.

One has the following diagram:

$$\begin{array}{ccc} T_2 \times I_2 \times T_3 \times I_3 & \xrightarrow{\mu} & \mathbb{R}^2 \times \mathbb{R}^2 \\ {\scriptstyle \pi}\downarrow & & \\ T_2 \times T_3 & & \end{array}$$

where π is the natural projection , and $\mu(t_2,x_2,t_3,x_3) = (t_2+x_2,t_3+x_3)$.

Lemma 3 : With the above notations, let $\Delta \subset \mathbb{R}^2 \times \mathbb{R}^2$ be the diagonal, $C'_1 = \underline{\pi(\mu^{-1}(\Delta))}$; then $C_1 = T_1 \times C'_1$.

Proof: Immediate.

I will not make a precise analysis of the complexity of this algorithm.

§3 SOME REMARKS ABOUT THE PLACEMENT PROBLEM UNDER TRANSLATIONS AND ROTATIONS.

The same method as above works for the general case of placement under translations and rotations (for any number of polygons).

The problem is now that the equations are not linear any more, and so that the projection operation becomes complicated.

Let us look at the case of one polygon I to be placed in E.

Let $R \simeq \mathbb{R}^2$ with coordinates (u,v), $T \simeq \mathbb{R}^2$ as above, $\psi : R \times T \times \mathbb{R}^2 \to \mathbb{R}^2$ defined by $\psi((u,v),t,x) = t+y$, with $x = (x_1, x_2)$ and $y = (ux_1 - vx_2, vx_1 + ux_2)$; if $\pi : R \times T \times \mathbb{R}^2 \to T \times \mathbb{R}^2$ is the natural projection, a method for the determination of the set of placements of I in E is:

a) Determine $B = \underline{\pi(A)}$, where $B = \psi^{-1}(E) \cap (R \times T \times I)$.

b) Intersect B with the cylinder $u^2 + v^2 = 1$.

The method generally employed for a) ([Schw-Sh]$_1$ or [A-B]$_2$) is to find the set of placements with <u>contacts</u> (of the boundaries of I and E). In our context, this means that we compute the restriction of A to the boundary of $\psi^{-1}(E) \cap (R \times T \times I)$ before doing the projection. This is achieved by taking the equations $f_i(x_1, x_2) = 0$ of I under the form $x_1 = s_i(x_2)$ for instance, and to report x_1 in the equations $g_j(t_1 + ux_1 - vx_2, t_2 + vx_1 ux_2) = 0$ which define the boundary of $\psi^{-1}(E)$.

We have now to eliminate x_2 beetween all the pairs ($g_j = 0$, $g_k = 0$), which give equations $l_k(t,u,v) = 0$ of degree ≤ 2 in u,v,t.

The intersection with the cylinder $C : u^2 + v^2 = 1$ gives then surfaces of degree ≤ 4 contained in $R \times T \simeq \mathbb{R}^4$.

BIBLIOGRAPHY

[A-B]$_1$ Avnaim F. , Boissonnat J-D. : Simultaneous containment of several polygons , 3rd ACM Symp. on Comp. Geometry , Waterloo (1987).

[A-B]$_2$ Avnaim F. , Boissonnat J-D. : Polygon placement under translation and rotation , Proc. STACS (1988).

[Schw-Sh]$_1$ Schwartz J. , Sharir M. : On the piano mover's problem 1 , Comm. Pure and Appl. Math. , 36, 345-398 (1983).

[Schw-Sh]$_2$ Schwartz J. , Sharir M. : On the piano mover's problem 2 , Adv. Appl. Math. 4 , 298-351 (1983).

The algorithm by Schwartz, Sharir and Collins on the piano mover's problem

Joël MARCHAND
Laboratoire L.I.T.P. / GRECO de Calcul Formel
Université Pierre et Marie Curie PARIS VI
4 Place Jussieu 75252 PARIS CEDEX 05 (FRANCE)
e-mail: jma@frunip11.bitnet

Abstract

A presentation of the quantifier elimination problem, and other problems connected to it, such as the "piano mover's problem", is followed by the state of the art on the question. Afterwards are explained Schwartz', Sharir's and Collins' algorithm, based on the cylindrical algebraic decomposition, and two applications to bring out its possibilities and limitations. The tracks and views of development in this research domain, which is a useful tool for algebraic geometry, are lastly indicated. Its limitation, due to a very high cost, is an actual obstacle, which is to be removed, for its use for many classes of concrete geometric and algebraic problems.

1. Presentation and motivation of the problem

The matter is the following problem:

Given $f_1,...,f_k$ polynomials with real coefficients, write the conditions on these coefficients so that a formula of the first order predicate calculus, with atomic formulas formed with equalities and inequalities between the f_i's, has a solution.

Michel Coste, in his article "Géométrie semi-algébrique effective" [DRR], explains in detail how many other geometric and algebraic problems are connected to this general problem. We will just enumerate them here, without specifying their causality relations.

The piano mover's problem or motion planning

We want to know if an object can move from a position to another in a real geometric space with obstacles, called walls and defined by semi-algebraic sets. This is formulated in terms of equalities and inequalities on the polynomials which define the object and the obstacles. We must then know if the two positions of the object belong to the same connected composant of the space.

The immediate application of this general problem is naturally robotic and automatic determination of the path of a robot, from the knowledge of the space in which it stands and of the object it moves. It is in fact a static computation a priori, which supposes that the space is constant in time.

Computation of intersections of algebraic surfaces and curves

We want to know the algebraic equations of the intersections of algebraic surfaces and curves given in a real plane or space. It is equivalent to searching variables values for which the polynomials which define these curves are null.

The application is CAD, which has a constant need for this kind of calculation. The idea is to replace the actual approximate numerical resolution by a correct formal solution. We would then no longer be dependent on a given numerical precision (sometimes small in the neighbourhood of singularities) and we could for example refine the display of an area at the will of the user, while keeping the precision of the plotting.

Resolution of systems of polynomial equalities and inequalities

These systems come from various domains. So examples have appeared from automatic (pole assignment by output feedback, a project in progress in CERT at Toulouse (France)) [CM], from automatic parallelisation (computation of the set of the points with integer coordinates in a n-polyedra, MASI laboratory at Paris VI University (France)) and from the term-rewriting systems (test on the sign of a polynomial formed according to a rewriting rule, CRIN laboratory at Nancy (France)) [BC].

Computation of the topology of an algebraic surface or curve

We want to answer the questions coming from algebraic geometry. Given a surface or a curve, what are its connected componants, its zeroes, or the signs of the associated polynomial function in any point of the space [R] ?

It is actually a backward problem, because this kind of question is solved by the knowledge of the **cylindrical algebraic decomposition** of the given surface or curve. This consists in knowing a decomposition of the real plane or space in disjoint cells, homeomorphic to a point, a line, a plan or the whole space, defined as a semi-algebraic set, on which the sign of the polynomial associated to the surface or the curve is constant.

The method, extended to a family with several variables polynomials, is in fact the only known algorithm for the complete resolution of the initial quantifier elimination problem and all the problems, which have been set out above. It is indeed easy to verify that all these previous problems are solved, as soon as we know all the topology of the given space. It is enough then to look at the cells, which verify the conditions assigned at the beginning.

2. State of the art

2.1 Arnon-Collins-Mac Callum's works

At the end of the 70's and at the beginning of the 80's, an American team around Arnon (Purdue University, Indiana), Collins and Mac Callum (University of Wisconsin - Madison, Wisconsin) elaborated and formalized the quantifier elimination problem and its resolution by the calculation of the cylindrical algebraic decomposition of a family of integral polynomials [Co] [ACM1] [ACM2] [ACM3] [MC] [Ar1]. Their works have brought a theoretical and algorithmical answer to the quantifier elimination. Today it is the principal and even the only global solution to all these problems of algebraic geometry.

They have in addition developed a language, ALDES, and then, thanks to it, a system, SAC2, doing all the classical manipulations in computer algebra.

The ALgorithm DEScription language has been designed to write computer algebra programs in a form, close to the one used by Knuth in his algorithm description [Lo]. It is a simple language, which is an arrangement between Pascal and Fortran on the one hand, and Lisp on the other hand. It borrows from the first ones a close syntax, the classical control structures and the imperative aspect. From Lisp it adopts the basic data, which are atoms (integers) and lists, but it has no type constructor.

The implementation of this language is double. At first it was made in Fortran IV with care of a large portability. A translator translates the language in an equivalent Fortran program, which is then compiled and executed.

In a second time, Lars Langemyr (a Swedish researcher) has made a translator of ALDES in Common-Lisp [Lan]. The semantics of the two languages being very near each other, this translation is more natural than in Fortran, because it allows to use again the management of memory, lists and integers from Lisp. Besides we now have a language working in an interactive Lisp environment.

From this language the computer algebra system SAC2 (Symbolic and Algebraic Computations) has been developed. It is made of a library of more than 700 functions. These are organized in modules around a same topic, modules which hierarchically depend one on the other. Thus they extend the basic language and implement elementary manipulations for arithmetical and polynomial calculation.

Collins' team has then implemented in the ALDES/SAC2 environment his quantifier elimination algorithm. Today this is the only software available in public domain, capable of solving a priori the formerly explained problems.

However, the theoretical complexity of the algorithm (doubly exponential in the number of the variables), the size of the handled objets, and the constant computation with algebraic numbers are such that this program is not very efficient. As we shall see later, it can only solve elementary examples. It doesn't implement many feasible theoretical optimisations and simplifications. It solves a complete general problem without taking into account the specific characteristics of some problems, which only need a partial solution.

2.2 Other works

After these works, James Davenport (professor at Bath University, Great-Britain) [Da1], the algebraic geometry team at Rennes University (France) around Marie-Françoise Coste-Roy and Michel Coste) [Pa] [CR] and researchers at Paris (France) like Jean-Jacques Risler [Ri], Daniel Lazard and Marc Giusti have brought various theoretical improvements and many ideas around this subject.

We must also remark that the piano mover's problem has been studied from other points of view, more specifically geometrical, in the case of the plane and obstacles with polygonal contours.

Sharir and Sifroni have thus proposed an explicit algorithm when the object is a rod [SSi]. The principle then is to join a wall situated near the initial position and to slip along the walls up to the goal. The critical curves related to the allowable positions of the rod with regard to the walls must be computed so that we might build the connexity graph. The complexity of this algorithm is in $O(n^5)$ (where n is the number of walls). An implementation has been made in Le-Lisp by François Ollivier (Ecole Polytechnique, France) [Ol]. He has noted a real saving of time with respect to the general algorithm, but has few hopes of implementing the generalisation proposed by Schwartz and Sharir, when the object is a polygon and even a polyedra [SSh] [ASSh].

Besides a more intuitive approach is formalized by the Voronoi diagrams [ODY]. The point is to try to move between the obstacles remaining at equal distance of the nearest ones. An explicit algorithm exists in the case of a disc and can be generalized to the case of a polygon [ODSY].

3. The cylindrical algebraic decomposition algorithm

This paragraph only wants to give a general idea of the algorithm. The reader is invited to refer to the article by Michel Coste "Géométrie semi-algébrique effective" [DRR]. He exposes in details the necessary mathematical tools and the way the algorithm works.

Definition

Let R be a real closed field, A a finite family of polynomials (p_i) in n variables with coefficients on R. A cylindrical algebraic decomposition (C.A.D) of R^n with respect to A is a decomposition of the space in disjoint cells which are homeomorphic to a sub-space R^j (j<=n), and on which all p_i have a constant sign (positive, negative or null).

Remarks

The decomposition is said to be:
- *algebraic*, because the cells are defined as semi-algebraic sets, by a finite number of polynomial disjunctions and conjunctions, called the associated formula of the cell.
- *cylindrical*, because it is built in a recursive way from a decomposition of R^{n-1} with respect to a family of polynomials obtained by projection from the p_i.

Notations

$deg(P)$ = degree of P with respect to its main variable
$res(P,Q)$ = resultant of P and Q with respect to their main variable.
$discr(P)$ = discriminant of P with respect to its main variable.
$ldcf(P)$ = leading coefficient of P.
$red(P) = red^1(P)$ = P minus its leading term.
$red^i(P) = red(red^{i-1}(P))$.

$PSC(P,Q)$ = the set of the non-zero principal subresultant coefficients of P and Q (cf [BCL]).

Principle

The decomposition is computed by successive projections of the family A on $R^{n-1}, \ldots, R$. We then obtain a C.A.D over R by isolating the zeroes of the polynomials in one variable x_1. Next we construct a cylinder over each cell with sections and sectors, which are separated by the zeroes of the polynomials of R^2. They are obtained by the projection and considered as polynomials in one variable x_2 (the value of x_1 is equal to a point of the cell on the basis of the cylinder). We iterate this method up to R^n and obtain the C.A.D associated to A.

In fact the projection is chosen to assure the climb: over each cell of R^i, the considerated polynomials in x_{i+1} have a constant number of zeroes. Thus we can represent a cell by its index (a i-tuple corresponding to its numeration at each climb), an example point of R^i and if necessary a formula built on the sign of the polynomials at this point.

Algorithm

(0) $A_n = A$

(1) Computation of a square-free basis of A_n noted B_n, of cardinal m.

(2) Computation of A_{n-1} by projection of B_n :

$$\text{PROJ}_1 = \bigcup_{1 \leq i \leq m \ \& \ G_i \in R_i} \{ \{ \text{ldcf}(G_i) \} \ ; \ \text{PSC}(G_i, G_i') \}$$

$$\text{PROJ}_2 = \bigcup_{1 \leq i < j \leq m \ \& \ G_i \in R_i \ \& \ G_j \in R_j} \{ \text{PSC}(G_i, G_j) \}$$

$$A_{n-1} = \text{PROJ}_1 \cup \text{PROJ}_2$$

where, for $i = 1,...,m$ and P_i of B_n,

$$R_i = RED(P_i) = \{ red^k(P_i) \mid 0 <= k <= deg(P_i) \ \& \ red^k(P_i) \neq 0 \}$$

(3) Iteration of (1) and (2) to obtain B_1, set of polynomials in one variable.

(4) Isolation of r zeroes of B_1 by Sturm's method (cf [BCL]) and formation of the C.A.D of R with r 0-cells (the zeroes) and $r+1$ 1-cells (the intervals between the cells), numbered from 1 to $2r+1$.

(5) For each cell, computation of the associated cylinder R^2. For the cell i represented by a, we have the polynomials of B_2 in the variable x_2 for $x_1 = a$. We compute other Sturm's sequences of these polynomials to isolate their t zeroes. We obtain t sections et $t+1$ sectors, numbered from $(i,1)$ to $(i,2t+1)$. Thus we obtain the C.A.D of R^2 with respect to B_2.

(6) Iteration of (5) to obtain the C.A.D of R^n with respect to A_n.

Example

On the unit circle of R^2, the algorithm unfolds as follows:

- $B_2 = A_2 = \{ x^2 + y^2 - 1 \}$
- $A_1 = \{ -4x^2 + 4, 1 \}$
- $B_1 = \{ x - 1, x + 1 \}$
- 2 zeroes (-1 and 1) and 5 cells on R.
- 13 cells on R^2 (2 of dimension 0, 6 of dimension 1, and 5 of dimension 2).

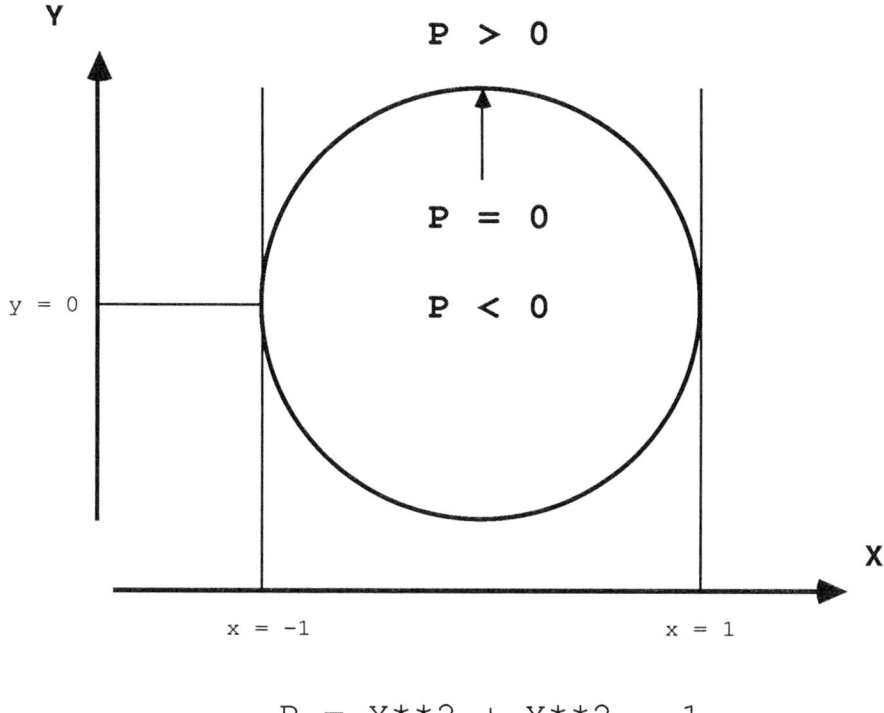

$$P = X**2 + Y**2 - 1$$

Remarks

The coefficients of the initial polynomials are rational in practice. The zeroes of the polynomials could be irrational. So from the construction of the C.A.D. in dimension one, the polynomials are viewed as having algebraic coefficients. This is very expensive in computation time and memory space. The representation used for these algebraic numbers is in fact a couple with the minimal polynomial and a framework of rationals isolating the zeroes of the polynomials. In case of double algebraic points during the successive climbs we must calculate the primitive element to represent the two numbers by the same polynomial.

If we want to compute a quantifier elimination, we join each cell to its own formula. In this way we can look at the set of these formulas and compare them with the initial equations and inequations.

In an improved version of Collins' team software, a computation is made to determine the adjacencies between cells. Two cells are adjacent if they belong to neighbouring cylinders and if they touch each other. This computation is correctly made in dimension 2 [ACM2], but partially in dimension 3 [ACM3]. Conjugate zeroes can also be taken into account and the cells gathered in clusters. This undoubtedly reduces the computation time.

We can see that the phase of successive projections is less expensive than the phase of climbs. This comes from the double exponential growth of the number of cells with the number of variables and from calculation with algebraic numbers.

4. Work on two quantifier elimination problems

4.1 The rod in the corridor

The first problem, explained by James Davenport, is the study of the possibility for a rod to turn in a corridor with a right angle [Da2]. This problem depends of course on the length of the rod and can be showed as a quantifier elimination: is there for a given length a position of the rod, for which both extremities are in the two branches of the corridor ?

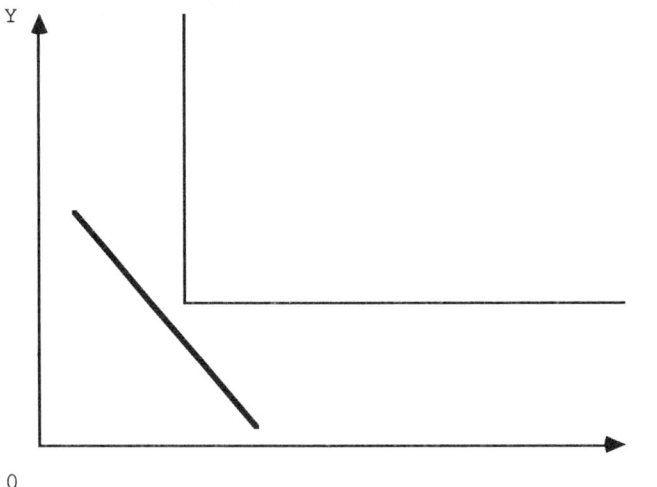

James Davenport had viewed the problem with the four variables describing the four coordinates of the two ends of the rod. After applying Collins' algorithm by hand (under Reduce), he had failed, because the polynomials were too long. He had to stop after one CPU-hour (on a DEC 20) and the computation of more than 300 zeroes of the univariate polynomials, which were obtained after three successive projections. Seeing that the projection phase is, in practice, insignificant in the total cost of the computation, this approach seems impossible.

On the other hand the problem could be expressed in algebraic terms with only three variables: the coordinates of an end and the tangent of the half angle between the x-axis and the rod.

Thus Collins' software answered - after seven and a half CPU-hours in Fortran on a Vax 11/780 and after three CPU-hours in Lucid Common-Lisp on a Sun 3/50 - that the problem had no solution for a length three times the width of the corridor.

It is probably the first automatic resolution of this problem, which is however mathematically very simple, by such a method. Finally we must note that the problem, in which the length of the rod is a fourth variable (to the purpose of obtaining the limit value of $2\sqrt{2}$), is totally unpracticable.

4.2 The four degree polynomial

The second example is the study of the necessary conditions on the coefficients for a generic univariate four degree polynomial ($x^4 + p\,x^2 + q\,x + r$) to be positive or strictly positive (without zero). This problem has been mathematically resolved by Daniel Lazard [Laz]. The result is that the discriminant of the polynomial must be positive and two simple conditions on p et r must be satisfied.

Collins' standard algorithm gives a list of clauses, which after a final work appear to be equivalent to the conditions found by hand. The study has here centered on the choice of the order of the variables, which have much effect on the faisability and on the cost of the computation, and on the logic deduction which is needed to simplify the clauses given by the program.

Arnon has also worked on this example: he has improved the basic program to obtain a condensed solution [Ar2].

4.3 Remarks

These examples and others have permitted to establish that the way of writing the problem has a great effect on the efficiency of the software and even on the termination of the computation. No systematic rule appears, but some remarks must be made:

- **the number of variables** must be minimal: most cases with two variables are practicable, some with three and almost any with four. The first example is a good illustration.

- **the order of variables** must if possible be such as to enable us to eliminate the most "involved" variables first. Thus the most efficient order - in the second example - is x,r,p and q. Not to begin by x makes the computation impossible.

- **the coordinates marks** must be chosen to minimize the number of zeroes of the respective polynomials. In the case of the rod in the corridor, taking the origin of the coordinates marks at the external corner of the corridor makes the polynomials very simple.

- **the computation with algebraic numbers** is very expensive. An example coming from the term-rewriting systems with four variables without algebraic zero has been solved after a few CPU minutes, whereas examples with three variables and a same scale-order number of cells require much more than ten CPU hours or even never ends.

All this agrees to say that the user must first simplify the problem. Unluckily we don't know what is a well-written problem and the user cannot be sure a priori of the practical ending of the computation.

5. Projects of developments

The computer science laboratories at Paris VI University have now an IBM 4381 with the experimental computer algebra language Scratchpad II [Je] & [G]. This powerful, modular, strong typed language seems ideal to implement complex computer algebra algorithms. The handling of real algebraic objects, the control of types, and the polymorphism of functions beyond the representation of the objects allow the programmer to think on a high level of abstraction and not to worry about low level problems, which are difficult to solve in ALDES/SAC2. The first attempts augur well for its efficiency.

So it is possible to rewrite Collins' algorithm. We can build an opened software with many possible improvements, because the actual software and ALDES/SAC2 are definitely not adapted to the necessary modifications.

Among the provided developments we can note:

- **the adequation of the projection method** to the problem which is considered. Following the cases the requested projection set is more or less great [MC] [R]. So it is enough - up to dimension three - to consider only the polynomials P_i and not the sets of their successive *red(P_i)* reductums. On the other hand the complete knowledge of the topology and the adjacencies can require us to consider the derivates of the polynomials P_i.

- **a new cell management.** For now we must compute all the cells of the considered space. It is preferable to restrict the computation to certain domains of the space. For example in the problem of the rod in the corridor, we can only consider points with positive coordinates, i.e. the quarter of the plane. In the same order the motion planning could be made easier, if it were split into several steps. The space could be first of all cut into sub-spaces and we could indicate points or areas of the space, where the robot would necessarily go along. The algorithm would also be applicated successively in these sub-spaces with minor complexity.

Note then that the movement of an object in a space of dimension n can be made only through areas of dimension n or $n-1$. So it is needless for the piano mover's problem to compute the cells of co-dimension less than $n-1$. This would reduce their number a lot from dimension 3. We can hope a saving all the greater as these little cells are the most "expensive" to calculate. They are those which introduce the most algebraic numbers and therefore computations of primitive elements, which are very costly.

For all this a more individualized and flexible management of these cells and of the work space is necessary, in opposition to current general and strict management (due to the numbering system).

- **the use of the adjacencies between cells.** The adjacency relation between adjacent cells is now partially used. Its computation could be improved, for the complete resolution of the piano mover's problem (calculation of the connected components and the way to move inside a same component).

A grouping of these adjacent cells could be made possible in the successive climb phases of the algorithm. For instance in the example of the unit circle, we would like to consider that the eight exterior cells of the circle form only one. This would reduce the work, if we had to do another climb phase.

- **the improvement of algebraic numbers management.** The great cost of computation with these numbers can probably be reduced by the use of techniques like that of the D5 system, which is developed in TIM3 laboratory at Grenoble (France) [DDDD]. D5 is in fact based on the recursive representation of algebraic numbers (probably less expensive than the use of primitive elements) and doesn't distinguish between the zeroes of a polynomial when it is not necessary.

Another way to isolate the zeroes of a polynomial, and then to represent these algebraic numbers, is also worth studying. It is based on Thom's lemma and consists in representing the zeroes, not by their values (in the form of an interval), but by the sign of the polynomial and its derivates in these zeroes.

Finally it seems absurd to consider a priori all the zeroes algebraic. We must consider them as rationals, when they are.

6 Conclusion

We must note that this project involves various aspects such as learning and using a new and yet experimental computer algebra language, computing on algebraic numbers, modelisation and management of large and complex geometric data, and possible interfaces with graphic softwares in order to display geometric objects.

Moreover, an important research is in progress on motion planning with numerical techniques. It also comes up against important difficulties due to the complexity of the problem and the impossibility to treat correctly the natural singularities. So it is interesting to study whether this algebraic approach brings an improvement or a help in some cases.

This quantifier elimination algorithm brings also a solution, today theoretical, to many badly solved practical problems (for example, the calculation of the intersection of two surfaces). This generality increases the usefulness of this algebraic and geometric tool.

The work which has already been accomplished leads us to think that the further improvements will bring an exponential gain on algorithm efficiency. In fact we have seen, with simple examples, that the theoretical complexity is not necessarily an insurmountable obstacle in practice. We can reasonably hope that the further improvements will have a notable influence on the practical complexity and will permit to divide the computation times by an important factor.

Thus a research effort must continue for many years to integrate and test all these ideas expressed by the community, in order to provide adapted and performed answers to the questions which are raised.

Bibliography

General books

[AB] D.S. Arnon, B. Buchberger, *Algorithms in Real Algebraic Geometry*, Reprinted from the Journal of Symbolic Computation, vol 5, number 1-2, Academic Press, 1988.

[DST] J. Davenport, Y. Siret, E. Tournier, *Calcul formel: systèmes et algorithmes de manipulations algébriques*, Masson, Paris, 1987.

[BCR] J. Bochnak, M. Coste, M-F. Roy, *Géométrie algébrique réelle*, Springer-Verlag, Berlin, 1987.

[BCL] B. Buchberger, G.E. Collins, R. Loos, *Computer algebra. Symbolic and Algebraic Computation*, Springer-Verlag, New-York, 1982.

[DRR] J. Della-Dora, M-F. Roy, J-J. Risler, *Calcul Formel et outils algébriques pour la modélisation géométrique*, Actes du Séminaire, CNRS, 1988.

Arnon-Collins-Mc Callum's works

[Co] G.E. Collins, *Quantifier elimination for real closed fields by Cylindrical Algebraic Decomposition*, Lecture Notes in Comp. Sc., vol 33, pp 134-183, Springer-Verlag, 1975.

[ACM1] D.S. Arnon, G.E. Collins, S. Mc Callum, *Cylindrical Algebraic Decomposition I: the basic algorithm*, SIAM J. Computations, vol 13, pp 865-877, 1984.

[ACM2] D.S. Arnon, G.E. Collins, S. Mc Callum, *Cylindrical Algebraic Decomposition II: an adjacency algorithm for the plane*, SIAM J. Computations, vol 13, pp 878-889, 1984.

[MC] S. Mc Callum, *An improved projection operator for Cylindrical Algebraic Decomposition*, Ph.D. Thesis, University of Wisconsin-Madison, 1984.

[ACM3] D.S. Arnon, G.E. Collins, S. Mc Callum, *An adjacency algorithm for Cylindrical Algebraic Decompositions of three-dimensional space*, Proc. EUROCAL 85 Vol. 2, Lecture Notes in Computer Science, vol 204, pp 246-261, Springer-Verlag, 1985.

[Ar1] D.S. Arnon, *A cluster-based cylindrical algebraic decomposition algorithm,* Proc. EUROCAL 85 Vol. 2, Lecture Notes in Computer Science, vol 204, pp 262-269, Springer-Verlag, 1985.

Works of the team of Rennes University

[Pa] A. Paugam, *Comparaison entre trois algorithmes d'élimination des quantificateurs sur les corps réels clos*, Thèse de troisième cycle, Université de Rennes I, 1986.

[CR] M. Coste, M-F. Coste-Roy, *Thom's lemma, the coding of real algebraic numbers and the computation of the topology of semi-algebraic sets*, Journal of Symbolic Computation, vol 5, number 1-2, pp 121-129, Academic Press, 1988.

[R] M-F. Roy, *Computation of the topology of a real curve*, IRMAR, Université de Rennes I.

[RS] M-F. Roy, Aviva Szpirglas, *Complexity of the computation on real algebraic numbers*, IRMAR, Université de Rennes I, to appear.

Others works on the piano mover's problem and the quantifier elimination

[Da1] J. Davenport, *Computer algebra for Cylindrical Algebraic Decomposition*, TRITA-NA-8511, NADA, KTH, Stockholm, 1985.

[Ri] J-J. Risler, *About the piano mover's problem*, Université de Paris VI, to appear.

[Me] M. Merle, *Le problème du déménageur*, Ecole Polytechnique, Palaiseau, 1987.

[SSh] J.T. Schwartz, M. Sharir, *On the piano mover's problem, I, II, III, V*, Comm. Pure Appl. Math, vol 36 & 37, 1983 & 1984 et Adv. Appl. math., vol 4, 1983.

[ASSh] E. Ariel-Sheffi, M. Sharir, *On the piano mover's problem, IV*, Comm. Pure Appl. Math, vol 37, 1984.

[SSi] M. Sharir, S. Sifroni, *A new efficient motion planning algorithm for a rod in polygonal space*, ACM, 1986.

[Ol] F. Ollivier, *Implantation d'un algorithme de M. Sharir et S. Sifroni pour le déplacement d'une tige dans un espace polygonal*, Université de Paris VI, Rapport de DEA-LAP, 1987.

[ODY] O'Dunlaing, Yap, *The Voronoï method for motion planning: I. The case of a disc*, Robotics Research, Technical Reports, Courant Institute, N.Y. University, 1983.

[ODSY] O'Dunlaing, M. Sharir, Yap, *Generalized Voronoï diagrams for motion a ladder: I, II*, Robotics Research, Technical Reports, Courant Institute, N.Y. University, 1984.

Applications and computer implementations

[Lo] R. Loos, *The Algorithmic Description Language: ALDES*, Universität 675, Kaiserslautern.

[Lan] L. Langemyr, *Converting SAC2-Code to Lisp*, SIGSAM Bulletin, vol 20.4, pp 11-13, 1986.

[Je] R.D. Jenks, *A primer: 11 keys for a new SCRATCHPAD*, Proc. EUROSAM 84, Lecture Notes in Computer Science, vol 174, pp 123-147, Springer-Verlag, 1984.

[G] M-C. Gontard, *Une première approche de la sémantique du langage de calcul formel Scratchpad II*, Université de Paris VI, Rapport de DEA-LAP, 1987.

[DDDD] J. Della-Dora, C. Dicrescenzo, D. Duval, *About a new method for computing in algebraic number fields*, Proc. EUROCAL 85 Vol.2, Lecture Notes in Computer Science, vol 204, pp 289-290, Springer-Verlag, 1985.

[Da2] J. Davenport, *A "piano mover's" problem*, SIGSAM Bulletin, vol 20.1&2, pp 15-17, 1986.

[Laz] D. Lazard, *Quantifier elimination: optimal solutions for two classical examples,* Journal of Symbolic Computation, vol 5, number 1-2, pp 261-266, Academic Press, 1988.

[Ar2] D.S. Arnon, *On mechanical quantifier elimination for elementary algebra and geometry: solution of a nontrivial problem*, Proc. EUROCAL 85 Vol. 2, Lecture Notes in Computer Science, vol 204, pp 270-271, Springer-Verlag, 1985.

[CM] C. Champetier, J-F. Magni, *Commande nodale par retour statique de sortie: méthode algébrique*, Rapport C.E.R.T, Toulouse, 1987.

[BC] A. Ben Cherifa, *Preuves de terminaison des systèmes de réécriture. Un outil fondé sur les interprétations polynomiales*, Thèse de doctorat, Université de Nancy I, 1986.

A practical exact motion planning algorithm for polygonal objects amidst polygonal obstacles

Francis Avnaim, Jean Daniel Boissonnat and Bernard Faverjon

INRIA
Centre de Sophia-Antipolis
Route des lucioles
06565 Valbonne

Abstract:

Let I be a 2-dimensional polygonal rigid object (with m edges) moving amidst polygonal obstacles E (with n edges) and let P_{init} and P_{end} be two free placements of I, where the interior of I does not intersect E. We investigate here the problem of finding a continuous motion of I from P_{init} to P_{end}, such that during this motion the interior of I does not intersect E, or to establish that no such motion exists. This problem is an instance of the well known "Piano Movers' Problem". We have shown in [2] that it is possible to compute an exact description of free space in time $O(m^3 n^3 \log(mn))$. We show in this paper that, using this description, a motion can be found in time $O(m^3 n^3)$. The actual complexity of our algorithm in many practical situations is much smaller. In particular, for the so called situation of local bounded complexity often encontered in robotics, the complexity of computing free space is $O(n \log n)$ and the complexity of planning a motion is $O(n)$. The method has been implemented and experimental results are discussed.

1   Introduction

We investigate here the problem of planning the motion of a 2-dimensional polygonal rigid simply connected object I (with m edges) which is free to move by translation and rotation amidst polygonal obstacles E (with n edges). More specifically, given two placements P_{init} and P_{end}, we want either to find a continuous motion connecting P_{init} and P_{end} during which the interior of I avoids collision with E, or else establish that no such motion exists.

This problem has been attacked from both practical and theoretical points of view in the literature. The only solutions that have been implemented use heuristic approaches: in [8] an approximation of the set of free placements (the so called free space) is computed and then a path is searched. Due to the approximation done, we are not guaranteed to always find a path if one exists. In [3] the motion of I is restricted to be a sequence of pure translational and pure rotational movements [3]. Here again, we are not guaranteed to always find a path if one exists. The first exact solution to the problem is due to Schwartz and Sharir [11]. The complexity of their algorithm is $O(n^5)$ in the case that I is a line segment (a *ladder*). This algorithm is rather involved and, according to the authors themselves, several technical delicate issues are ignored. Recently, Kedem and Sharir [7] improved on this result in the case that I is convex. The complexity of their algorithm is $O(mn\lambda(mn)\log(mn))$, where $\lambda(q)$ is an almost linear function of q. The improvement is obtained by only computing a judicious subset of the set of free placements, namely a set of edges on its boundary. This is sufficient to find a path if one exists but the computed path may be very unsatisfactory in practice since, during the motion, I keeps at least two points in contact with E.

In this paper, we propose a rather simple method to handle the general case where I may be non convex. Our algorithm makes use of the exact description of the boundary of free-space obtained in [2]. This description consists of a set of "faces" that are portions of ruled surfaces. There are $O(m^3 n^3)$ faces in the worst case. The faces and their adjacency relationships can be computed in time $O(m^3 n^3 \log(mn))$ in the worst case and much faster in many practical situations [2]. For example, when the number of vertices of I is small (and thus can be considered as a constant) and when, in addition, the edges of E are not concentrated near each other compared with the diameter of I –a situation refered to as of local bounded complexity [12]– the boundary of free space is computed in $O(n \log n)$ time. In this paper, we show that this description can be used to solve the motion planning problem. The time complexity is in the worst case proportional to the number of faces which is $O(m^3 n^3)$. For situations of local bounded complexity, the time complexity is only $O(n)$. An important aspect of our method is that the motions produced are searched in a 2-dimensional variety thus allowing to locally modify and optimize them. The method has been implemented and experimental results are discussed.

2 Representation of the boundary of free-space

Let I be a 2-dimensional polygonal rigid simply connected object with m edges moving amidst a set E of polygonal obstacles with a total number of n edges. We assume that the complement of E is a bounded polygonal region (this is always possible by enclosing obstacles in a sufficiently large rectangle) (see Fig. 1).

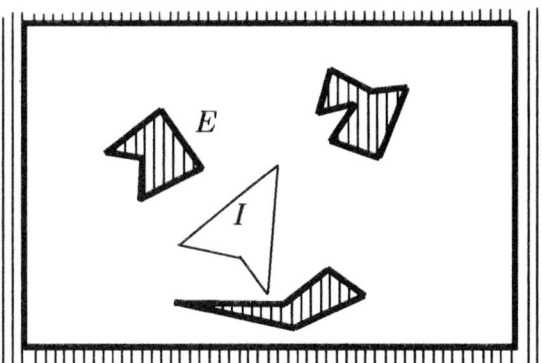

Figure 1: Polygon I and obstacles E

Free-space FP is defined to be the closure of the subset of $[0, 2\pi[\times R^2$ consisting of all placements $(\theta, \vec{u})$ satisfying $R_\theta \circ T_{\vec{u}}(I) \cap E = \emptyset$ where R_θ denotes rotation with center at the origin and angle θ and $T_{\vec{u}}$ denotes translation by vector $\vec{u}$. A placement of FP is called a *free placement*. An algorithm has been described in [2] which computes a complete description of the boundary BFP of FP. This description consists of a disjoint union of "faces" and of their adjacency relationships. Each face is a portion of a ruled surface generated by a line segment $P(\theta)Q(\theta)$ when θ ranges in a subinterval $[\theta_{min}, \theta_{max}]$ of $[0, 2\pi[$ (see Fig. 2).

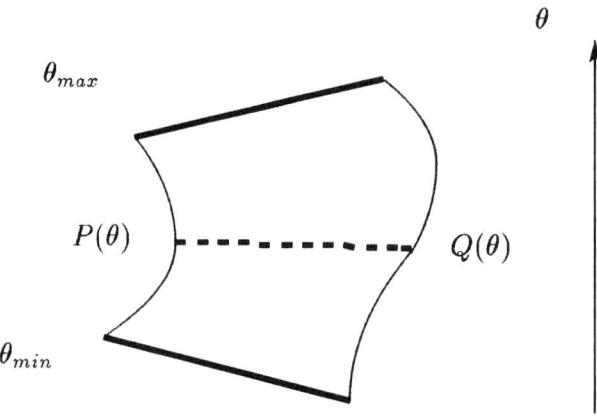

Figure 2: A face of BFP

Points $P(\theta)$ and $Q(\theta)$ are of type $(\theta, f(\theta), g(\theta))$ where $f(\theta)$ and $g(\theta)$ are analytic functions of θ. Each face is a set of placements involving a given contact between I and E (i.e., a given vertex of I in contact with a given edge of E or a given vertex of E in contact with a given edge of I). There are $O(m^3n^3)$ faces in the worst case.

Each face is bounded by at most four edges; two edges are (portions of) the curves $P(\theta)$ and $Q(\theta)$ for $\theta \in [\theta_{min}, \theta_{max}]$. The two others are the line segments $P(\theta_{min})Q(\theta_{min})$ and $P(\theta_{max})Q(\theta_{max})$, which may be reduced to points. Two faces are adjacent if they share an edge or a portion of an edge. In the sequel, we will represent BFP by a graph called the *boundary graph* of FP and denoted by $\mathcal{G}$. The nodes of $\mathcal{G}$ are the faces of BFP and its edges join adjacent faces. It is shown in [2] that the size of this graph is $O(m^3n^3)$ and that it can be computed in time $O(m^3n^3\log(mn))$ in the worst case.

In many practical situations, the number of faces composing the boundary of FP is much smaller than in the worst case. For example, when the number of vertices of I is small (and thus can be considered as a constant) and when, in addition, the edges of E are not concentrated near each other compared with the diameter of I –a typical situation in robotics refered to as a situation of local bounded complexity– the boundary of FP has $O(n)$ faces and can be computed in time $O(n\log n)$.

The intersection FP_{θ_0} between FP and the plane $\theta = \theta_0$ (for any fixed orientation θ_0) is a polygonal region which is the set of free placements when I can only move by translation with fixed orientation θ_0. It is shown in [1] that such a polygonal region has $O(m^2n^2)$ edges in the worst case and can be computed in $O(m^2n^2\log(mn))$ time. Moreover, any of the connected components of FP_{θ_0} has $O(mn\alpha(mn))$ edges where $\alpha(mn)$ is the functionnal inverse of Ackermann's function, and thus is extremely slowly growing [9]. It is plain to compute the boundary of FP_{θ_0} from graph $\mathcal{G}$ in time $O(m^3n^3)$ (a better result will be given in Section 4.3).

3 Computing a free path

Let $P_{init} = (\theta_{init}, X_{init}, Y_{init})$ and $P_{end} = (\theta_{end}, X_{end}, Y_{end})$ be two free placements of I. We want to find a continuous obstacles avoiding motion of I between P_{init} and P_{end}. This is equivalent to searching a curve inside FP joining P_{init} and P_{end}. Such a curve is called a *free path* from P_{init} to P_{end}. We successively study the three following instances of the problem (with increasing difficulty):

1. P_{init} and P_{end} belong to the same face of BFP.

2. P_{init} and P_{end} belong to the same connected component of BFP.

3. P_{init} and P_{end} are in general position.

Case 1: Let f be a face of BFP and A and B two points of f. As f is a ruled surface swept by a line segment $P(\theta)Q(\theta)$, A (resp., B) is completely defined by an orientation θ_A (resp., θ_B) and a real α_A (resp., α_B) such that $\overrightarrow{P(\theta_A)A} = \alpha_A \overrightarrow{P(\theta_A)Q(\theta_A)}$ (resp., $\overrightarrow{P(\theta_B)B} = \alpha_B \overrightarrow{P(\theta_B)Q(\theta_B)}$). It is plain to observe that the curve Γ defined by

$$\Gamma = \{M(\theta), \overrightarrow{P(\theta)M(\theta)} = (\alpha_A + (\alpha_B - \alpha_A)\frac{\theta - \theta_A}{\theta_B - \theta_A})\overrightarrow{P(\theta)Q(\theta)}, \theta \in [\theta_A, \theta_B]\}$$

passes through A and B and lies entirely inside face f (see Fig. 3).

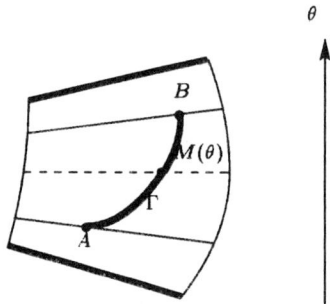

Figure 3: Curve Γ joining two points on a face of BFP

Thus Γ is a free path between A and B. Notice that if face f corresponds to contact C, this contact will be maintained all along Γ.

Case 2: When P_{init} and P_{end} belong to the same connected component of BFP (but not to the same face), we can search in the boundary graph of FP a sequence S of faces $f_1, \cdots, f_k$ such that f_1 contains P_{init}, f_k contains P_{end} and f_i is adjacent to f_{i+1} ($i = 1, \cdots, k-1$). Let us consider two faces adjacent in the sequence S, say f_i and f_{i+1}. We associate to f_i and f_{i+1} a point P_{ii+1} belonging to the two faces. If f_i and f_{i+1} are adjacent by a segment, P_{ii+1} is simply the middle of the segment. If they are adjacent by a portion of curve ranging from θ_1 to θ_2, P_{ii+1} is the point on this curve corresponding to $\frac{\theta_1+\theta_2}{2}$ (see Fig. 4).

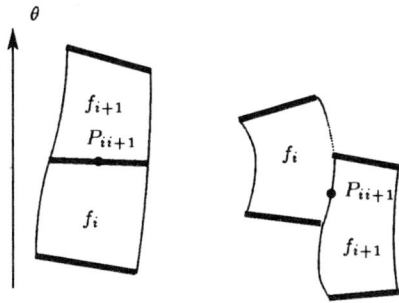

Figure 4: Common point of two adjacent faces

Thus the sequence S yields a sequence of points $P_{init}, P_{12}, \cdots, P_{k-1k}, P_{end}$ such that P_{init} and P_{12} belong to f_1, P_{i-1i}, P_{ii+1} belong to f_i (for $i = 2, \cdots, k-1$), P_k and P_{end} belong to f_k. As any two consecutive points in that sequence belong to the same face it is possible to compute a curve between them lying on that face (Case 1). Thus the concatenation of theses curves is a free path between P_{init} and P_{end}. This path is entirely contained in BFP and thus corresponds to a motion where I remains in contact with E. Clearly, the dominant step in the above procedure is the search of a sequence of faces of BFP. This search can be done in time proportional to the size of the graph, which is in turn proportional to number of faces of BFP (see Section 2).

Case 3: In that case, P_{init} and P_{end} belong to the interior of FP. Due to the fact that the complement of E is bounded (see Section 2), FP is also bounded. Let us denote by FP^{init} (resp., FP^{end}) the connected component of FP containing P_{init} (resp., P_{end}). The boundary of FP^{init} (resp., FP^{end}) has several connected components, one of them enclosing the others. We call it the *external boundary* of FP^{init} (resp., FP^{end}) and denote it by FP_{ext}^{init} (resp., FP_{ext}^{end}). There exists a free path from P_{init} to P_{end} iff P_{init} and P_{end} belong to the same connected component of FP, i.e., iff $FP^{init} = FP^{end}$. As in Section 2, let FP_θ be the set of free placements for a fixed value θ of the orientation of I. Let us call U^{init} (resp., U^{end}) the union of the boundary of FP^{init} and of the polygonal region $FP_{\theta_{init}}$ (resp., the union of the boundary of FP^{end} and the polygonal region $FP_{\theta_{end}}$). As FP^{init} (resp., FP^{end}) is bounded and connected, there exists a continuous path in U^{init} (resp., U^{end}) which joins point P_{init} (resp., P_{end}) and the external boundary of FP^{init} (resp., FP^{end}). Indeed, let D be any half line lying in $FP_{\theta_{init}}$ with P_{init} as its endpoint. Let $S = \{S_1 = P_{init}, S_2, \cdots, S_{k-1}, S_k\}$ be the sequence of the intersection points between D and the boundary of FP^{init}, sorted along D. Note that S_k belongs to FP_{ext}^{init}. The interior of any segment $S_i S_{i+1}$ is either inside FP^{init} or outside FP^{init}. Let $S_i S_{i+1}$ be a segment such that its interior is outside FP^{init}. As FP^{init} is connected, there exists a path α_{ii+1} contained in the boundary of FP^{init} joining S_i and S_{i+1}. Thus the concatenation C_{init} of $S_1 S_2, \alpha_{23}, S_3 S_4, \cdots, S_{k-3k-2}, \alpha_{k-2k-1}, S_{k-1} S_k$ is a continuous path joining P_{init} and S_k. Similar arguments show that there exists a continuous path C_{end} in U^{end} and a point $S'_{k'}$ belonging to FP_{ext}^{end} such that C_{end} joins P_{end} and $S'_{k'}$. If there exists a path in FP joining P_{init} and P_{end} then $FP^{init} = FP^{end}$, thus FP^{init} and FP^{end} have the same external boundary. Moreover, as this external boundary is connected, there exists a path C_{ext} contained in it that joins S_k and $S'_{k'}$. The concatenation of C_{init}, C_{ext} and C_{end} is a continuous path from P_{init} to P_{end} contained in the union of $BFP, FP_{\theta_{init}}$ and $FP_{\theta_{end}}$. Figure 5 illustrates such a construction (for clarity, D and D' have been taken to be coplanar).

In conclusion, if there exists a path in FP joining P_{init} and P_{end}, then there exists a path in the union of the boundary of $FP, FP_{\theta_{init}}$ and $FP_{\theta_{end}}$ joining P_{init} and P_{end}. Reciprocally, any path in the union of the boundary of $FP, FP_{\theta_{init}}$ and $FP_{\theta_{end}}$ joining P_{init} and P_{end} is clearly a path in FP joining P_{init} and P_{end}.

We can deduce from the above discussion a method to compute a free path joining P_{init} and P_{end}. First we triangulate $FP_{\theta_{init}}$ and $FP_{\theta_{end}}$. Let $T_{\theta_{init}}$ and $T_{\theta_{end}}$ be the adjacency graphs of the two triangulations. Each edge of the boundary of one of the two triangulations belongs to a face of BFP. Let t be a triangle of a triangulation having an edge e belonging to a face f of BFP. We create an adjacency relation between f and t. Doing so for all possible triangles, we merge the graphs $\mathcal{G}, T_{\theta_{init}}$

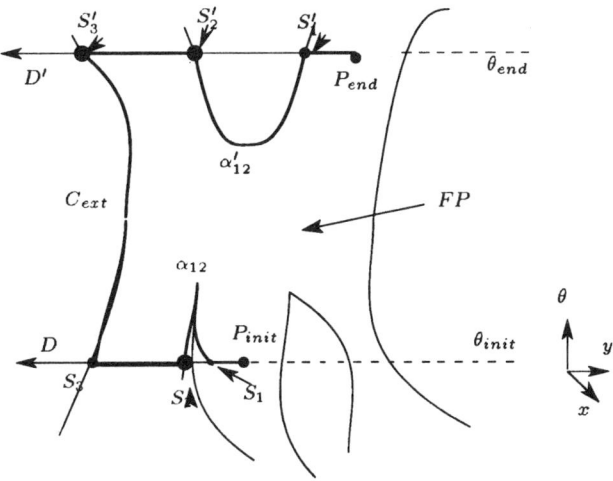

Figure 5: A path from P_{init} to P_{end}

and $T_{\theta_{end}}$ and achieve a new graph $\mathcal{G}^*$. As any triangle is a portion of a ruled surface (a plane) which accepts exactly the same description as a face of $\mathcal{G}$, the resulting graph $\mathcal{G}^*$ is a graph of faces such that P_{init} and P_{end} belong to faces of it. Moreover the existence of a free path from P_{init} to P_{end} is equivalent to the existence of a path in $\mathcal{G}^*$ from the triangular face containing P_{init} to the one containing P_{end}. Thus searching a free path when P_{init} and P_{end} are in general position reduces to Case 2 using graph $\mathcal{G}^*$.

We briefly describe the computation of a motion:

1. Deduce from graph $\mathcal{G}$ the polygonal region $FP_{\theta_{init}}$. Compute a triangulation of $FP_{\theta_{init}}$ yielding a graph T_{init} of triangular faces. Merge T_{init} and $\mathcal{G}$ by making an adjacency relation between a face t of T_{init} and a face f of $\mathcal{G}$ iff an edge of t belongs to f.

   The same is done for P_{end} yielding a final graph $\mathcal{G}^*$.

2. Search the two triangular faces f_{init} and f_{end} containing respectively the points P_{init} and P_{end}.

3. Search in $\mathcal{G}^*$ a sequence S of faces $f_1, \cdots, f_k$ such that $f_1 = f_{init}$, $f_k = f_{end}$ and f_i is adjacent to f_{i+1} ($i = 1, \cdots, k-1$). Compute the corresponding sequence of points $P_{init}, P_{12}, \cdots, P_{k-1k}, P_{end}$. If this search is unsuccessful, return "no path".

4. Compute the k curves $\Gamma_1, \cdots, \Gamma_k$ such that Γ_1 is a curve inside f_1 joining the points P_{init} and P_{12}, Γ_i is a curve inside f_i joining the points P_{i-1i} and P_{ii+1} ($i = 2, \cdots, k-1$) and Γ_k is a curve inside f_k joining the points P_{k-1k} and P_{end}.

5. Return the path obtained by concatening $\Gamma_1, \cdots, \Gamma_i, \cdots, \Gamma_k$.

Complexity analysis Let K_{init} (resp., K_{end}) be the number of edges composing the boundary of $FP_{\theta_{init}}$ (resp., $FP_{\theta_{end}}$) and F be the number of faces of BFP. Let K be the sum $K_{init} + K_{end}$. The complexity analysis is done in function of K and F.

1. $FP_{\theta_{init}}$ is computed in time $O(F)$ using graph $\mathcal{G}$. As $FP_{\theta_{init}}$ has at most $O(K_{init})$ edges, T_{init} can be computed in $O(K_{init} \log K_{init})$ time. Similarly, T_{end} can be computed in $O(K_{end} \log K_{end})$ time. Merging the graphs T_{init} and $\mathcal{G}$ takes $O(K_{init})$ time and merging the graphs T_{end} and $\mathcal{G}$ takes $O(K_{end})$ time. Therefore, the final graph $\mathcal{G}^*$ is computed in time $O(K_{init} \log K_{init} + K_{end} \log K_{end} + F) = O(K \log K + F)$.

2. The localisation of P_{init} and P_{end} in their respective triangulation takes $O(K)$ time since they are $O(K)$ triangles in each triangulation. The final graph $\mathcal{G}^*$ has $O(F)$ edges, thus a searching a path in $\mathcal{G}^*$ takes $O(F)$ time.

3. The computation of curves Γ_i takes $O(F)$ time.

We sum up the above results in the following proposition :

Proposition 1 *A free path between P_{init} and P_{end} can be computed in time $O(K \log K + F)$.*

In the worst case $K = O(m^2n^2)$ and $F = O(m^3n^3)$ thus the worst case complexity of the computation of a free path between P_{init} and P_{end} is $O(m^3n^3)$. For situations of local bounded complexity $K = F = O(n)$ thus the complexity of our motion planning algorithm is $O(n \log n)$.

4 Final remarks

4.1 Computing pseudo-optimal motions

Computing a free path from P_{init} to P_{end} has been reduced to a simple search in a graph. If the edges of these graphs are valuated by some real numbers, we can find a shortest path according to these values. The value associated to an edge joining two nodes can be, for example, the euclidean length of the corresponding curve. Depending on the choice of the unit respective lengths on the axis x, y and on axis θ, the resulting path will minimize preferably the translational movements or the rotational movements of I. This technique does not yield theoretic optimal paths but these paths appear to be reasonable in practice.

Searching such a pseudo optimal motion takes $O(M + N \log(N))$ where M is the number of edges and N is the number of nodes in the graph [6]. As $M = N = F$, a pseudo optimal motion can be found in time $O(F \log F + K \log K) = O(F \log F)$ which is $O(m^3n^3 \log(mn))$ in the worst case.

Furthermore, an important aspect of our method is that the produced motions can be locally modified and optimizied. Indeed, we have at our disposal a 2-dimensional variety (represented by graph $\mathcal{G}^*$). Although we have only used a finite set of curves in this variety (a 1-dimensional subvariety) to search a motion. Once a motion has been computed, we can locally improve this motion by relaxing the positions of the points $Pii + 1$ on their respective edges (see for analogous point of view [4].

4.2 Motion along a reference trajectory

In some applications, a point of reference on I is required to move along a given polygonal line while I may rotate around this point. Let k be the number of segments composing the polygonal line. In that case, we need only to compute the faces corresponding to the vertex-edge contacts involving the point of reference (which plays the same role as a vertex of I) and the k segments of the polygonal line. Refering to [2], it is easy to see that the computation of these faces takes $O(km^2n^2 \log mn)$ time and thus searching a free motion takes $O(km^2n^2)$ time.

4.3 Repeated queries for translational motion planning

Our description of FP can also be used to find repeated translational motions. This is done by constructing an appropriate data structure which allows efficient queries. To each face f of BFP corresponds an interval $[\theta_{min}, \theta_{max}]$ denoted by I_f. We denote by $\mathcal{I}$ the set of all the intervals corresponding to the faces of BFP.

We want to compute from boundary graph $\mathcal{G}$ of FP the boundary of the polygonal region FP_{θ_0}. Let K be the number of edges composing FP_{θ_0}. Any edge e belonging to the boundary of FP_{θ_0} belongs to exactly one face f of BFP. Moreover, the interval I_f contains orientation θ_0. Reciprocally, let f be any face in BFP generated by the segment $P(\theta)Q(\theta)$ when θ ranges in I_f, and assume that θ_0 belongs to I_f. It is clear that $P(\theta_0)Q(\theta_0)$ is an edge belonging to the boundary of FP_{θ_0}. Thus the set of edges composing the boundary of FP_{θ_0} is trivially deduced from the set L_{θ_0} of faces defined by:

$$L_{\theta_0} = \{f \in BFP, \theta_0 \in I_f\}$$

Note that L_{θ_0} has exactly K elements. We store the intervals of L in a segment tree [10]. This takes $O(F \log F)$ time as there are F intervals in L (note that this does not increase the time complexity of the computation of graph $\mathcal{G}$). Then using this segment tree, we compute in time $O(K + \log F)$ the set L_{θ_0} and thus the set of edges composing the boundary of FP_{θ_0}. To get a complete description of the boundary of FP_{θ_0}, it remains to compute the adjacency relationships between these edges. This is done in time $O(K \log K)$ as follows. Each endpoint of an edge of the boundary of FP_{θ_0} is labelled by a double-contact (see [2]). Thus, after sorting these edges with respect to their endpoint's labels –which takes $O(K \log K)$ time– we can easily compute in time $O(K)$ the adjacency relationships between the edges of the boundary of FP_{θ_0}. In conclusion, the boundary of FP_{θ_0} is computed in time $O(K \log K + \log F)$. Searching a translationnal motion between two points of FP_{θ_0} can then be done within the same time bound.

4.4 Implementation

The search of a motion has been implemented in C on a Sun workstation. Examples of free spaces and motions are shown in Figure 6a, 6b.

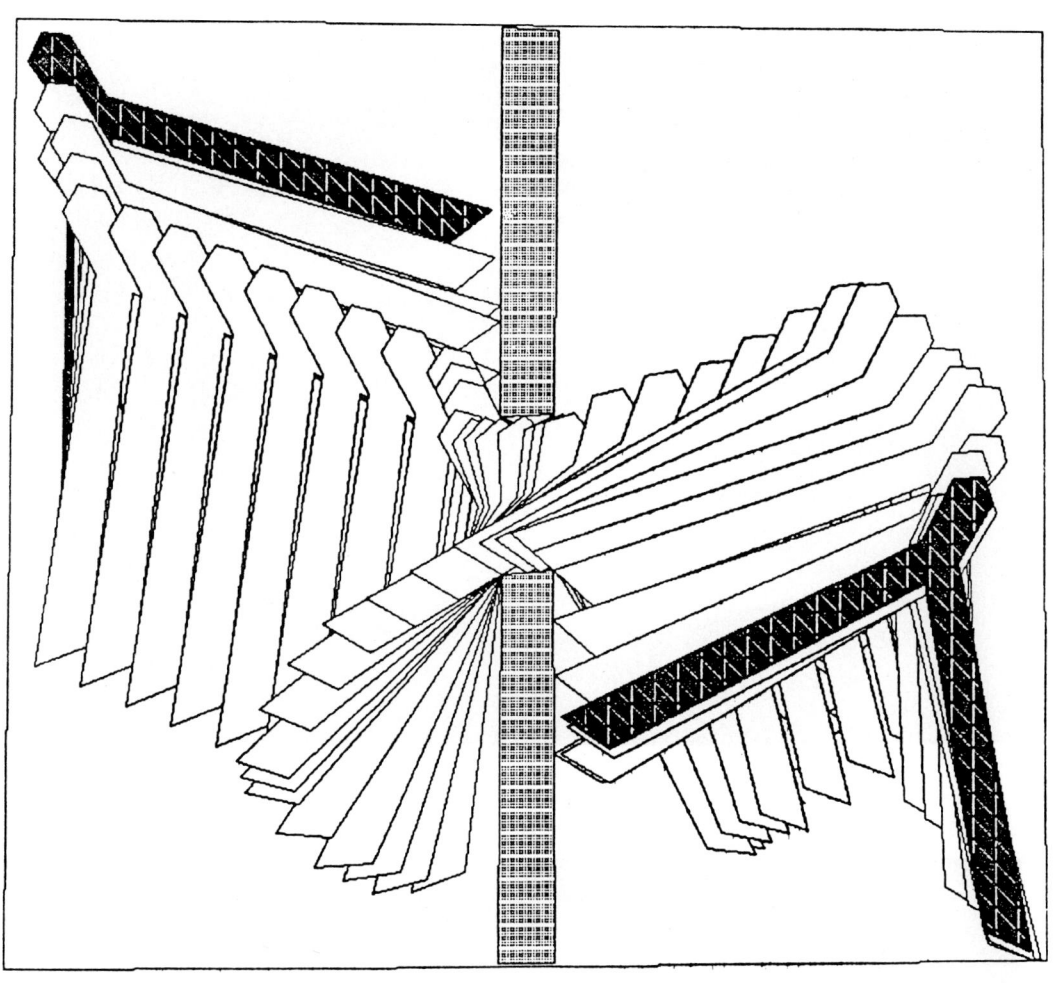

Figure 6a: Free space (perspective and top view), free motion

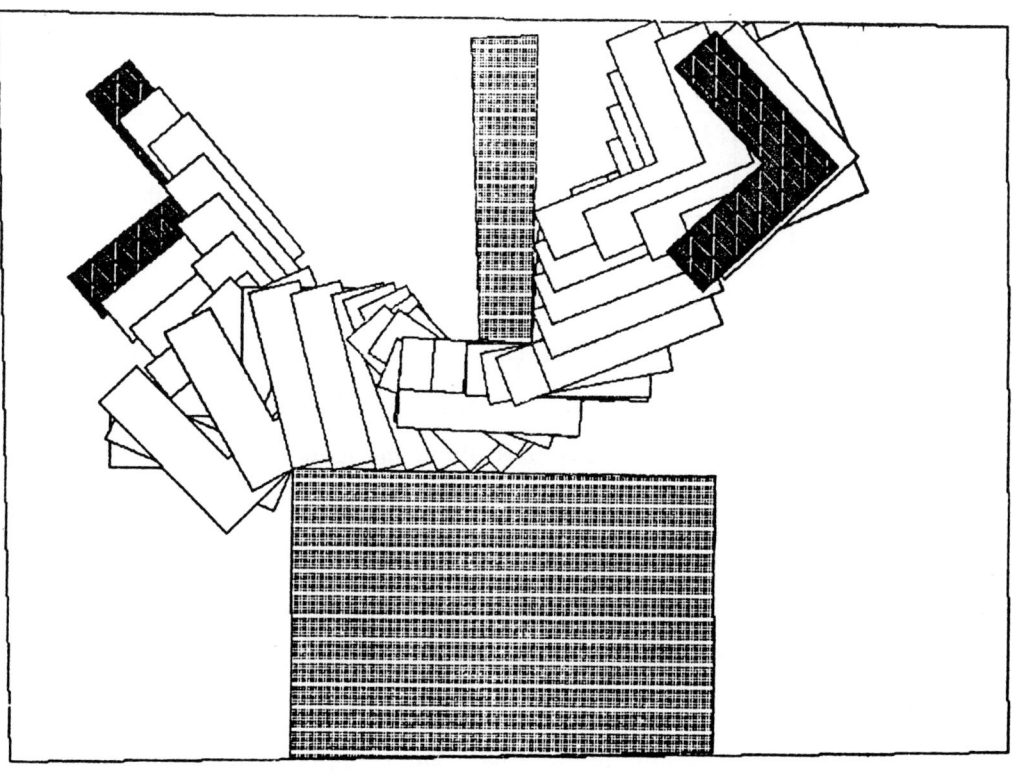

Figure 6b: Free space (perspective and top view), free motion

The table below sums up the experimental results:

Exp. results	I	E	Faces	Computing FP (s)	Computing a free path (s)
Fig. 6a	10	12	613	270	0.5
Fig. 6b	6	12	285	92	0.3

The program consists of approximatively 15000 lines of C, including the computation of FP and the computation of a motion. First experiments have shown that computing FP is the most costly part of our method. It can be considered as a pre-processing.

Compared to the traditional approaches discretizing FP, our method has the important advantage of being exact. This means that very "difficult" paths (see Fig. 6a, 6b for examples) can be found. Moreover, the program is rather simple and the computing times and the complexity of the description of FP (number of faces) compare favourably with others approaches. For example, to solve the problem shown in Figure 6b, we have stored a graph with 270 nodes while Faverjon's method [5] requiers to store a graph of 10000 cells. It seems impossible to solve the problem shown in Figure 6a with approximate or heuristique methods.

5 Conclusion

We have shown that, using a complete description of the boundary of free space FP(i.e., the set of all free placements for a polygonal object I (with m edges) which is free to translate and to rotate but not to intersect another polygonal object E), it is possible to compute a free motion for I between two free placements, if such a motion exists. We compute first and once for all the boundary of FP. This preprocessing takes $O(m^3 n^3 \log(mn))$ [2]. Then each free motion is computed in $O(m^3 n^3)$ time. As we get a complete exact description of the boundary of free space, it is possible to compute many different motions between two free placements, according to different criteria. Last but not least, the algorithm has been implemented and first experimental results are very hopeful. The immediate application of our algorithm is the motion planning for a planar mobile robot moving amidst polygonal obstacles.

References

[1] AVNAIM F., BOISSONNAT J.D., Simultaneous containment of several polygons, 3rd ACM Symp. on Computational Geometry, Waterloo (June 1987).

[2] AVNAIM F., BOISSONNAT J.D., Polygon placement under translation and rotation, LNCS N. 294 Springer–Verlag pp.322-333 (1988). To appear also in RAIRO Informatique théorique et applications 1988.

[3] BROOKS R.A., Solving the find-path problem by good representation of free space, IEEE Trans. on Systems, Man and Cybernetics, Vol. SMC-13 pp.190-197, (March-April 1983).

[4] CHANDERJIT BAJAJ, MOH T.T., Generalized unfoldings for shortest paths. The international journal of Robotics Research vol. 7 number 1 ISSN 0278-3649 MIT press (Feb. 1988).

[5] FAVERJON B., Obstacle avoidance using an octree. In Proceedings of IEEE Int. Conference on Robotics and Automation (March 1988).

[6] FREDMAN M., TARJAN R.E., Fibonacci heaps and their uses uin improved network optimization problems. In Proc. 25th IEEE FOCS, pp.338-346, (1984).

[7] KEDEM K., SHARIR M., An efficient motion planning algorithm for a convex polygonal object in 2-dimensional polygonal space, Tech. Rept. No 253, Comp. Sci. Dept., Courant Institute, (Oct. 1986).

[8] LOZANO-PEREZ T., BROOKS R.A., A subdivision algorithm in config uration space for findpath with rotation, IEEE Trans. on Systems, Man and Cybernetics, Vol. SMC-15 No 2, pp.224-233 (1985).

[9] POLLACK R., SHARIR M., SIFRONY S., Separating two simple polygons by a sequence of translations, Discrete and Computational Geometry 3:pp.123-136 (1988).

[10] PREPARATA F.P., SHAMOS M.I., Computational geometry: an introduction, Springer-Verlag, (1985).

[11] SCHWARTZ J.T., SHARIR M., On the piano movers' problem: I. The special case of rigid polygonal body moving amidst polygonal barriers, Comm. Pure Appl. Math., Vol. XXXVI, pp.345-398 (1983).

[12] SIFRONY S., SHARIR M., A New Efficient Motion Planning Algorithm for a Rod in Two-Dimensional Polygonal Space, Algorithmica 2, pp. 367-402 (1987).

Motion Planning for Manipulators in Complex Environments

B. Faverjon * and P. Tournassoud **

(*) INRIA Sophia-Antipolis, Av. Emile Hugues, 06565 Valbonne, France
(**) INRIA, Domaine de Voluceau, BP 105, 78153 Le Chesnay Cedex, France

1 Introduction

1.1 Why is Path-planning for Manipulators Difficult ?

The problem of motion-planning for manipulators is that of finding a path from an initial to a goal configuration, avoiding collisions with obstacles in the workspace. This ability is crucial in order to achieve *task-level programming*, this is, being allowed to describe a task in terms of geometric relationships between objects, rather than purely joint angle commands. We will attack in this paper the general problem of motion-planning for manipulators with prismatic and revolute joints (eventually co-operating robots) in a three-dimensional environment.

Searching for a path involves representing the geometry of the problem in the *Configuration Space* of the system, chosen among the spaces of independent parameters describing the position of all moving bodies (e.g., joint angles for a manipulator robot). Conceptually, planning motions for the robot is then reduced to planning motions for a point representing the configuration. Path-planning algorithms will be composed of two critical steps. The first one is the computation of the transform of obstacles in Configuration Space, this is, the set of configurations for which there is collision. The latter consists in structuring *Free Space*, the complement of the transform of obstacles in Configuration Space, so that the search for a path can be performed efficiently.

This construction is difficult, for two main reasons. First, we are faced with the difficulty of representing rotations in three-dimensional space, which makes it quite irksome to compute transforms of obstacles for kinematic chains with many revolute joints. The second reason is the inherent exponential complexity of the problem with the number of degrees of freedom of the system. Many instances of the general motion-planning problem are indeed provably *NP-hard* or *PSPACE-hard* with the number of degrees of freedom, even if there exist general polynomial algorithms to answer the problem with a fixed number of degrees of freedom (Schwartz and Sharir [SS83]).

The first point has found a solution with the idea of discretizing Configuration Space, giving approximate, conservative solutions. Answering the latter point will always be difficult. A tricky answer has been to reduce the number of degrees of freedom taken into account to describe the search space (e.g., Brooks [Bro83]). Another -treacherous- answer is to forget the requirement of good termination of the motion (reach the goal) and make the system incrementally move towards the goal using minimization technique. This second class of approaches, illustrated by the *Potential Field* method (Khatib [Kha86]), more inspired by control theory, has proved to be very powerful for describing a large family of robotic tasks.

1.2 Bases of our Approach

Our approach deeply relies on the ideas of Lozano-Pérez to represent the problem as a graph search in a discretized Configuration Space [Loz87]. The graph represents adjacency between elementary cells of safe configurations.

Other tools that we introduce are original. These are essentially hierarchical geometric models of fixed and moving solids, to accelerate the computation of critical distances by use of thresholding, general techniques to compute distances between solids and their variations when the configuration of the system changes, and structuration of Free Space in Configuration Space as a hierarchical grid (2^n-tree).

These tools are used to attack the problem under a number of angles.

Local Planning: A first order model of the variation of distances, mapped into Configuration Space, is used to build a simple local model of Free Space. This model is first used as a basis for minimization techniques by a local planner. As opposed to the standard *Potential Field* approach, the idea is to separate the realization of the task, described by the minimization of a set of measures of the problem, and eventually some geometric constraints, from the constraints of anti-collision between the robots and the environment. This makes it possible to respect accurately both the geometric constraints on the task, and anti-collision constraints that remain very close to the geometry of the problem.

Global Planning: Similar local models are used to build incrementally a representation of Free Space as a connectivity graph between free cells. The amount of memory required is important, as small cells are needed to describe the boundaries of the transforms of obstacles. Due to memory limitations and construction cost, this can only be acheived up to three or four degrees of freedom, for example, to represent motions of the arm of a manipulator (its first three links). For planning the path of a manipulator carrying a small load, we propose to have local and global techniques collaborate, local six-dimensional models of Free Space being used at both ends of the trajectory to generate *fine motions* close to the obstacles, and during *gross-motions* of the arm as heuristics to influence the search of a path for the first three links.

Mixed Approach: For systems with over three or four significant degrees of freedom (e.g., a six-joint manipulator with a bulky load), we propose to systematically decompose the motion-planning problem into two levels. We build a graph whose nodes represent relatively large cells of Configuration Space. Transitions between adjacent cells are weighted by the probability for the local planner to succeed in moving the system from one cell to the other. These probabilities are used by a minimum cost path finding algorithm to generate subgoals for the local planner. They are updated using Bayesian rules, from the results of the execution of trajectories planned locally. As we deal with the high complexity of graph searching only at the relevant level of the description, it is possible to apply this technique to robotic systems with more degrees of freedom.

2 Geometric Models

2.1 A Hierarchical Description of Solids

A key element in motion-planning algorithms is the geometric representation of objects they use. In our implementation solids are approximated by unions of simple convex volumes that we call *primitives*. These are 3D-rectangles, prisms, cylinders, truncated cones and spheres.

Each solid is represented by a hierarchical structure that we call an *Assembly Tree*. This is a binary tree with a primitive attached to each node. This primitive contains both primitives attached to its sons. The object is most precisely described by the unions of leaf primitives, but any cross-section of the tree provides a *conservative* approximation (that contains the object).

This hierarchical structure is computed automatically by the CAD system. When we assemble two objects, a new node is created with the smallest volume primitive containing both of them attached to it. Figure 1 illustrates this construction.

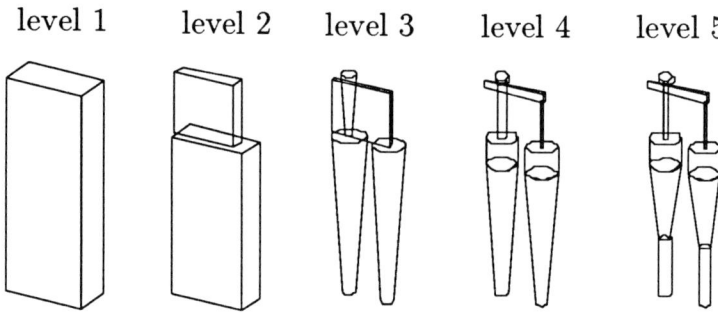

Figure 1: An Assembly Tree representing part of the environment. Any cross-section of the tree is an approximation containing the exact description.

A straightforward application is to test efficiently the intersection of two solids, starting at the roots and performing a balanced descent in both Assembly Trees.

2.2 A Hierarchical Description of Manipulators

We describe a robot by an open articulated chain of $n+1$ solid bodies $B^i, i = 0, \ldots, n$, each body being described by an Assembly Tree. The i-th joint, of parameter q^i, links bodies B^{i-1} and B^i. It is translational or revolute.

A n degree of freedom manipulator is represented by a pyramid with $n+1$ levels. Level k, $0 \leq k \leq n$, represents a virtual manipulator M^k with k degrees of freedom, the last links being replaced by their swept volumes when joints $k+1$ to n vary freely. More precisely, M^k is the union :

- of bodies $B^1, \ldots, B^k$,

- and of the Assembly Tree representing volumes $SV^{i,k}$ swept by bodies B^i, $i > k$, when $q^1, \ldots, q^k$ are fixed and $q^{k+1}, \ldots, q^i$ vary freely.

We call the root of this tree *Swept Volume* of level k, and denote it SV^k. Figure 2 illustrates this pyramidal structure, for the first three links of a vertical manipulator. Leaf primitives only are represented at each level. A node of this structure contains the node directly below it, and, if it is not a leaf or a body, two sons on the same level.

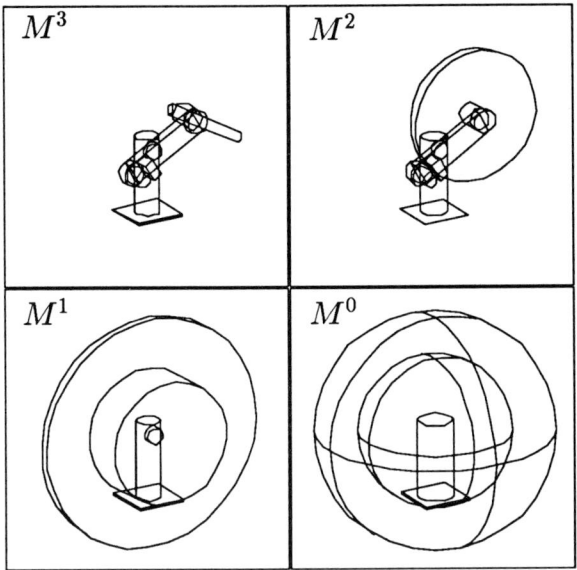

Figure 2: Hierarchical model of a 3 dof manipulator.

3 On Local Planning

In this section we propose a local, memory-less method to generate trajectories for one or more manipulators. If it has many features similar to the *Potential Field* method, it also makes use of different tools and exhibits different behaviours.

At each time increment, we will perform the following computations.

- From the geometric models of the robots and their environment, we compute the list of pairs of primitives that need to be separated : a pair consists in two primitives lying at less than a conservative influence distance σ, function of the maximum Cartesian displacement for points on these solids. One primitive is part of the description of a body of a manipulator, the other part of a fixed obstacle. Both of them also can be mobile primitives, either on the same kinematic chain, or on two different robots in a multi-robot application.

- From the list of interactions we compute a local view of obstacles as seen in the Configuration Space of the system, that translates the control of critical distances between pairs of interacting solids.

- Finally, we will use minimization techniques to get closer to the realization of the task, while remaining in the local, conservative model of Free Space we have built.

3.1 The Local Model of Free Space

3.1.1 Derivative of the distance between two Convex sets

In the sequel, S_1 and S_2 denote two convex solids in Cartesian space, x_1 and x_2 two points belonging respectively to S_1 and S_2, and n a unit vector. Notation $(u|v)$ refers to the inner product of vectors u an v.

The Euclidean Distance between S_1 and S_2 is equal to $d(S_1, S_2) = \min_{x_1 \in S_1, x_2 \in S_2} \|x_1 x_2\|$.

We now consider that S_1 and S_2 are moving solids. Their position and orientation are described by vectors of configuration parameters q_1 and q_2. An important property of the distance $d(S_1, S_2)$ between two non overlapping moving objects, is that it can be simply differentiated:

Proposition: *The time derivative of the distance between moving solids S_1 and S_2 writes:*

$$\dot{d} = (n|J_2\dot{q}_2 - J_1\dot{q}_1), \tag{1}$$

where

- *J_1 and J_2 are the Jacobian matrices for S_1 and S_2 calculated respectively at points x_1 and x_2 realizing the minimum distance,*

- *n is the unit vector on the line $x_1 x_2$.*

Writing $J = (-J_1 \; J_2)$, and $q = (q_1^t, q_2^t)^t$ the global configuration vector, equation 1 also writes:

$$\dot{d} = (n|J\dot{q}) = (J^t n|\dot{q}). \tag{2}$$

We call J the "relative Jacobian" at points x_1 and x_2.

3.1.2 Translating Constraints in C-Space

In this section we describe how to translate an interaction between two convex primitives into a constraint on the displacements of the manipulator in Configuration Space. Between a moving primitive and an obstacle, or between two moving primitives, we impose a constraint that we call *velocity damper* : it expresses that the distance between them must not decrease too fast when it is less than the influence distance σ. It is based on the following proposition :

Proposition: *There cannot be any collision if we impose :*

$$\dot{d} \geq -\xi \frac{d-\varepsilon}{\sigma-\varepsilon}, \qquad (3)$$

where ε is the security distance at which the robot must stop and ξ a positive coefficient for adjusting convergence speed.

We recall (equation 2) that $\dot{d} = (\mathbf{n}|\mathbf{J\dot{q}}) = (\mathbf{J^t n}|\dot{\mathbf{q}})$, where $\mathbf{n}$ is the unit vector on the line of minimum distance and $\mathbf{J}$ the relative Jacobian matrix at points $\mathbf{x}_1$ and $\mathbf{x}_2$ realizing the minimum. Thus inequality 3 can be translated into a simple linear constraint on configuration parameters increments $\delta\mathbf{q}$ between time t and $t+\delta t$, namely with $\hat{\mathbf{n}} = \mathbf{J^t n}$:

$$(\hat{\mathbf{n}}|\delta\mathbf{q}) \geq -\xi \frac{d-\varepsilon}{\sigma-\varepsilon} \delta t \qquad (4)$$

A nice property of these constraints is that they allow to go arbitrarily near from the limit $d = \varepsilon$ in a finite time, though the component of the velocity normal to the obstacle gets very small.

As illustrated by figure 3, combining those constraints and bounds on configuration parameter increments (standing for maximum joint velocities) results in building a local model of Free Space in Configuration Space. This simplified model is convex and conservative (included in the current connected component of Free Space). A positive point is that the generated constraints remain close from the original geometry of the problem. As the minimization we perform will be initiated with zero joint increments, which clearly respect the constraints, redundant constraints will not be examined in the process and will not have any influence on the result.

This local model of Free Space is used in next subsection as a basis for minimization techniques. It can have many other applications. In section 4, similar computations are used for incrementally building a global model of Free Space ; in section 5, for doing some naive geometric reasoning to accelerate the learning of safe displacements.

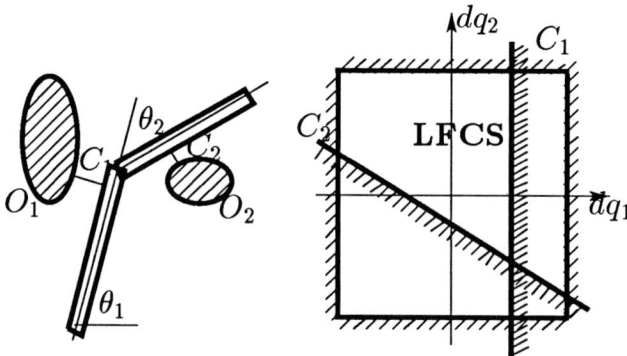

Figure 3: A local model of Free Space in Configuration Space : the simple case of a planar manipulator with two revolute joints.

3.2 Description of a Task

The realization of a task is expressed by the minimization of a set of measures on the system, and eventually the respect of some geometric constraints. Anti-collision constraints are expressed separately by means of the local model of Free Space introduced above.

We will use homogeneous configuration parameters $\mathbf{q} = (q^1, \ldots, q^n)$, such that $q^i = \theta^i/\omega^i$ for $i \in \{1, \ldots, n\}$, where ω^i is the maximum value of the i-th configuration parameter derivative. Hence, with the norm $\|\mathbf{q}\| = \max_{i=1,\ldots,n} |q^i|$ for the vector of configuration parameters, we always have $\|\dot{\mathbf{q}}\| \leq 1$.

A task is described by the control of a set of measures of the problem, a vector $\mathcal{T}(\mathbf{q})$ of a p-dimensional space. We again write its norm $\|\mathcal{T}(\mathbf{q})\|$ for simplicity. We suppose that we want to reach a state $\mathbf{q}$ such that $\mathcal{T}(\mathbf{q}) = 0$. Such a state will be called a solution of the problem. For example, if the task simply consists in reaching a goal configuration $\mathbf{q}_g$, we write $\mathcal{T}(\mathbf{q}) = \mathbf{q} - \mathbf{q}_g$.

The first step consists in translating this task into an optimization problem.

We write :

$$\mathbf{J}_\tau = (\frac{\partial \mathcal{T}}{\partial q^1}, \ldots, \frac{\partial \mathcal{T}}{\partial q^n}) \quad (5)$$

the *Jacobian of the task* at time t, and $\|\mathbf{J}_\tau(\mathbf{q})\|$ its algebra norm for configuration $\mathbf{q}$, equals to $\max_{\delta \mathbf{q} \neq 0} \|\mathbf{J}_\tau(\mathbf{q}) \delta \mathbf{q}\| / \|\delta \mathbf{q}\|$. Let j_τ be the maximum value of $\|\mathbf{J}_\tau(\mathbf{q})\|$ for all configurations. In practice, we make use of a conservative approximation of j_τ.

Writing that we try to follow a straight line in the space where the task is expressed, we formulate the problem :

$$(\mathcal{P}) \quad \text{minimize} \quad \frac{1}{2}\|\dot{\mathcal{T}}(\mathbf{q}) - \tilde{\mathcal{T}}\|^2, \quad \text{with}$$
$$\tilde{\mathcal{T}} = -j_\tau \frac{\mathcal{T}(\mathbf{q})}{\|\mathcal{T}(\mathbf{q})\|}$$

Note that $\tilde{\mathcal{T}}$ is not the derivative of the measure along a reference trajectory, but only the current desired value of $\dot{\mathcal{T}}(\mathbf{q})$. In particular, we do not memorize eventual deviations caused by an obstacle.

In the case we simply want to reach a goal configuration, this translates easily in terms of desired configuration parameters increments :

$$\delta \mathbf{q}^* = \frac{\mathbf{q}_g - \mathbf{q}}{\max_{i=1,\ldots,n} |q_g^i - q^i|} \delta t \tag{6}$$

In the absence of any obstacle, the trajectory then follows a straight line in Configuration Space.

For a given time increment δt, optimal variations $\delta \mathbf{q}^*$ of configuration parameters are computed by solving minimization program $(\mathcal{P}')$, discrete expression of $(\mathcal{P})$:

$$(\mathcal{P}') \quad \text{minimize} \quad \frac{1}{2}(\mathbf{J}_\tau^t \mathbf{J}_\tau \delta \mathbf{q}|\delta \mathbf{q}) - \delta t(\mathbf{J}_\tau^t \tilde{\mathcal{T}}|\delta \mathbf{q})$$

We add the constraints of anti-collision generated by velocity dampers :

(i) $(\hat{\mathbf{n}}_c|\delta \mathbf{q}) \geq \alpha_c \, \delta t, \quad \text{for } c \in \{1,\ldots,n_c\}$,

n_c being the total number of constraints, and bounds on the variations of joint angles :

(ii) $\|\delta \mathbf{q}\| \leq \delta t$ this is $-\delta t \leq \delta q^i \leq \delta t$ for $i \in \{1,\ldots,n\}$.

In the case we simultaneously want to realize k sub-tasks $\mathcal{T}_1(\mathbf{q}),\ldots,\mathcal{T}_k(\mathbf{q})$, the composed task is simply the vector $\mathcal{T}(\mathbf{q}) = (\mathcal{T}_1(\mathbf{q}),\ldots,\mathcal{T}_k(\mathbf{q}))$ written in the space cross-product of the spaces of the sub-tasks. We can easily prioritize tasks by adding weighting coefficients p_i and write $\mathcal{T}(\mathbf{q}) = (p_1 \mathcal{T}_1(\mathbf{q}),\ldots,p_k \mathcal{T}_k(\mathbf{q}))$.

Once a subtask is realized, we can impose it remain as a geometric constraint included in the definition of the task. It is then translated into an equality constraint $\mathcal{T}_e(\mathbf{q}) = 0$, that we linearize around the current value :

$$\mathcal{T}_e(\mathbf{q}) + \mathbf{J}_{\tau_e} \delta \mathbf{q}^* = 0. \tag{7}$$

This describes such a subtask as following a line with the tip of a tool, or keeping contact with the surface of an object.

We have thus reduced the control of the measures $\mathcal{T}(\mathbf{q})$ to a simple minimization problem with quadratic criterion and linear constraints. Strong points of this technique is that it enables to control independently relevant measures of the problem, to respect precisely given geometric constraints, and to asymptotically approach an obstacle, for example to generate grasp trajectories.

We must outline again that this method is limited in its applications by eventualities of dead-locks before the goal is reached. This limitation is inherent to any local, memory-less approach.

3.3 An Example

Figure 4 shows an example of a trajectory obtained with this algorithm, for a redundant 10 link 8 degree of freedom arm carrying a part to be welded, in the complex environment of a nuclear plant reactor. Tasks that can be performed consist in reaching a goal configuration, or in achieving geometric constraints on the final position of the part, described in Cartesian space. Note that in the example illustrated in figure 4, it was necessary to provide one sub-goal along the trajectory.

Hierarchical descriptions introduced in section 2 are used for efficiently updating the current model of the environment when the robot is moving. This model consists in a list of nodes from the Assembly Tree describing the environment. When a primitive of the robot moves towards a given node, a more precise description is obtained by replacing this node by its sons in the tree. On the contrary, the robot may move away from an obstacle, so that two nodes with the same father now lie at a distance greater than the threshold σ from the robot. Then they are replaced by their father in the list.

In the application illustrated in figure 4, the environment is composed of about 500 primitives and the robot of 20. A straightforward use of the hierarchical structures enables to deal with only about 100 interactions at a time, compared with the 10000 potential ones. Furthermore, most of the corresponding primitives of the environment lie at a distance superior to the threshold σ from the primitive on the robot. We can perform a worst case updating of the distance using a conservative approximation of the maximum velocity of points on the moving primitive. Then only nodes lying at an estimated distance lower than σ from the robot need to be examined precisely at next step, typically 10 at each time increment. This enables real-time computation of collision-free trajectories for this particular manipulator, which moves only at a maximum velocity of 20 cm per second at the tip of the hand.

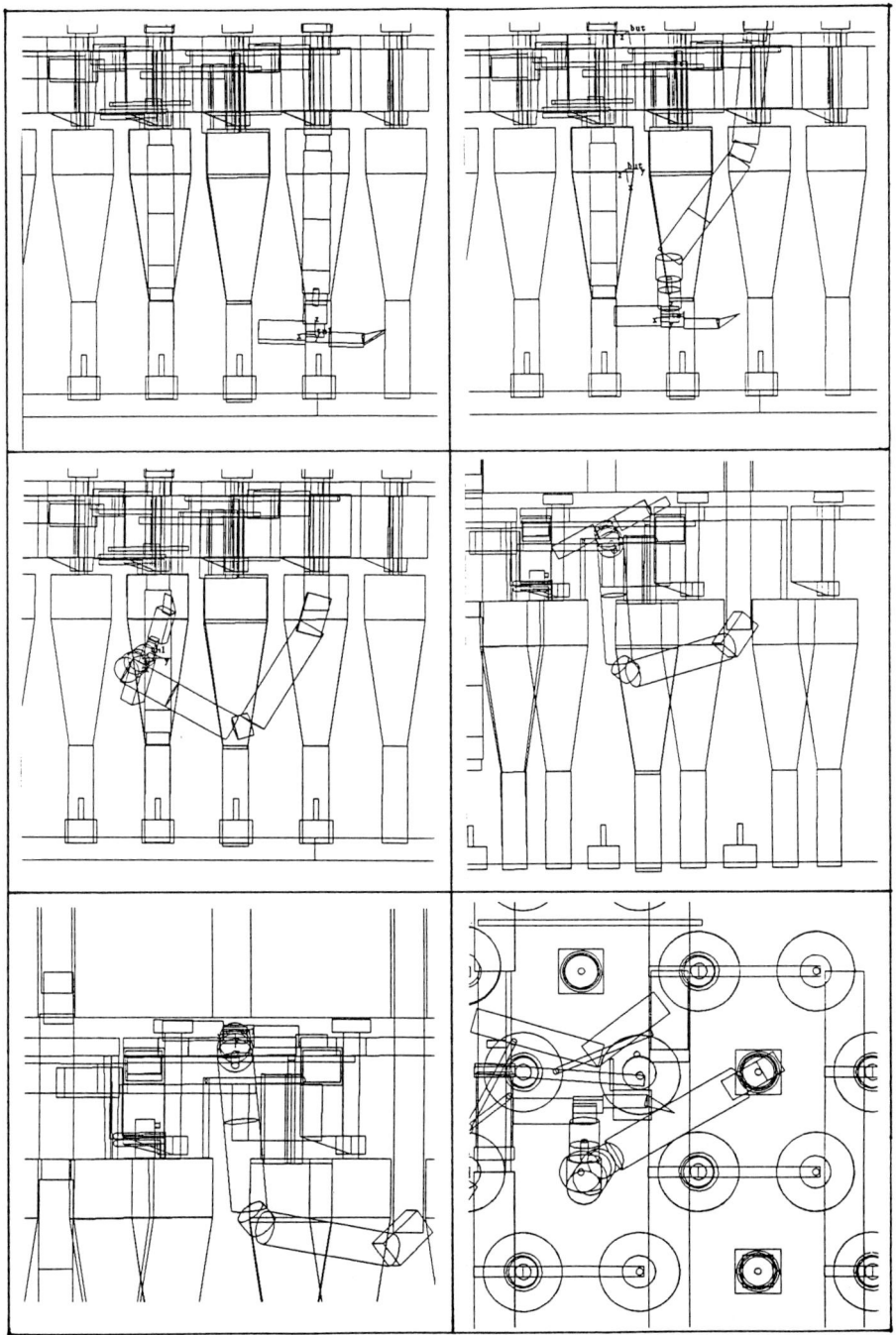

Figure 4: A trajectory of a 10 link robot in the reactor.

4 On Global Planning

In this section we apply general tools defined in sections 2 and 3 to the construction of a global model of Free Space. The technique we propose is general, and the model we obtain not too conservative, as the way we "grow" obstacles to ensure safety does take into account the various configurations of the system.

4.1 The Generic Construction of Free Space

We begin by defining a generic construction of Free Space induced by the natural order of the joint variables along the articulated chain. It is based on the hierarchical description of a manipulator introduced in section 2, as a pyramid of $n+1$ virtual manipulators M^k, $0 \le k \le n$. Virtual manipulator M^k is made of the k first bodies $B^i, 1 \le i \le k$, and of a virtual solid SV^k, describing the volume swept by following bodies when the last $n-k$ joint variables vary freely. The volume occupied by SV^k depends only on the k first joint variables $\mathbf{q}^k = (q^1, \ldots, q^k)^t$.

For each of those virtual manipulators M^k, we define the corresponding Free Space FS^k in the k-dimensional Configuration Space $CS^k = Q^1 \times \ldots \times Q^k$: it consists in the configuration vectors $\mathbf{q}^k$ such that M^k does not collide with any obstacle in 3D Cartesian space. We also define $\widetilde{FS}^k$ as the subset in CS^k of "possibly free configurations", the set of configuration vectors $\mathbf{q}^k$ such that none of the k first bodies collide with any obstacle in 3D Cartesian space (that is, virtual bodies are not taken into account). Any set $FS^k \times Q^{k+1} \ldots \times Q^n$ provides a conservative approximation of FS. Obviously, $\widetilde{FS}^k$ strictly contains FS^k for $k < n$, and $\widetilde{FS}^n = FS^n = FS$. Furthermore, these sets can be iteratively computed using the relationships :

$$FS^1 = I^1$$
$$\widetilde{FS}^1 = \tilde{I}^1$$
$$FS^{k+1} = FS^k \times Q^{k+1} \cup \{\mathbf{q}^{k+1} | \mathbf{q}^k \in \widetilde{FS}^k \setminus FS^k \text{ and } q^{k+1} \in I^{k+1}(\mathbf{q}^k)\}$$
$$\widetilde{FS}^{k+1} = FS^k \times Q^{k+1} \cup \{\mathbf{q}^{k+1} | \mathbf{q}^k \in \widetilde{FS}^k \setminus FS^k \text{ and } q^{k+1} \in \tilde{I}^{k+1}(\mathbf{q}^k)\}$$

where I^k and $\tilde{I}^k$ are monodimensional subsets of Q^k defined for obstacles O by :

$$\tilde{I}^k(\mathbf{q}^{k-1}) = \{q^k \in Q^k | B^k(\mathbf{q}^k) \cap O = \emptyset\}$$
$$I^k(\mathbf{q}^{k-1}) = \tilde{I}^k(\mathbf{q}^{k-1}) \cap \{q^k \in Q^k | SV^k(\mathbf{q}^k) \cap O = \emptyset\}$$

Thus only monodimensional free spaces need to be computed for building FS. Sets I^k and $\tilde{I}^k$ are indeed finite unions of intervals whose end-points correspond to configurations for which there is contact between a body (real or virtual) and an obstacle. For each body, the list of obstacles it can collide with is initialized using the corresponding leaf of Assembly Tree M^0, that is, its whole swept volume. The list is pruned during the iterative process by removing the obstacles that do not collide with $M^k(\mathbf{q}^k)$ when computing I^{k+1} and $\tilde{I}^{k+1}$ for a given configuration vector $\mathbf{q}^k$. Thus only the obstacles possibly yielding a collision are considered when computing monodimensional free spaces.

In practice, parameter ranges are cut into slices : monodimensional free spaces are computed only for a finite number of discrete configuration vectors lying at the center of cuboids of the type $\prod_{i=1}^{k-1}[q_c^i - \Delta q^i, q_c^i + \Delta q^i]$, where Δq^i is half the discretization step for the i-th joint. The sets I^k and $\tilde{I}^k$ must be such that for any value q^k in the set, and for any configuration vector $\mathbf{q}^{k-1}$ in the corresponding cuboid, we obtain a free configuration $\mathbf{q}^k$. Next section explicits the computation of a monodimensional free space.

4.2 Computation of a Monodimensional Free Space

We now consider a body B^k and one obstacle O, both convex objects. The position of body B^k, either real or virtual, depends on k configuration parameters, and we compute the monodimensional free space in q^k for a discrete configuration vector $\mathbf{q}_c^{k-1}$ at the center of the cuboid $\prod_{i=1}^{k-1}[q_c^i - \Delta q^i, q_c^i + \Delta q^i]$. The method we propose to compute the discrete monodimensional free space is based on a local model of Free Space similar to that introduced in section 3, the only difference being that we use exact distances between objects in place of conservative bounds generated by damper equations. The idea is to count how many elementary cuboids one can stack in the k-th direction while remaining within the local model of Free Space, and then recompute this model at the center of the top and bottom cuboids until no extension of the free interval is possible any more.

More precisely, from equation 2, the variation of the distance between B^k and O writes :

$$\delta d = (\mathbf{n}|\mathbf{J}\delta\mathbf{q}^k) \qquad (8)$$
$$= (\mathbf{J}^t\mathbf{n}|\delta\mathbf{q}^k) \qquad (9)$$
$$= \sum_{i=1}^{k} \nu^i \delta q^i \qquad (10)$$

If the distance to the obstacle equals d for configuration vector $\mathbf{q}_c^k$, we deduce that all configurations inside the cuboid are safe as long as q^k verifies inequation :

$$\nu^k q^k \geq \nu^k q_c^k - d + \sum_{i=1}^{k-1} |\nu^i|\Delta q^i. \qquad (11)$$

Depending on the sign of ν^k this yields either an upper or a lower bound on q^k. In the case of a revolute joint, another bound must be imposed in order to insure the validity of the first order approximation of the distance. A new distance computation is performed for the more constraining of the two bounds, and the process is iterated until the remaining free interval becomes smaller than Δq^k.

Strong points of this technique are, first, that it is applicable to all types of articulated chains, second, that the way we "grow" moving bodies to ensure safety of the model of Free Space is not too conservative : coefficients ν^i are computed for the exact value of the Jacobian, thus taking into account the various configurations of the system. This method is as well applicable to the computation of configuration obstacles generated by collisions between two moving bodies, either part of the same kinematic chain, or of two different robots.

4.3   Structuration of the Data in a 2^n-tree

This generic construction yields a tessellation of Free Space in very thin cuboids, some of their sides being necessarily as small as the discretization step in that direction. The structuration will aim at reconstructing a more homogeneous representation (cf. [Fav86]). The structure we use is a 2^n-tree, generalization of the *quadtree* introduced by Samet to hierarchically represent binary arrays [Sam84]. It is a tree of degree 2^n whose nodes at level l are n-dimensional cuboids of size 2^{n-l}. Each node is labelled as *empty*, *full* or *mixed*. Mixed nodes only are split up into their 2^n sons, until they become of unit size.

In our context, such a structure presents several advantages :

- *Isotropy* : That is, all joints are treated equally.

- *Hierarchy* : Big cells will be sufficient far from configuration obstacles, small ones being only needed to describe in details their boundaries. Thus, when looking for a gross motion, one can use big cells only to obtain quickly a path along which the robot remains far from obstacles. The 2^n-tree data structure also enables efficient boolean operations : this is useful to merge the structures computed for different parts of the environment of the robot.

- *Regularity* : This enables to have access to the neighbors of a cell without explicitly storing the connectivity graph, which is most important considering memory limitations.

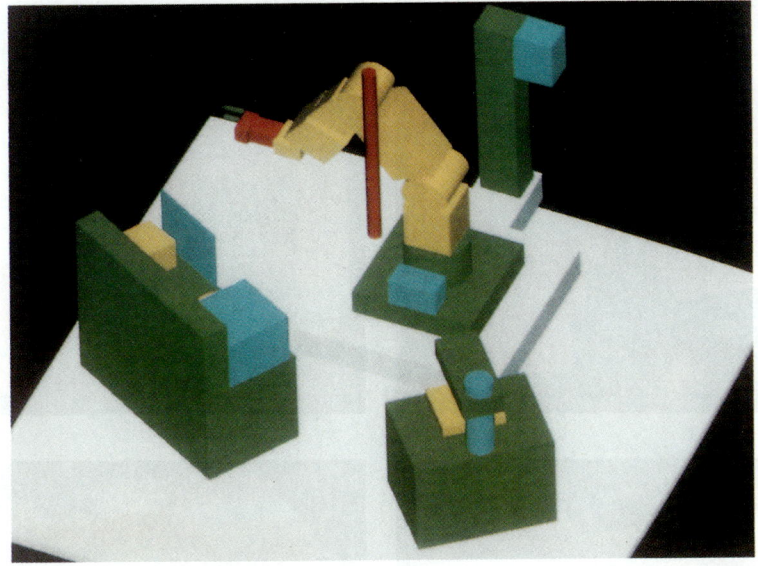

Figure 5: Robot and obstacles in Cartesian Space.

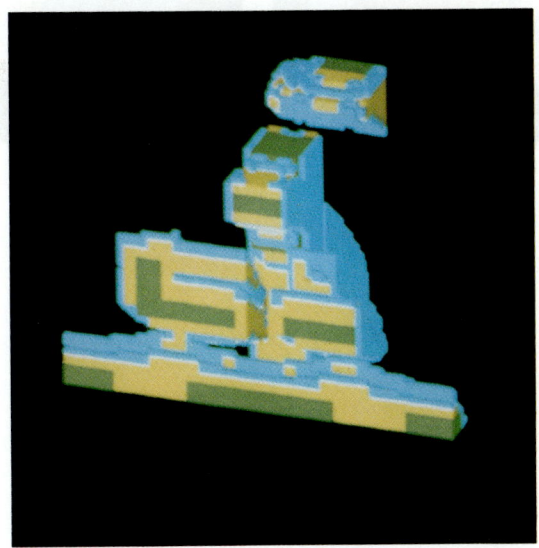

Figure 6: Transforms of obstacles in the Configuration Space of the first three links.

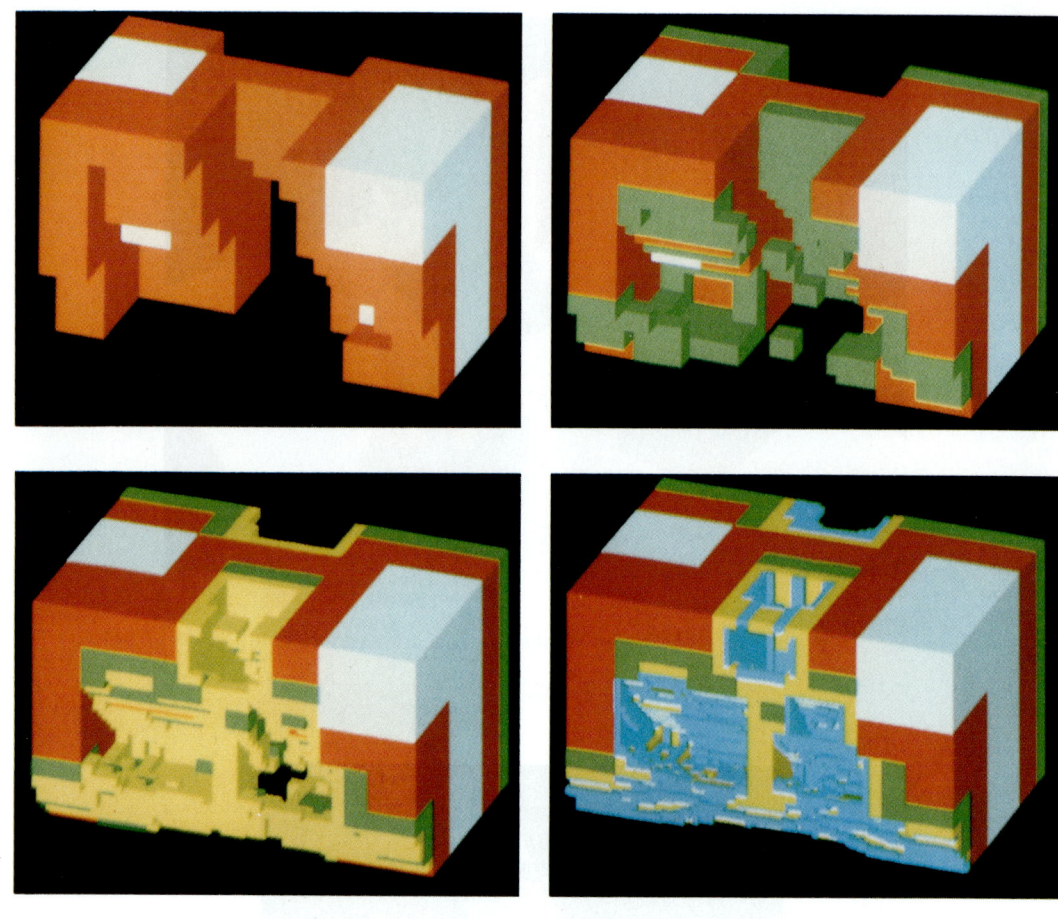

Figure 7: *Octree* representation of Free Space at levels 4, 5, 6, 7. The freeway between the front obstacles appears at level 6.

level	3	4	5	6	7	Total
% nodes	0.06	0.93	5.08	13.55	80.38	100%
% Free Space	19.0	29.0	26.5	19.0	6.5	100%

Table 1: Repartition of nodes in Free Space.

Figure 6 shows the transforms of obstacles in the 3D Configuration Space representing the first three links of the manipulator in the environment of figure 5. Figure 7 gives the corresponding *octree* representation of Free Space, at levels 4, 5, 6 and 7 respectively. The total number of nodes for representing Free Space is 10591, which would correspond to 129694 cells of level 7. Table 1 gives the percentage of nodes at each level, and the corresponding percentage of Free Space volume. Note in particular that 93.5% of Free Space is described by nodes up to level 6, with 20% of the nodes only.

4.4 Search for a Collision Free Trajectory

When computing a trajectory between an initial and a goal configuration, we first search for a path in the connectivity graph of the 2^n-tree, which yields a list of cells to be traversed. The search is performed with an A^* algorithm, the criterion used being usually the Manhattan distance between the centers of the cells, and the heuristic function h^*, estimate of the exact cost function h between a node and the goal, the Cartesian distance to the goal.

As mentioned earlier, a lower bound on the size of the cells to be examined during the search can be imposed. The number of nodes explored is then much smaller, and the search faster, but some solutions in constrained regions of the environment can be lost. However, in most practical applications, one will prefer a safer but longer trajectory than a shorter but very constrained one.

Once a list of free cells has been found, we must still compute a trajectory joining the initial and the final configurations, that remains within those cells. We have chosen to search a trajectory among polygonal lines, using the following recursive procedure :

1. Hypothesize the line segment joining the two end-configurations as a trajectory. If it intersects all the faces common to two adjacent cells, it is a feasible trajectory.

2. Else, on each face not intersected by the line-segment, compute the point that minimizes the sum of the length of the two segments to the end-points.

3. Choose the point yielding the largest sum as an intermediate configuration, and call recursively the procedure on both line segments.

In general, this procedure yields a trajectory with few intermediate points. The speed along each line segment can be set according to the size of the corresponding empty cells : the larger the cell is, the faster one can move safely.

4.5 Heuristics for the hand

Although in theory there is no bound in the dimension of the Configuration Space, one is limited to three or four degrees of freedom in practice, because of the combinatorial explosion due to the discretization. However, most of the industrial robots can be decomposed into an arm with three joints, and a hand articulated at the wrist. The hand is usually much smaller than the arm, and does not play an essential part in the search for a gross motion. It can then be replaced by virtual body SV^3, the volume swept by the last three links when joints q^4, q^5, q^6 vary freely. As the hand is usually close to obstacles at both ends of the motion, a local method is nevertheless needed to compute a motion that pushes the hand away from obstacles. This yields a three step procedure for planning the motion of a six joint manipulator :

1. Use a local method to compute new initial and final configurations away from obstacles. The task consists in maximizing the distance to obstacles, under the collision avoidance constraints.

2. Search for a gross motion for virtual manipulator M^3. This step yields a list of cells in the *octree*.

3. Search for a collision free trajectory, using the procedure of subsection 4.4. The trajectory in the six dimensional space is obtained by simple linear interpolation for the last three joints.

This algorithm works quite well for manipulators having small hands and moving small objects. If it is not the case, one can no more replace the hand by its real swept volume, as in many cases no trajectory at all would be found. However, we still influence the construction of the *octree* by using a virtual body to represent the hand, in general a sphere centered at the wrist, smaller than the real swept volume. We then modify the two last steps of the algorithm above as follows :

- In step 2, we add an additional cost to the criterion, that measures the violation of the constraints of the 6D Free Space local model, for an interpolated joint vector. Different orientations of the hand, for the same start and goal cells, may now yield different lists of cells in the *octree*.

- In step 3, we add the constraints generated by the local model of Free Space, to compute intermediate points and test collisions along line segments.

This algorithm may fail in the last step because no correct orientation can be found for the hand in the proposed list of cells. However, it does yield correct trajectories in many practical cases. Figures 8 and 9 show some intermediate configurations computed for two problems with the same initial and final positions of the wrist, but with different orientations of the hand.

For this example, computation time is of the order of five minutes for the construction of Free Space and its structuration, and ten seconds for the search of a collision-free trajectory.

Figure 8: A collision-free trajectory.

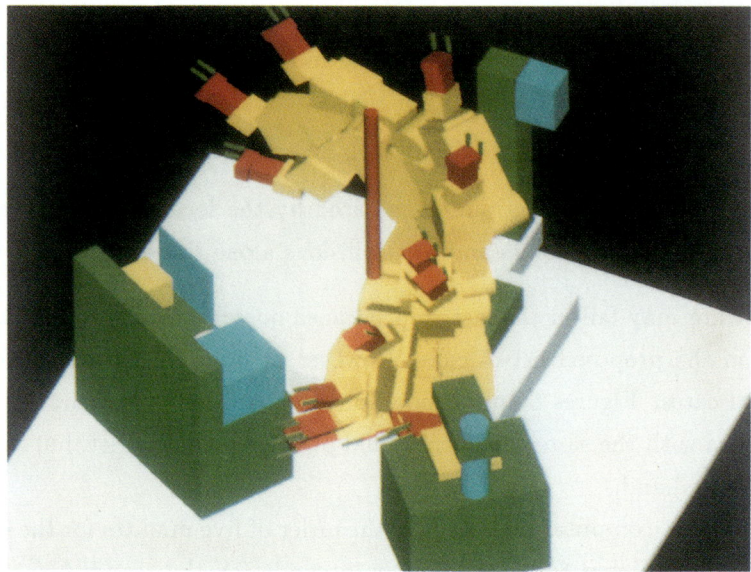

Figure 9: Another trajectory with the same initial and final configurations of the arm, but with different orientations of the hand.

5 A Mixed Approach to Planning

On one hand, global constructions of Free Space are limited by combinatorial explosion with the number of degrees of freedom. On the other, local methods are limited in their scope of applications by the eventualities of dead-locks. One obvious solution would be to give the local method "good" subgoals so that it avoid dead-ends. In this section, we propose to decompose the general problem of path-planning for manipulators with many degrees of freedom into :

- a *low complexity local planner*,

- and a *higher complexity global planner* working on a graph of cells representing relatively large regions of Configuration Space.

We will describe in details the interactions between the two levels, which result in learning automatically good subgoals for the local planner.

The main idea underlying this approach is to take advantage of the power of the local method to follow configuration obstacles boundaries, so as to deal with the high complexity of global planning only at the relevant level of the description. At the global level, no geometric description of the obstacles is stored, but only weights indicating the probability that a trajectory computed locally does not lead to a dead-end. Let us emphasize that this approach is relevant only when it is performed using a coarse quantization of Configuration Space, otherwise it is not better than a classical global approach in terms of computational cost. Dealing with a loose graph is made possible only because of the intrinsic power of the local planner that produces long segments of collision free trajectory.

5.1 A Probabilistic Model of Safe Displacements

We assume that an algorithm that locally computes segments of safe trajectories is available. We write $\mathbf{q} = (q^1, \ldots, q^n)^t$ a n dimensional configuration vector describing the state of the system, and partition Configuration Space into cells of the type :

$$\mathcal{C}_k = \prod_{i=1}^{n} [q_k^i - \Delta q^i, q_k^i + \Delta q^i].$$

Again $(q_k^1, \ldots, q_k^n)$ are the configuration parameters of the center of cell $\mathcal{C}_k$, and Δq^i equals half the width of a cell in the i-th direction. Starting with this description, we build a graph whose nodes stand for the cells themselves, each node having $2n$ neighbors. It is called the *State Graph* in the sequel. When the robot configuration lies inside a cell $\mathcal{C}_k$, it is said to be in state $\mathcal{C}_k$.

As opposed to a standard Configuration Space approach, we do not label nodes as *free* or *intersecting configuration obstacles*. This would not be relevant as the grid is based on a coarse discretization of parameter ranges, and we want to be allowed to navigate in partially occupied cells.

For each oriented transition between two neighboring cells $\mathcal{C}_k$ and $\mathcal{C}_l$ in the State Graph, we only memorize a weight that estimates the difficulty for the robot to enter cell $\mathcal{C}_l$ when coming from cell $\mathcal{C}_k$. More precisely, we define p_{kl} as the probability for the local planner to succeed in making the robot enter cell $\mathcal{C}_l$ from neighboring cell $\mathcal{C}_k$, when aiming at some point located inside $\mathcal{C}_l$.

We call any connected sequence of nodes of the State Graph a *path*. The probability for a path to yield a successful trajectory (with no dead-lock) is defined as the product of probabilities $p_{k_m k_{m+1}}$ along the path. This implies a hypothesis of independence, namely that the probability of realizing a transition is independent of the sequence of nodes we followed so far. In practice, we attribute to each arc of the graph a weight equal to

$-\log p_{kl}$, and we minimize the cost

$$g = - \sum_{m=0}^{M-1} \log(p_{k_m k_{m+1}}) \qquad (12)$$

for traversed transitions along the path.

Initially, in absence of any information on its environment, the robot initializes the graph with given *a priori* probabilities $p_{kl} = \overline{p}$, $0 < \overline{p} < 1$. Hence the path we compute first is a path of minimum length in terms of the number of cells traversed from the initial configuration to the goal. Later as probabilities vary to reflect the robot's accumulated knowledge of its environment, the path we compute realizes a compromise between minimum distance and the assurance that we will reach the goal.

5.2 The Interface Between the Local and Global Planners

In the simplest implementation of the algorithm, the interactions between both levels are quite limited : the global planner sends subgoals to the local planner, and observes successive configurations of the robot to run the learning process.

5.2.1 Choosing Subgoals on the Path

When executing a path, a simple choice of a subgoal is the center of next cell on the path. However, this may yield a trajectory similar to a drunkard's walk because of the coarse discretization. In order to avoid this reprehensible behavior, we propose another choice that gives a smoother trajectory. The idea is to locally optimize the length of the trajectory. So, the subgoal is chosen as the point on the face common with next cell that minimizes the sum of :
– the distance from the current position to this point,
– and the distance from this point to the center of one of the cells farther on the path.

The farther this cell is chosen, the more we anticipate future displacements. In the case the robot is blocked, nearer cells are used in order to provide new subgoals.

5.2.2 Updating Probabilities During Path Execution

Updating of the transition probabilities is first performed whenever the motion supervisor detects a transition in the State Graph. If *the robot performed the transition ordered by the global planner*, the corresponding probability is increased. Else, *the robot has been deflected from its path by an obstacle,* and we decrease the probability of the desired transition. In this case, we must also modify the path so that it remains connected. This is done by adding to the path the current node, and the common neighbor of the current node and next cell on the path (see Fig. 10).

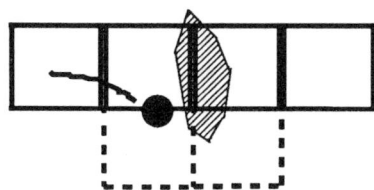

Figure 10: Modification of the path when the actual transition is not the desired one : additional cells are drawn with dotted lines.

Updating the transition probabilities is also performed whenever *the robot is blocked*. If it is not inside the goal node, we decrease the probability of the desired transition and invoke the global planner with the updated State Graph. As the probability of the problematic transition has been decreased, it will eventually be avoided by the newly computed path.

If a dead-lock occurs inside the goal node, the goal is declared not reachable. This event may have several causes :
– the goal lies inside an obstacle (or another connected component of Free Space).
– the local method failed to reach the goal because of a difficult layout of obstacles inside the goal node.

In the latter case, a solution would be to move the robot outside of the goal node, setting a subgoal in a neighbor chosen from local information, before aiming again at the goal. This strategy might succeed in making the robot pass round the obstacle. But the main problem in fact is to distinguish between the two cases above. Elements of the answer are given in subsections 5.5 and 5.7.

5.3 More Information from the Local Model

The idea outlined in this section is to acquire more information from the local model, to accelerate the updating of the graph when a dead-lock occurs. Recall that we work in a high dimensional space : in the case of a failure of the local planner, several transitions with neighbors of the current node may yield similar costs, and more than one call to the global planner is needed in general to make the system move again significantly.

Before searching for the best global path, we thus execute *virtual moves* in all the directions of the $2n$ neighbors, except the one we were aiming at that led to a dead-lock. Probabilities for each of those transitions are then updated by increasing by 1 the number of events, and the "number of successes" by the ratio $\delta q^i / \delta q^i_{max}$, where δq^i is the increment effectively computed by the local planner on the i-th parameter, for desired increments $(0, \ldots, \delta q^i_{max}, \ldots, 0)$.

The search algorithm now tends to favor those transitions around the current configuration that are *a priori* safer for the local planner, and one or few calls to the global planner are sufficient for the system to start moving again.

5.4 Heuristics for the Global Planner

When searching for a path from an initial to a goal cell, we use an A^* algorithm maximizing the product of probabilities p_{kl} along the path. This translates into minimizing the cost $g = -\sum_m \log(p_{k_m k_{m+1}})$ for traversed transitions along the path from initial cell $\mathcal{C}_0$ to goal cell $\mathcal{C}_g$. The heuristic cost we use to guide the search is for a node $\mathcal{C}_k$ the number of cells N_k that are traversed by a straight line to the goal, weighted by the log of the average probability $\bar{p}$ for all transitions of the State Graph :

$$h_k^* = -N_k \log \bar{p}, \tag{13}$$

This heuristic function is generally not admissible, but gives good results in our experiments. It can be seen as an estimate of the minimum cost we expect if we suppose that the difficulty to move around is about the same everywhere.

It must also be noticed that since all weights are finite in our model and the graph is connected, a path is always returned by the global planner, even when the goal intersects an obstacle, or lies in another connected component of Free Space. As we do not want the robot to roam around indefinitely while trying to reach the goal, we have to decide *a priori* about the feasibility of the trajectory from the cost returned by the search algorithm.

We impose a threshold on the ratio of the optimal cost and the heuristic cost, when we perform the planning starting from some cell $\mathcal{C}_k$:

$$\frac{g_k^*}{h_k^*} < \chi. \tag{14}$$

The heuristic function can be seen as the expected cost in the case of a uniform distribution of the difficulty to move in the environment. Thus, if the optimal cost actually computed is much higher than this estimate, it means that the optimal path is much longer than the expected one, or that there exists on the optimal path one or more transitions with a low probability of success. So, the constant χ measures the relative length of the detour we tolerate in order to have a reasonable chance to succeed. The more the environment resembles a labyrinth, the higher χ has to be set.

5.5 When the Goal is Described by a Task Function

Isotropy of the representation of Free Space by a grid makes it possible to adapt the search for problems described by a task function (as defined in subsection 3.2).

First, recall that the A^* algorithm works as well with a *list* of goal nodes, if for all of them the heuristic cost $h^*(n)$ equals 0. We valuate a cell $\mathcal{C}$ of the State Graph with the norm of the task vector computed at the center of the cuboid, $\|\mathcal{T}(\mathbf{q}_c)\|$, and we evaluate the distance to be traversed to reach the goal by : $d(\mathbf{q}_c) = \frac{1}{\lambda}\|\mathcal{T}(\mathbf{q}_c)\|$.

It is easy to see that in this evaluation is lower than the exact value for :

$$\lambda \geq \max_{\mathbf{q} \in CS} \|\mathbf{J}_\tau(\mathbf{q})\|, \tag{15}$$

with $\|\mathbf{J}_\tau(\mathbf{q})\|$ the algebra norm of the Jacobian of the task for configuration $\mathbf{q}$. This distance is used to compute the number of cells to be traversed in the heuristic search function.

A last problem is that of the practical termination of the algorithm. When we reach during the search a node of the graph such that the estimated number of cells to be traversed from the center of the cell to the goal is less than one, a more precise test is performed :

$$\|\mathcal{T}(\mathbf{q}_c)\| < \Delta \|\mathbf{J}_\tau(\mathbf{q}_c)\|, \tag{16}$$

with Δ of the order of half the width of a cell, $\max_{i=1,\ldots,n} \Delta q^i$. If the condition is satisfied, this node is retained as a goal node and the A^* algorithm terminates.

5.6 Adaptation of the Size of the Grid

In order to decrease the number of nodes in the graph, it may be useful to adapt the discretization of the Configuration Space to the planning difficulty in the various regions. This can be done by analyzing the variation of the estimated transition probabilities. Indeed, if a cell is too large, its transition probabilities will not converge towards 0 or 1, because of alternate success and failure. Such transitions are characterized more precisely by a high value of the entropy, measured by $-\overline{p}_e \log \overline{p}_e - (1 - \overline{p}_e) \log(1 - \overline{p}_e)$, or, more

simply, by :
$$\overline{p}_e(1 - \overline{p}_e) = \frac{(s+1)(e-s+1)}{(e+2)^2}. \tag{17}$$

To solve this problem, it is necessary to split the corresponding node into several new cells, thus refining the description of this part of the environment at the global level.

Following section 4, we propose to make use of a 2^n-tree in place of a regular grid. Using such a structure in this context may proceed as follows. Exploration starts with a regular grid corresponding to a relatively high level in the 2^n-tree. When an event occurs for a given transition, we first look at the corresponding entropy and compare it to a threshold (a decreasing function of time). If the entropy is smaller than the threshold, updating is performed as usual. Else, the node we aim at is split into its sons and new transitions are initialized as detailed below.

- Transitions between sons and neighbors of the father are initialized assuming that the events that happened to the former transition are uniformly distributed on the 2^{n-1} new transitions.

- Transitions between two sons are initialized with the *a priori* uniform law.

This process can be used similarly for solving the problem of a dead-lock occurring inside the goal node.

Splitting is performed until the size of a node is smaller than a threshold, function of the ability of the local method to reach a subgoal at a distance corresponding to that threshold.

A question naturally arising is whether one should now influence the cost function considering the width of a cell. We argue that we should not, as long as the hypothesis of independence of the various transitions stand, which implicitly means that we will use smaller cells only in cluttered areas.

5.7 A Six Degree of Freedom Example : a Manipulator with a Bulky Load

To compute trajectories of a manipulator robot carrying a large object, one cannot simply consider the first three degrees of freedom to describe Free Space. We propose to build a State Graph that takes into account the whole six degrees of freedom. The State Graph is initialized from the *octree* computed for the 3 degrees of freedom of the arm with a global method. The *octree* is a structure of about 210 Kbytes, which corresponds to a discretization of configuration parameters in respectively 64, 64 and 44 intervals, for ranges of 270° (the idea being that the contribution of each discretization interval to the precision of a Cartesian displacement of the wrist be roughly equal). Due to memory limitations, the 6D graph is based on a discretization 4 times coarser. The probability of entry in a 3D cell is initialized with the ratio of its free successors at the lowest level in the *octree*. Joint ranges of the arm are thus divided into 16, 16 and 11 intervals, and those of the hand into 6, 4 and 6 (for ranges of 270, 150 and 270 degrees respectively). The total cost of the representation is then about 1.6 Mbytes. The State Graph is updated from results of execution of the local trajectory generator, which uses an exact geometric model of the arm, the hand and the load.

Before learning, the robot is aware of the interactions between its arm and the obstacles. The learning process essentially discriminates between different configurations of the hand, for given positions of the wrist.

Trajectories obtained are illustrated by figure 11 and 12. Computation time is of the order of a few minutes for the first execution of a trajectory.

6 Conclusion

Difficulties we are faced with in order to solve the motion-planning problem for manipulators are twofold : 1) the practical difficulty of dealing with rotations in a three dimensional space, 2) the inherent exponential complexity of the Free Space construction with the number of degrees of freedom.

To answer the first point, we have given general tools to translate the control of critical distances between solids in Cartesian Space into constraints in Configuration Space. This results in building a *local model of Free Space*, that we use both to locally generate safe trajectories by minimization techniques, and to build global discretized models of Free Space.

To answer the second point, we have proposed to decouple the search for the global shape of the trajectory from the fine motion computations. The two levels operate with very distinct time-scales : a fast one for the computation of displacements from local information, a slow one for the learning of global strategies from execution of a path. This method can also be adapted to various applications, such as the exploration of an unknown, eventually changing, environment by a mobile robot.

In addition, we have extended the motion-planning problem to a broader class of examples, defined by the control of a set of measures on the system.

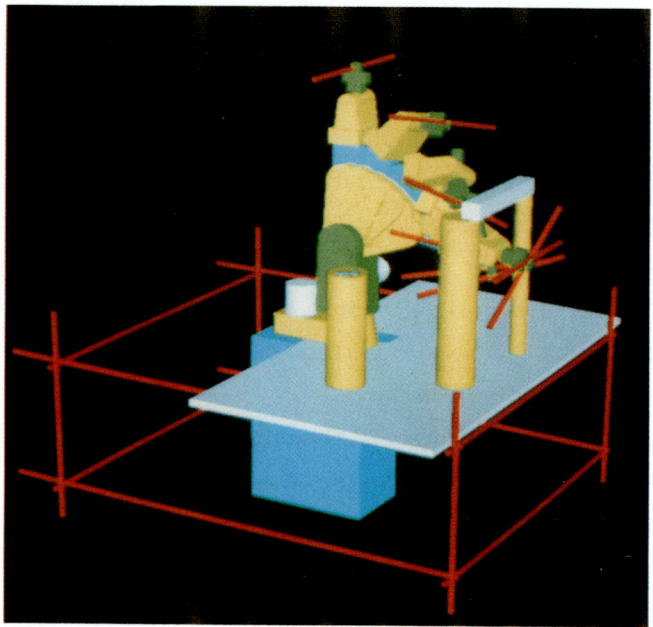

Figure 11: Moving a rod between two columns: Nine calls to the global planner were needed on the first trial, three on the second trial and one on the third one.

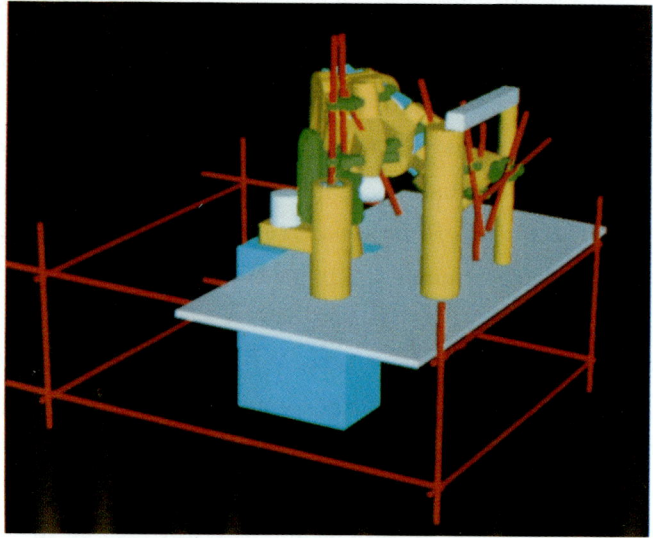

Figure 12: Moving a rod into a cylindrical hole. The goal is defined in the cartesian space and correspond to a variety of dimension 1 in the Configuration Space.

References

[Bro83] R.A. Brooks. Planning collision-free motions for pick-and-place operations. *Int. Journal of Robotics Research*, 2(4):19–44, 1983.

[Fav86] B. Faverjon. Object level programming of industrial robots. In *Proceedings of IEEE Int. Conference on Robotics and Automation*, pages 1406–1412, San Francisco, April 1986.

[FT86] B. Faverjon and P. Tournassoud. A hierarchical CAD system for multi-robot coordination. In *Proceedings of the NATO Advanced Research Workshop on Languages for Sensor Based Control in Robotics*, Pisa, Italy, September 1986.

[Kha86] O. Khatib. Real time obstacle avoidance for manipulators and mobile robots. *Int. Journal of Robotics Research*, 5(1):90–99, Spring 1986.

[Loz87] T. Lozano-Pérez. A simple motion planning algorithm for general robot manipulators. *IEEE Journal of Robotics and Automation*, 3(3):224–238, June 1987.

[Sam84] H. Samet. The quadtree and related hierarchical data structures. *ACM Computing Surveys*, 16(2):187–260, June 1984.

[SS83] J.T. Schwartz and M. Sharir. On the piano movers' problem II: general techniques for computing topological properties of real algebraic manifolds. *Advances in Applied Mathematics*, (4):298–351, 1983.

Planning Collision Free Trajectories by a Configuration Space Approach

Thierry Siméon
Groupe Robotique et Intelligence Artificielle
LAAS-CNRS
7, Avenue du Colonel Roche
31077 Toulouse cedex

Abstract

This paper addresses the motion planning problem. We present a simple and efficient algorithm for generating a approximate representation of the free Configuration Space for a robot moving in a workspace with obstacles. The algorithms presented in this paper have been implemented and can deal with all robots made of revolute or prismatic joints. We present an exemple of collision-free trajectories computed by the system for a manipulator with 4 degrees of freedom.

1 Introduction

Designing automatic programming systems for robots is now a major issue in robotic research.

The objective of programming systems is to describe tasks involving a fairly high level of abstraction. However, they raise serious concerns, some of which have not yet been satisfactorily addressed.

In recent years, programming aid systems have nonetheless been devised for robotics tasks, SHARP [8], HANDEY [16], NNS [3] [1].

The first two generate a program based on a sequence of orders to be followed by the system's components, e.g. robots, sensors, auxilary devices.

On the other hand, the NNS system structures manipulation knowledge into an execution model (set of programs and rules for these programs). An on-line system featuring decisional capabilities utilizes this model according to the asynchronous events that affect the progress of the task.

Both types of systems utilize different dedicated modules to tackle specific issues, the interdependencies between modules being handled by a coordinator. Usually, these modules rely on geometric reasoning. Hence, algorithms featuring, structuring and reasoning capabilities based on the geometric information supplied by a perception system or an environmental model, are needed to plan and control system motions.
In particular, the following issues must be addressed:

- Planning ollision-free trajectories
- Study of fine motions (local motions in the task space)
- Effects of uncertainties on sensor information, models used, or robot control
- Determination of stable grasp positions

This paper is concerned with collision-free trajectory planning. Although algebraic methods have permitted construction of a general purpose algorithm [18], their complexity makes them impractical.

Together with these results which indicate an insight into the level of difficulty of this problem, some methods designed to address subclasses of problems have been devised. Among these we note:

- *local approaches* which rely on partial environmental knowledge to generate incremental robot motions toward the envisaged goal [17][19].
- *global methods* which rely on a characterisation of the regions belonging either to the Cartesian Space or to Configuration Space of the system.

While local methods cannot guaratee an optimal solution and may even fail when a solution exists, global methods are time-consuming in terms of computation.
This paper deals with the implementation of an obstacle avoidance system based on a discretisation of the Configuration Space (CSpace).

2 Global methods in CSpace

2.1 Principle

The obstacle avoidance problem can be viewed as the collision-free motions (with possible contact) of one or several objects in the same environment.

Methods based on the analysis of CSpace formalized in [12] [14], avoid the geometry of links, the latter being regarded as subsets A of the real space $\Re^m$, whose position and orientation are defined by the n-tuple of independent parameters $Q = (q_1, \ldots, q_n)$. In this space, two cases can be envisaged as to whether Q defines a position and orientation leading to a collision with the subset $OBST = \bigcup_{i=1}^{nobst} O_i$ of $\Re^m$ representing the *nobst* environmental obstacles:

- $OCS = \{Q \in CS / A(Q) \cap OBST \neq \emptyset\}$.
- $FCS = \{Q \in CS / A(Q) \cap OBST = \emptyset\}$.

Figure 1 derived from [15] represents the configuration space of a revolute robot in $\Re^2$ for which $CS = [0, 2\pi[\times [0, 2\pi[$ (OCS corresponds to the shaded areas).

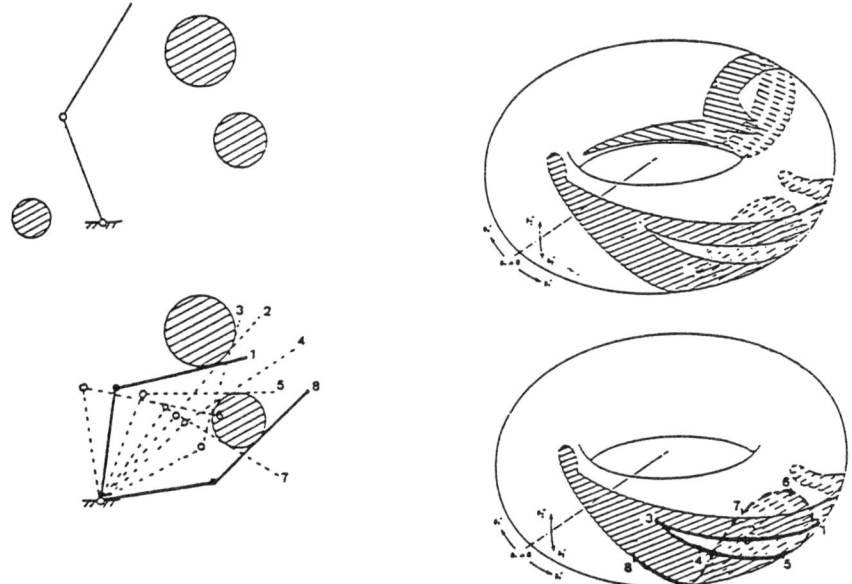

Figure 1: An example of Configuration Space

These algorithms therefore compute the obstacle transforms in the CSpace in order to address the simpler problem of the motion of a point Q in the subset FCS.
This requires embedding FCS in a graph describing connectivity between free regions. The existence of a trajectory between Q_i and Q_f is therefore analogous to knowing whether they belong to the same connected component of FCS.
Trajectory generation can be decomposed into two steps:

- finding a path in this graph that furnishes a trajectory envelope

- finding a specific trajectory within this envelope according to some specific criteria (eg. the optimal distance solution or a safe distance relative to obstacles).

2.2 Approximation methods

In the general case of a system with n degrees of freedom, the CS transforms are complex volumes of $\Re^n$ whose characterization has been envisaged for some particular cases only, namely:

- polyhedra in translation for which the polyhedric structure of the transforms is maintained.

- polygons in translation and rotation

The methods that have led to actual, more general implementations (able to handle several rotations with $n \geq 3$) are based on a discretization of CS that leads to a conservative approximation FCS_{app} of FCS [7][9][4][13].
They simply require that the type of space quanta be determined from computations made at discrete points of CS and that the discretized space be filled in the most economical manner.
The major differences between these methods arise from the representation used for the manipulated geometric entities, the selected space filling algorithm and the latter's structuring.
In the remainder of the article, an implementation based on [7] is presented.

3 CS analysis algorithm

The algorithm involves the following:

- hierarchical modeling of the robot workspace

- definition of a function to compute the distance between two links. This distance is defined as the value of the shortest translation permitting contact (positive) or separation (negative) of the two links.

- a mechanism for storing and propagating information supplied by this function

From the robot's hierarchical description, subspaces of dimension 1 to n are analyzed recursively to produce a tree structuring of the space whose regions are modeled as hyperparallelepipeds (products of intervals).

3.1 Robot's hierarchical modeling - CS tree structure

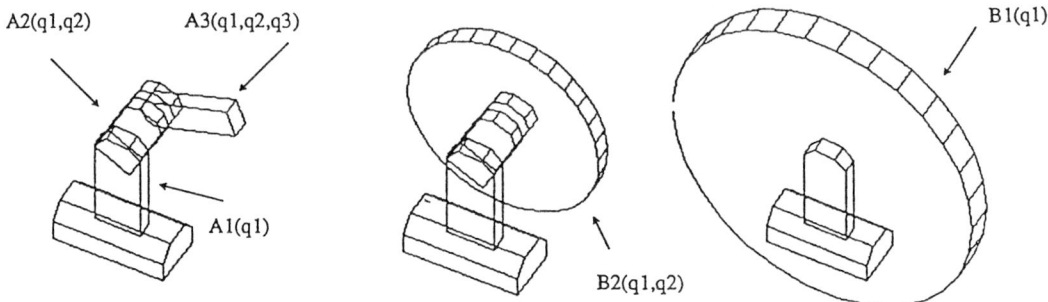

Figure 2: Hierarchical structure of a robot manipulator

Let $A_i(q_1, \ldots, q_i)$ be the space of $\Re^m$ occupied by the robot's links whose position depends on the first i configuration parameters only. A set of fictitious links $(B_1, \ldots, B_{n-1})$ is defined by (see Figure 2):

$$B_i(q_1, \ldots, q_i) = \bigcup_{i < j \leq n} \left\{ A_j(q_1, \ldots, q_i, \ldots, q_j) / q_j \in [q_j^{min}, q_j^{max}] \right\}$$

This description is tantamount to representing the working volume of n sub-robots B_i ($i = 1, n-1$) with i degrees of freedom. The tree structure of the configuration space is directly derived from this representation by defining the following sets (see Figure 3):

- $OCS_i = \{Q \in MCS_{i-1} / A_i \cap OBST \neq \emptyset\}$

- $FCS_i = \{Q \in MCS_{i-1} / (A_i \cap OBST = \emptyset) \wedge (B_i \cap OBST = \emptyset)\}$

- $MCS_i = \{Q \in MCS_{i-1} / (A_i \cap OBST = \emptyset) \wedge (B_i \cap OBST \neq \emptyset)\}$

($MCS_0 = CS$)

The sets OCS_i and FCS_i correspond to subspaces of dimension i whose analysis enables us to know whether regions of the complete space belong to the free space.

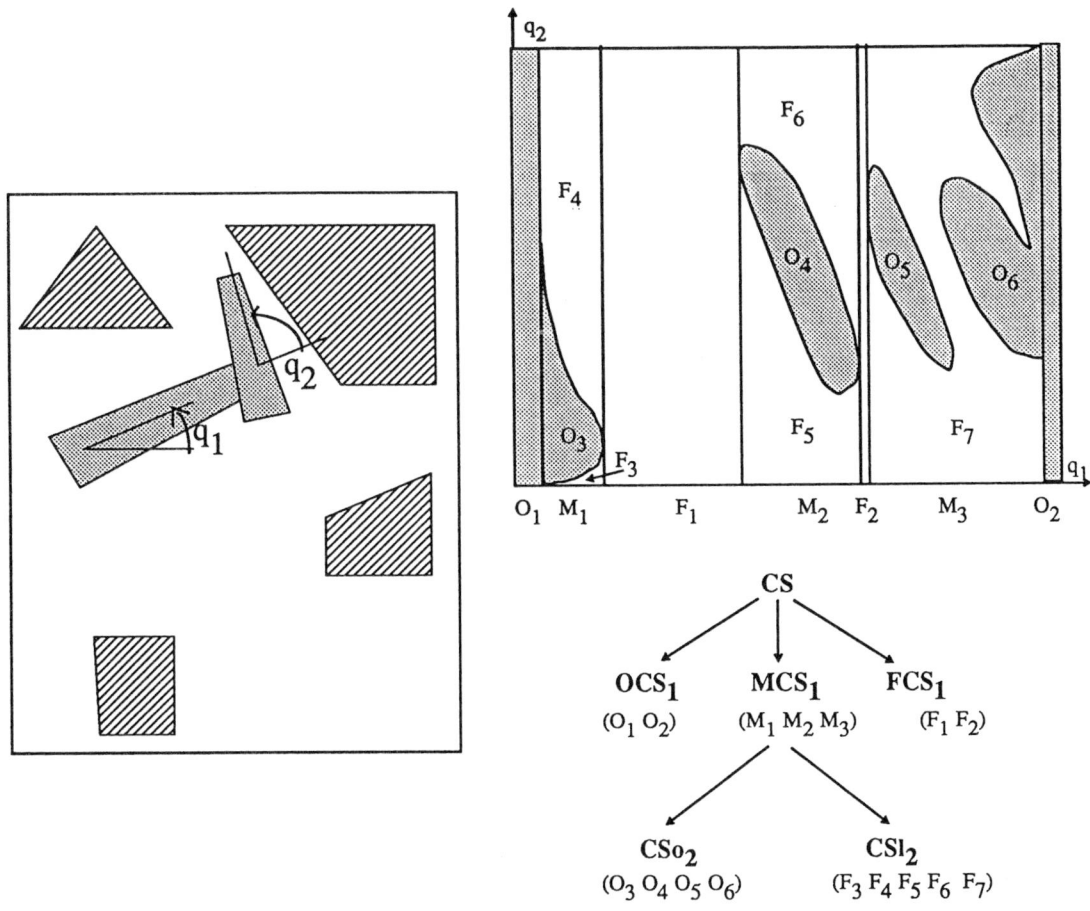

Figure 3: CS tree structure

3.2 Algorithm

The analysis algorithm of $CS = I_1 \times \ldots \times I_n$ will therefore consist of computing recursively the type of space quanta resulting from the discretization of subspaces $CS_{i(i=1,n)} = L_1 \times \ldots \times L_i$.

Distance computations of the type $(A_i, OBST)(B_i, OBST)$ are effected on discrete positions of $Q_i = (q_1, \ldots, q_i)$ (the quanta centers). The results are then propagated into a "ball" of this space according to this distance and its maximal variation:

$\epsilon = Max_{Q_i \in CS_i} \|J_i(Q_i) DQ_i\|$.

with $J_i(Q_i)$: Jacobian of the i first joints

and DQ_i: increments on the i first joints

The procedure can be described by the subspace analysis function:

Subspace-Analysis(CS,i,CS_{i-1},DQ_i)
   $pave = CS_{i-1} \times I_i$
   $cdis = $ discretize-space($pave$,DQ_i)
   $\epsilon = $ compute-dep-max($pave$,DQ_i,A_i,B_i)
   For all the quanta of $cdis$
     test-quantum($cdis$,$quantum$,ϵ,A_i,B_i,$OBST$)
   For all CS_i of merge-space($cdis$)
     If CS_i is occupied by A_i
        $OCS_i = OCS_i \cup (CS_i \times I_{i+1} \ldots \times I_n)$
     If CS_i is free with A_i and B_i
        $FCS_i = FCS_i \cup (CS_i \times I_{i+1} \ldots \times I_n)$
     Else
        $MCS_i = MCS_i \cup CS_i$
        If $i < n$
           Subspace-Analysis(CS,$i+1$,CS_i,DQ_{i+1})

discretize-space creates a structure accounting for the discretization of a hyperparallelepiped HP.

compute-dep-max computes the maximal motion of links A_i and B_i according to this portion of space and the discretization step.

test-quantum determines the type of space quantum (free or occupied) by computing the distance and the propagation around the quantum (distances are computed only if the type has not already been determined by propagation).

merge-space selects quanta of the same type in order to regenerate hyperparallelepipeds of dimension i.

3.3 Distance computations

These algorithms depend on the models used for representing the volumes describing the robot and the obstacles.
The selected models must allow for a tradeoff between a reliable description of these volumes and efficient distance computation.

3.3.1 Object modeling

In its present version, the modeling system possesses the following primitives:

- Planes
- Spheres
- Seg-Spheres { points located at a distance $< R$ from a segment }
- Dis-Spheres (same as above for a disk)
- Pol-Spheres (same as above for a polygon)
- Polyhedra

The advantage of the first 5 types of models is that the geometric computation of volumic primitives are limited to straightforward calculations of distance between points, circles and segments. For the polyhedrical primitives (which allow a more reliable description of the objects), we have implemented an algorithm for computing the distance between convex polyhedra which is quasi-linear in the total number of vertices.
The following section will briefly describe its principle (for more details see [5]).

3.3.2 Gilbert's algorithm

Let C_1 and C_2 be two sets of $\Re^3$ describing the space occupied by the objects whose distance is sought.

This distance can then be expressed as:

$d(C_1, C_2) = min\{\|z\|, z \in C_1 \ominus C_2\}$ if $C_1 \cap C_2 = \emptyset$
$d(C_1, C_2) = -min\{\|z\|, z \in cl\overline{C_1 \ominus C_2}\}$ if $C_1 \cap C_2 \neq \emptyset$ ("cl" indicates closure).

It corresponds to the shortest distance from the origin to a point situated on the boundary of the Minkowsky difference C_{12} between these two sets.
In the case where these sets are convex, the computation of C_{12} can be effected in time $O(n_1 + n_2 + s)$ where s can be $O(n_1 n_2)$ [2].

The principle of the algorithm consists of generating a sequence of polytopes with less than four vertices, whose distance to the origin converges towards d. The algorithm enables us to obtain these elementary polytopes from the vertices of C_1 and C_2 with a complexity proportional to $n_1 + n_2$, and to calculate their distance to the origin by an efficient procedure.

In addition, the algorithm furnishes the C_1 and C_2 boundary points that satisfy this minimal distance and the elementary sets of vertices which permitted this computations. In the case of collision-detection along a trajectory, this information can then be employed to initialize the algorithm, thus accelerating the search for the following set. Notice however, that for intersecting objects, the algorithm only provides a lower bound of $|d|$.

3.4 Complexity

As stated in the previous section, distance computations are made in constant or linear time according to the type of primitive involved.

On the other hand, complexity depends on the level of accuracy required, but this is inherent in any discretization method.

With respect to the analysis of an n dimensional space, if K represents the mean number of values to be tested per dimension, the complexity is $O(K^n)$, i.e. an order of magnitude above that of [13][8] which involves a complexity $O(K^{n-1})$.

Thanks to the propagation mechanism, the distance computation will seldom be effected in the CS regions that are sufficiently far (inner or outside) from the obstacle transform. These computations become more frequent when space portions, close to the transform boundaries, are analyzed.

On the other hand; when the environment includes several obstacles, they can be sorted so that these distance tests are performed only on certain ones.

These remarks highlight the advantage of the method whose complexity decreases when these mechanisms are efficient.

4 Building and using the CS structure

4.1 Creation of the adjacency graph

The graph can be easily obtained from the tree structure of CS and the modeling of the space regions in the form of interval products.

Two nodes corresponding to regions R_1 and R_2 of FCS_i and FCS_j are adjacent if the

$min(i,j)$ first intervals of these regions possess a non-empty intersection area (exept for the particular case of rotations $[0, 2\pi[$).
Finding a collision-free trajectory envelope between Q_i and Q_f can then be decomposed into:

- finding the source and destination nodes with the help of the tree structure.

- finding a path connecting these nodes in the graph with a A* algorithm.

Remark: The algorithm described in the preceding paragraph can also be used to consider fictitious obstacles by adding obstacle-typed leaves to the tree.
These fictitious obstacles can be used to represent space regions of the robot (camera field of view, workspace of another robot ...) requiring region allocation in order to traverse them.
During the search in the graph, these nodes are regarded as free only when the robot can allocate the corresponding region.

4.2 Application to motion planning for manipulators

To generate a trajectory contained in the free space envelope, the simplest approach consists of employing a segment to connect all the boundary centres between consecutive hyperparallelepipeds. This method leads to unnecessary motions when large regions are traversed.
Thus, optimization of the trajectory in terms of distance requires an algorithm adapted from [6]. It consists of recursively tightening a trajectory obtained by selecting the Q_{i+1} of the boundary between R_i and R_{i+1} closest to Q_i (see Figure 4).

Figures 6 and 7 show a trajectory furnished by the obstacle avoidance system for a manipulator with six revolute joints. In this example, only four degrees of freedom have been taken into account (with discretization steps of 3^0, 3^0, 4.5^0 and 12^0).
Figure 5 shows a slice of the corresponding four dimensional CSpace. The pyramidal object represents a camera field of view. For the computation of FCS, this object was considered as a ficticious obstacle i.e. the $CSpace$ regions which cause a collision between the robot and this object were labelled by this obstacle.

The two trajectories were thus computed from the same $CSpace$ representation. In the first case, the camera is not considered (the regions labelled by the camera field of view

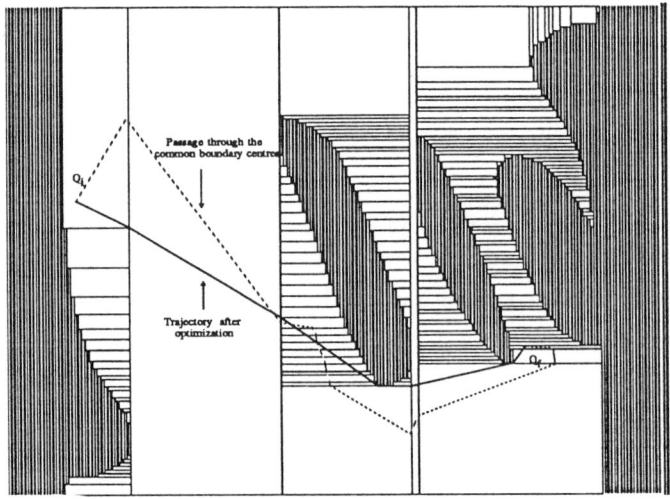

Figure 4: Trajectory optimization in terms of distance

are regarded as free during the search in the adjacency graph). In the other case the robot can not allocate these regions (which are considered as occupied). The computed trajectory is then different though the initial and final configurations are the same.

The system is written in Common Lisp. The computation time required about 50 minutes on a Sun 3/60 workstation for the global free space description, and less than ten seconds for generating a particular trajectory.

4.3 Mobile robots

The method is equally applicable to the case where the configuration space of the system is (x,y,θ). This has resulted in the validation of a first approach which takes into account kinematic constraints while allowing for the generation of trajectories for non-holonomic robots [11].

For such systems all trajectories in FCS are not necessarily feasible. From results established in [10] and from a particular procedure adapted to the free space modelization by a set of hyperparallelepipeds, this approach enables us to produce trajectories with maneuvers.

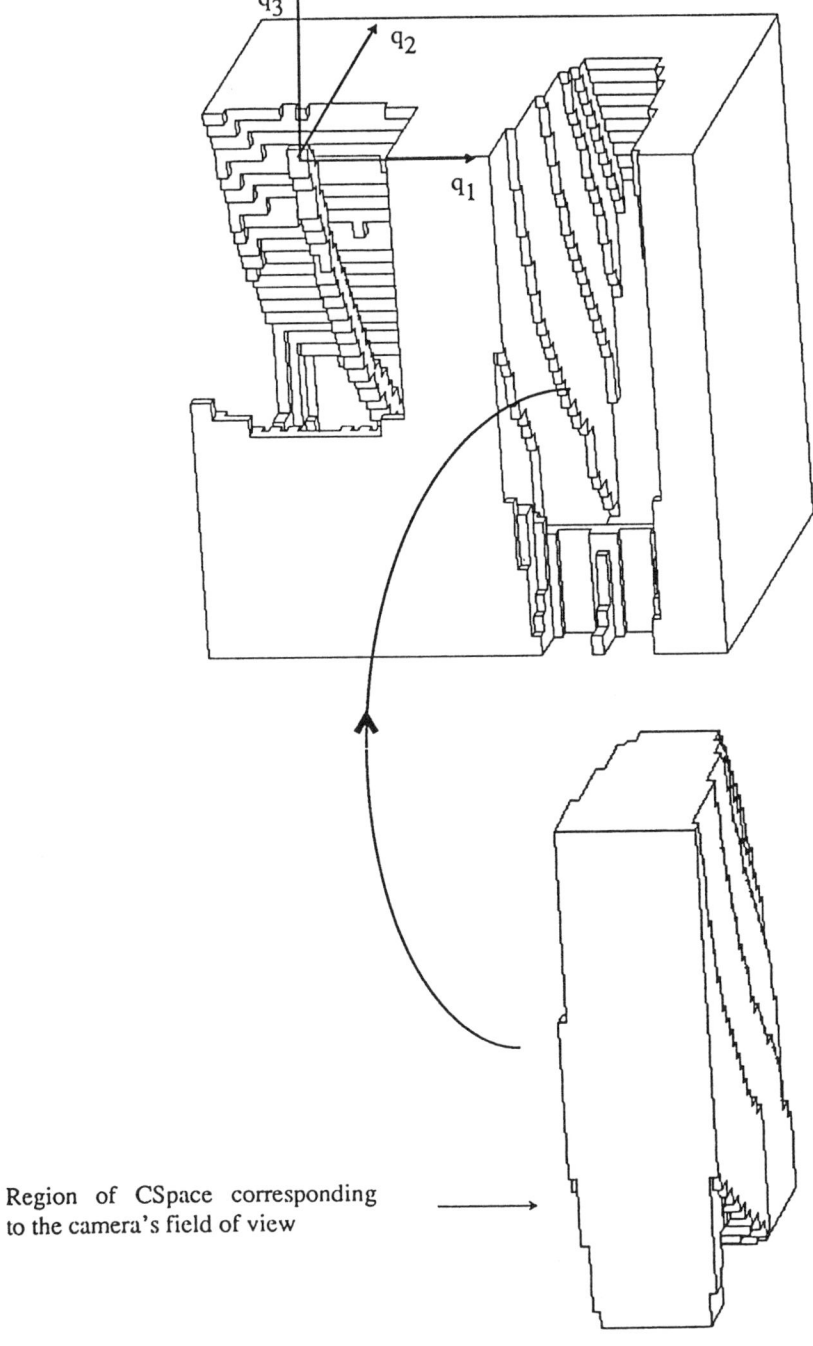

Figure 5: A three dimensional slice $(\theta_1, \theta_2 \theta_3)$ of the free space description

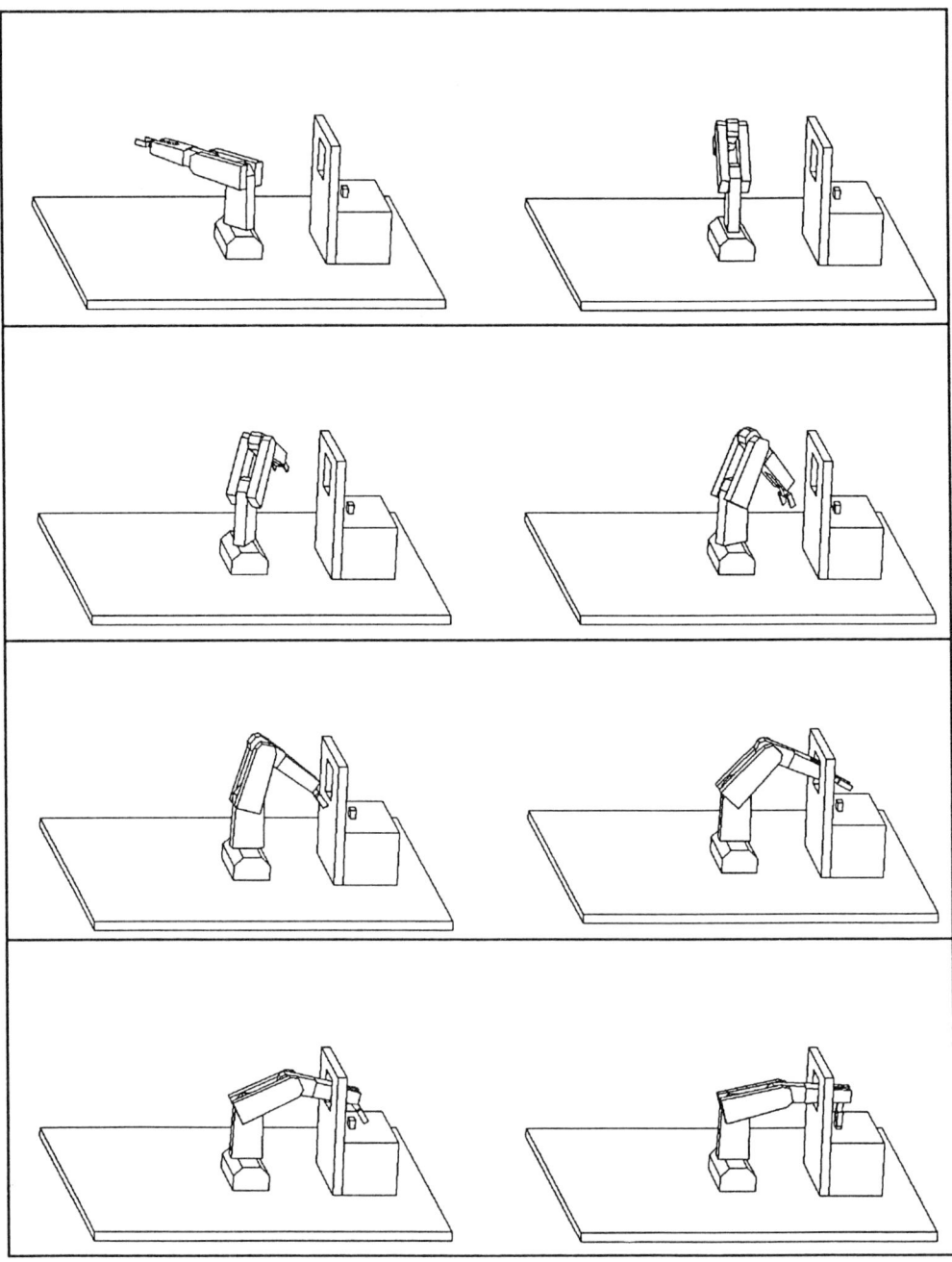

Figure 6: A path requiring 4 degrees of freedom

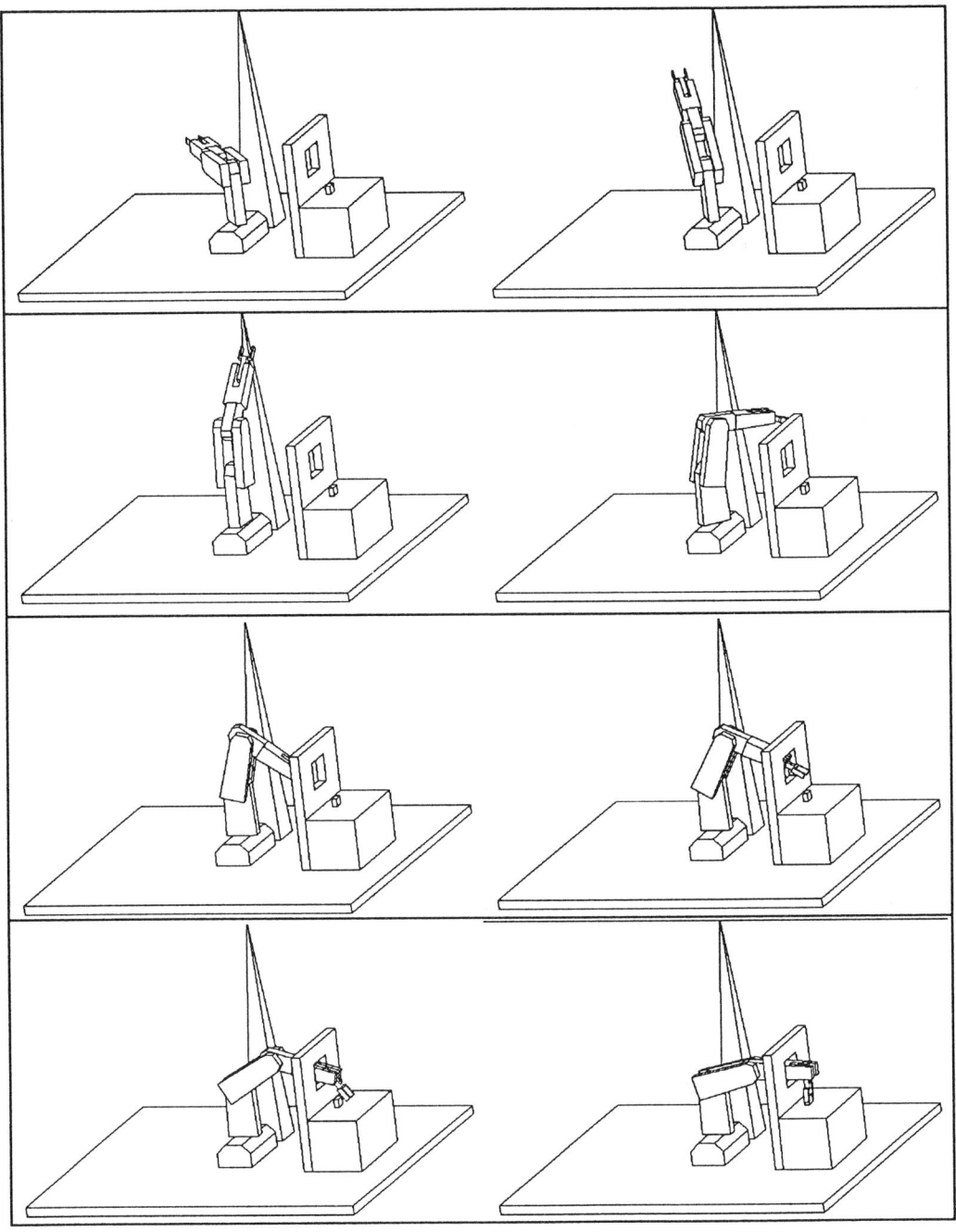

Figure 7: Another path computed by the system when the camera is active

5 Conclusion

The present version of this system is written in Common Lisp (except for distance computations that are C functions which have been termed Lisp).
This explains the relatively long time needed for the computation of the free space description.
Because global methods are usually time-consuming, the robot manipulator's six degrees of freedom cannot be taken into account. However, efficiency can be dramatically enhanced by customizing the algorithms for the geometric and kinematic structure of a particular robot.
On the other hand, the effect of a fine discretization on computational times limits the scope of global methods to transfer motions. This does not mean that complete space structuring is not interesting. Without it, interdependencies with other modules (e.g. for planning grasp position) could only be achieved through the use of "generate and test" methods.
These global methods can be combined with local methods for handling a certain number of degrees of freedom or for assisting a local planner in its search.

References

[1] R. Alami and H. Chochon. Programmation et contrôle d'exécution d'une cellule flexible d'assemblage. In *Journées Robotique, INRIA, Sophia-Antipolis (France)*, June 1987.

[2] J.D. Boissonnat. Complexité géométrique et robotique. In *Journées Robotique, INRIA, Sophia-Antipolis (France)*, June 1987.

[3] H. Chochon and R. Alami. A knowledge-based system for programming and execution control of multi-robot assembly cells. In *'87 International Conference on Advanced Robotics (ICAR), Versailles (France)*, October 1987.

[4] B. Faverjon. Object level programming of industrial robots. In *IEEE, International Conference on Robotics and Automation, San Francisco (USA)*, 1986.

[5] E.G. Gilbert, D.W. Johnson, and S.S Keerthi. A fast procedure for computing the distance between objects in three space. In *IEEE, International Conference on Robotics and Automation, Raleigh (USA)*, 1987.

[6] L. Gouzènes. *De la perception à l'action en robotique d'assemblage: contribution à l'algorithmique de base*. Thèse de docteur ingénieur, Université Paul Sabatier, Laboratoire d'Automatique et d'Analyse des Systèmes (C.N.R.S.), Toulouse (France), November 1984.

[7] L. Gouzènes. Strategies for solving collision-free trajectories problems for mobile or manipulator robot. *International Journal of Robotics Research*, 3(4), Winter 1984.

[8] C. Laugier. *Raisonnement géométrique et méthodes de décision en robotique. Application à la programmation automatique des robots*. Thèse d'état, Institut National Polytechnique de Grenoble, December 1987.

[9] C. Laugier and F. Germain. An adaptative collision-free trajectory planner. In *'85 International Conference on Advanced Robotics (ICAR), Tokyo (Japan)*, September 1985.

[10] J.P. Laumond. Feasible trajectories for mobile robots with kinematic and environment constraints. In *International conference on autonomous systems*, Amsterdam, Netherland, 1987.

[11] J.P. Laumond, T. Siméon, R. Chatila, and G. Giralt. Trajectory generation and motion control of mobile robot. In the same book.

[12] T. Lozano-Perez. Automatic planning of manipulator transfer movements. *IEEE Transactions on System, Man and Cybernetics*, 11(10), 1981.

[13] T. Lozano-Perez. A simple motion-planning algorithm for general robot manipulators. *International Journal of Robotics Research*, 3(3):224–238, June 1987.

[14] T. Lozano-Perez. Spatial planning : a configuration space approach. *IEEE Transaction Computer*, 32(2), 1983.

[15] Lumelsky. *Adaptative and Learning Systems*, chapter Continous robot motion planning in unknown environment. Picnum Press, 1986.

[16] E. Mazer. *HANDEY : un modèle de planificateur pour la programmation automatique des robots*. Thèse d'état, Institut National Polytechnique de Grenoble, December 1987.

[17] O. Khatib. Real-time obstacle avoidance for manipulators and mobile robots. In *IEEE, International Conference on Robotics and Automation, St Louis (USA)*, 1985.

[18] J. T. Schwartz and M. Sharir. On the piano mover II : general techniques for computing topological, properties of real algebraic manifolds. *Applied Math.*, 4, 1983.

[19] P. Tournassoud and B. Faverjon. Cooperation of two manipulators. In *'87 International Conference on Advanced Robotics (ICAR), Versailles (France)*, 1987.

Trajectory planning and motion control for mobile robots *

J. P. LAUMOND, T. SIMEON, R. CHATILA, G. GIRALT
Groupe Robotique et Intelligence Artificielle
LAAS-CNRS
7 avenue du Colonel Roche, 31077 Toulouse Cedex, France

Abstract

This paper addresses two aspects of the navigation problem for a two d.o.f mobile robot (non holomic system): trajectory planning and motion control. Trajectory planning concerns the existence and the generation of a feasible collision-free trajectory, and motion control the actual execution of this trajectory.

The problem has to be solved in constrained and non-constrained environment. We summarize some results previously obtained in non constrained space and develop a general approach for finding feasible trajectory in constrained space. This method is based on a result which characterizes the existence of a feasible trajectory by means of the existence of a connected open component in the admissible configuration space. Its current implementation, based on a configuration space structured into hyper-parallelepipeds, is described.

The trajectory is then analyzed in order to smooth it when possible, using clothoid curves. Its execution is controlled by means of comparing sensor readings with the local environment model along it.

1 Introduction

Over the last decade, robotics researchers have had to address the problems of planning and control of robot motions, including issues that range from geometric reasoning to the study of control. To accomplish this, they have developed their own tools [3].

Over the same period, an important research effort in the field of geometry has primarily focused on the design and analysis of efficient algorithms relying on various approaches ranging from real algebraic geometry to computational geometry [26].

---
*This paper has also been presented at the I.U.T.A.M. / I.F.A.C. Symp. Dynamics and Controlled Mechanical Systems, June 1988, Zürich

Recent developments tend to establish a fruitful synergy between the techniques involved in these two fields. Notice that the desire to build actual physical systems gives rise to novel and challenging issues, as in the case of the problem dealt with in this paper.

The work described has been conducted within the HILARE mobile robot project developed at LAAS. It deals with the navigation of a mobile robot subject to major environmental and kinematic constraints. The problem is the following :

"How to plan and control collision-free trajectories for mobile robots for which the dimension of the configuration space (three) is larger than the number of d.o.f (two) ?"

After a brief review of existent methods for trajectory planning and motion control for mobile robots, we especially investigate the geometric aspects of the question. We mention in the last section the techniques which are in development in order to control robot motion from the sketch of trajectory provided by geometric reasoning.

2 Mobile robot motion planning

Trajectory planning is only one aspect of the global navigation problem that includes also environment perception and modeling, accounting for inaccuracies, real-time decision-making, spatial structure learning... An overall synthesis of such issues is given in [7].

Even if we restrict ourselves to the geometric and control aspects, collision-free motion planning for a mobile robot still remains an open problem, in spite of important partial results. There are four classes of methods to deal with this problem; according as the geometric constraints of the environment are more or less strong, the methods integrate more or less motion control aspects.

The first kind of approaches is applied in highly structured environments. The better known systems concern the road-following problem. [29] [22] study the global architecture needed by a trajectory planning and control system that uses vision for guidance. The most relevant issues in such systems are the real time processing of road feature extraction, and visual feature tracking in order to control vehicle motion.

The second class concerns the local methods. Their principle consists in using only local and poor but quickly acquired information on the environment, in order to plan a trajectory in real time. The potential fields based methods are the most commonly used [15]: the robot is supposed to be moving in a fictive potential field wherein obstacles are associated with a repulsive field and the goal with an attractive one. This method is efficient in numerous situations (convex obstacle avoidance for instance) but not in very constrained space, where the goal and the obstacles are very near. [10] palliates this last drawback by using an approach wherein collision-free constraints appear as linear constraints in a quadratic criterium minimization problem associated to the goal. Because of a local view of the environment these methods are not complete (i.e. they do not guaranty to find a solution if it exists).

Several methods can be gathered in a third class. They deal with unstructured environments, and are based on a structuring of the euclidian free-space using particular approximation shemas. If the robot is assumed to be circular [23] [17] propose as a structure, respectively the Voronoi diagram of the polygonal environment, and a generalized visibility graph in a more general environment; trajectories thus produced are smooth (they do not have angular points nor cusps). Other methods decompose the free-space in elementary places (convex polygons [6], generalized cones [4] ...), which are structured into a graph whose adjacency relation indicates the possibility (and the associated way) of moving from a place to another. [10] associates a local method developed for corners with a decomposition of the free-space into lanes based on its Voronoi diagram. All these methods are only applicable when free-spaces is large with respect to robot geometric and kinematic constraints.

Motion planning in a very constrained environment needs to consider the formalism of the configuration space (CS) [21]. This space is the space of independant parameters that characterize the position and the orientation of a mobile body ($R^2 * S^1$ in our case, where S^1 is the unit circle). It is divided into the admissible space (ACS) in which the mobile body does not intersect the obstacles, the free space (FCS), defined as the closure of the ACS interior, the occupied space (OCS), defined as complementary to ACS (for an analysis of the connectivity and topology of such spaces, see [18]).

There are two types of methods for configuration space exploration, viz:

- The methods [25] [2] [1] that lead to an exact partitioning of either FCS or its boundary which is constituted by quadratic surface patches [24]. Notice that the most efficient (in $O(n^3 \log n)$) is [1] and has been implemented.

- The numerous methods of *"paving"* CS into *"space quanta"*: cells [5], hyperparallelepipeds [12] [27], one-dimensional slices structured into regions [20], cubes structured into octrees [9]. All these methods have been implemented.

These last class of methods solves (completely or partially) the classical piano-mover problem that assumes the piano to be holonomic. The aim of the next section is to take into account kinematic constraints in such formalisms.

3 Trajectory planning for non-holonomic robots using a configuration space approach

3.1 Position of the problem

The last approaches characterize the existence of a trajectory by means of the existence of a connected component of ACS including the initial and final configurations. Such characterization is a priori valid only for holonomic systems. Let us recall briefly some fundamental concepts of analytical mechanics [30].

The joints expressing the relations between the velocities of the configuration parameters, which cannot be reduced to relations between these parameters (and which therefore cannot be integrated) are called non- holonomic joints. The number d, of degrees of freedom of a system is defined by $n - r$, n representing the number of configuration parameters and r the number of independent non-holonomic joints. A holonomic system is a system without non-holonomic joint, i.e., $d = n$. For such systems, any infinitesimal motion (*i.e.*, any infinitesimal variation of the configuration parameters) can be achieved. This property does not hold for non-holonomic systems.

Let us consider a mobile robot whose locomotion system consists of two independent driving wheels located on a common axis (see Fig. 1). Let (x, y, θ) be the three configuration parameters.

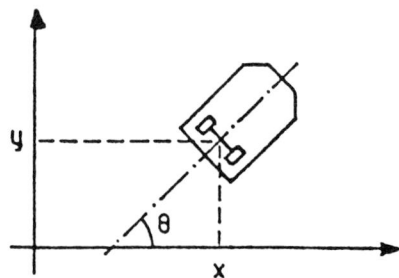

Figure 1: Configuration parameters

The state equations characterizing the system motion are defined by:

$$\begin{aligned} dx &= \frac{1}{2}(v_1 + v_2)\cos\theta \\ dy &= \frac{1}{2}(v_1 + v_2)\sin\theta \\ d\theta &= \frac{1}{2}(v_1 - v_2) \end{aligned} \qquad (1)$$

where v_1 and v_2 stand for the velocities of both driving wheels. From these equations we deduce that there exists one (and only one) non-holonomic joint :

$$dy - dx\tan\theta = 0.$$

For such a system all trajectories in ACS are not necessarily feasible (see Fig 2). A *feasible trajectory* is a function of time, piecewise continuous and differentiable (the robot's linear speed vector determines its orientation), the points where the linear speed is zero corresponding to "pure" rotations. In order to distinguish forward and backward motions [13] uses the notion of *tracing* that retains only the topological and geometrical characteristics of the trajectory. With respect to this terminology our problem consists in defining an algorithm for planning polygonal tracings.

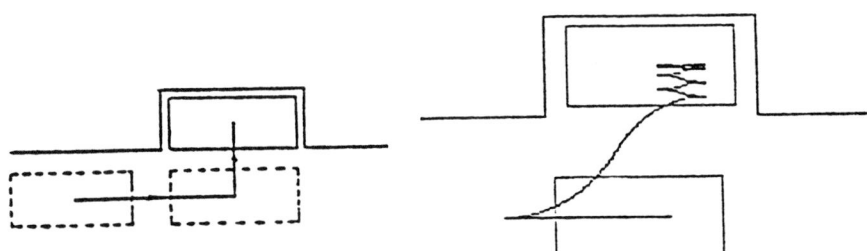

Figure 2:   Non feasible trajectory   Feasible trajectory

3.2   An algorithm schema

[18] establishes that:

Property : If c and c' are two configurations contained in a single connected domain of the interior of ACS, then there exists a feasible collision-free and contact-free trajectory between c and c'.

Remark: this result is established in the more general case where the gyration radius is lower bounded (as for a car).

The proof of this property is based on the existence of a feasible trajectory between any two configurations of an elementary open set of $R^2 * S^1$. This existence proves that any configuration resulting from a motion consequent to an infinitesimal variation of the configuration parameters can be reached in an open set. Several procedures for searching feasible trajectories between two configurations of an open set can be defined according to the type of open set considered. A detailed proof is given in [18]. It is constructive and leads to the following algorithm:

Input data:

- A contact-free trajectory $\mathcal{T}$ (i.e., in an open connected domain $\mathcal{D}$ of ACS) between two configurations c and c'.

- A procedure $\mathbf{P}(c_1, c_2)$ which produces a feasible trajectory between any two configurations c_1 and c_2 in an open set of given type O.

Output data: A feasible contact-free trajectory $\mathcal{T}'$ between c and c'.

Algorithm:

Cover $\mathcal{T}$ by a finite sequence of open sets $\mathcal{O}_1, \ldots, \mathcal{O}_p$ of type O such that:
$\mathcal{O}_i \subset \mathcal{D}, \mathcal{O}_i \cap \mathcal{O}_{i+1} \neq \emptyset, c \in \mathcal{O}_1, c' \in \mathcal{O}_p.$
$i \leftarrow 1$
$c_1 \leftarrow c$
While $(i < p)$
    Let c'' be a configuration of $\mathcal{O}_i \cap \mathcal{O}_{i+1}$
    $\mathcal{T}_i \leftarrow \mathbf{P}(c_i, c'')$
    $i \leftarrow i + 1$
    $c_i \leftarrow c''$
$\mathcal{T}' \leftarrow (\bigcup_{1 \leq i < p} \mathcal{T}_i) \cup \mathbf{P}(c_i, c')$

The implementation of this algorithm requires :

- A procedure for computing ACS or FCS.

- A procedure for searching a contact-free trajectory.

- The definition of a type of open sets of $\mathcal{R}^2 * \mathcal{S}^1$ and the associated procedure **P** for searching a feasible trajectory.

Notice that the data structures used by the methods representing the configuration space by means of discretization offer the advantage of directly providing a path in the space quantum adjacency graph.

To adapt this algorithm schema, it suffices to define a procedure for searching feasible trajectories within these quanta. (Remark: these space quanta are closed but it can easily be shown that the algorithm holds).

3.3 Planning of polygonal tracing

In this section we present an implementation based on a general software described in [27] and resumed in section 3.3.1, that structures FCS into hyperparallelepipeds (parallelepideds in our case). The procedure for searching feasible trajectories in these parallelepipeds is described in 3.3.2. The results, the extensions currently under study and the elements used for analyzing its complexity are discussed respectively in 3.3.3, 3.3.4 and 3.3.5.

3.3.1 FCS Computation and Exploration

The algorithm is based on some principles established in [12]. It receives as input:

- A mobile body $A_n(q_1, \ldots, q_n)$ (or an articulated system $A(q_1) \ldots A_n(q_1, \ldots, q_n)$) and a set of obstacles O_i, described by assemblies of elementary surfaces or volumes (polygons in this application).

- A CS to be analyzed (interval product $I_i = [q_{i_{min}}, q_{i_{max}}]$).

- A discretization step on each dimension.

From a hierarchical description of the mobile by means of different volumes

$$B_i(q_1, \ldots, q_i) = \bigcup_{i < j \leq n} \{A_j(q_1, \ldots, q_i, \ldots, q_j)/q_j \in [q_{j_{min}}, q_{j_{max}}]\}$$

one gets as output a tree structuring of CS of the form:

- OCS_i = subspace of CS occupied whatever $q_{i+1}, \ldots, q_n$.

- FCS_i = subspace of CS free whatever $q_{i+1}, \ldots, q_n$.

- $MCS_i = OCS_{i+1} \cup FCS_{i+1} \cup MCS_{i+1}$, subspace of CS for which a subspace of dimension $j > i$ had to be analyzed recursively to determine its belonging to either OCS or FCS.

Each component is represented by a set of hyperparallelepipeds (HP) of dimension i. The principle used to analyze a discretized subspace of dimension i relies on :

- The computation of a function $Distance(Q)$ (minimal translation allowing $A_i(q_1, \ldots, q_i)$ to be either put in contact or removed from an obstacle), for discrete values of Q.

- A function for propagating the results on a ball centered on Q and of radius $\mathcal{F}(Distance(Q), dQ)$.

- The use of diverse heuristic techniques permitting reduction of the number of calls of the distance function.

A tree representation of CS under the HP form permits easy superposition of a graph structure whose vertices are the elements of FCS_i and whose arcs reflect a connectivity relationship between two nodes.

For Q_{start} and Q_{end} given, the search carried out in this graph with an algorithm A^* provides a trajectory hull. In the case of a mobile robot with kinematic constraints, the heuristic used involves a weighting between the distance to the goal and a criterion characterizing the robot's maneuverability to traverse the HP.

3.3.2 Procedure P

The trajectories produced by this procedure are polygonal tracings *i.e.*, consisting of rotations and of line segments going either forward or backward.

The input data of **P** are a parallelepiped $Pa = [X_1, X_2] * [Y_1, Y_2] * [\theta_1, \theta_2]$, an initial configuration $c = (p, \theta) \in \mathcal{R}^2 * \mathcal{S}^1$ and a window Wd on the boundary of Pa allowing passage to the adjacent parallelepiped. Wd can be of three types according to whether the adjacency is for a constant x, y or θ_i.

The procedure **P** furnishes a feasible trajectory between c and a configuration c' of Wd. We denote by Pa^r and Wd^r the projections of Pa and Wd on $\mathcal{R}^2$ (see Fig. 3). δPa^r stands for the Pa^r boundary rectangle.

Given a point p of the plane and a sub-interval $[\theta, \theta']$ of $[\theta_1, \theta_2]$, we call Sec(p,$[\theta, \theta']$) the domain swept by the lines passing through p and of orientation $\theta'' \in [\theta, \theta']$. $Rot(p, [\theta, \theta'])$ refers to a rotation at point p, allowing to reach (p, θ') from (p, θ) by a segment included in $\{p\} * [\theta_1, \theta_2]$. $Line(p, p')$ stands for a translation of vector $\vec{pp'}$.

Procedure P

$\tau \leftarrow \emptyset$
$c_i \leftarrow c$ /*$c_i = (p_i, \theta_i))$*/
While $(Sec(p_i, [\theta_1, \theta_2]) \cap Wd^r = \emptyset)$
    Compute $Int = \{p_1, p_2, p_3, p_4\} = Sec(p_i, [\theta_1, \theta_2]) \cap \delta Pa^r$
    Choose $p_j \in Int$ such that $Dist(Line(p_i, p_j), Wd^r)$ is minimal
    /* Let θ_j the orientation of the line (p_i, p_j) in $[\theta_1, \theta_2]$ */
    $\tau \leftarrow \tau \cup Rot(p_i, [\theta_i, \theta_j]) \cup Line(p_i, p_j)$
    $c_i \leftarrow (p_j, \theta_j)$
Choose a point $p' \in Sec(p_i, [\theta_1, \theta_2]) \cap \delta Wd^r$
/* Let θ' the orientation of the line (p, p') in $[\theta_1, \theta_2]$, $c' = (p', \theta')$ */
$\tau \leftarrow \tau \cup Rot(p_i, [\theta_i, \theta']) \cup Line(p_i, p')$

Figure 3 illustrates this algorithm on an example.

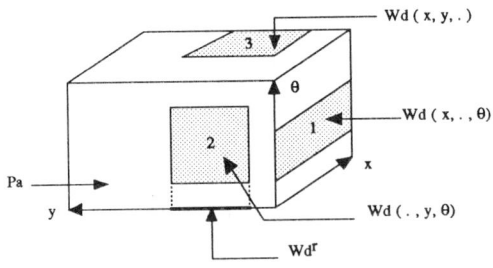

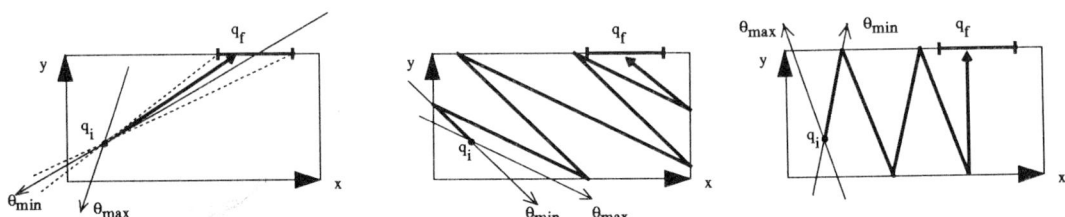

Figure 3: Trajectory planning in a parallelepiped of CS

3.3.3 An example

Figure 4 shows an environment with a corridor and a door, and the associated FCS. Figure 5 shows two results furnished by the algorithm starting with the same data, but with two distinct heuristics (in the path search step) : the first one minimizes the angular gap between two adjacent parallelepipeds in $\mathcal{R}^2 * \mathcal{S}^1$, the second chooses the parallelepipeds whose dimension on $\mathcal{S}^1$ is maximal.

3.3.4 Complexity

The complexity of the global algorithm is governed by the representation and exploration of CS. It is in $O(n/\varepsilon^3)$ where n represents the total number of obstacle vertices and ε the size of the elementary parallelepiped.

The complexity of the procedure **P** is difficult to assess since it depends on the number of maneuvers (defined as the configurations in which the robot's speed is zero). Evidently, there exists some cases where **P** is optimal (i.e., where no "better" trajectory exists in terms of number of maneuvers). However, in general, one would have to compare this number to the optimum number. Evaluating such optimum is a difficult task which, as

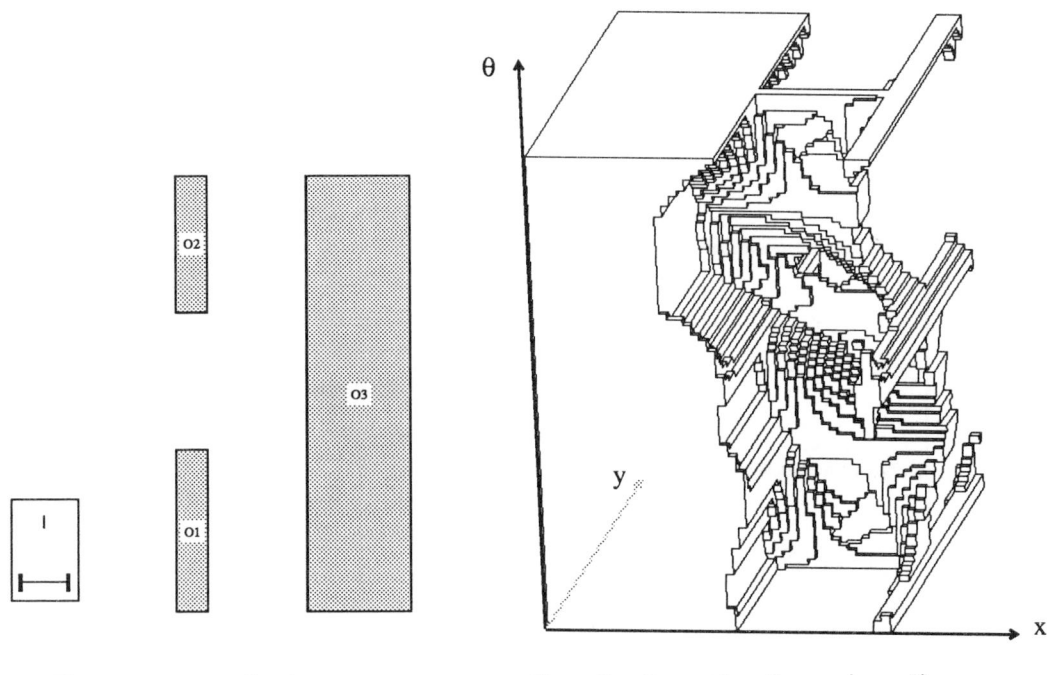

Figure 4:   Environment   Free Configuration Space (x, y, θ)

yet, has not been performed.

Initial results have been obtained : [19] proposes an algorithm sketch for searching maneuver-free trajectories for a non-holonomic circular robot with a lower bounded gyration radius. [11] shows that the problem for a point in a polygonal environment is decidable in $2^{O(poly(n))}$ where *poly* designates a polynomial function.

3.3.5   Extensions

The extensions under study concern with:

- Extending the scope of **P** to parallelepipeds adjacent to the hull provided by algorithm 3.3.1, in order to decrease the number of maneuvers when it is possible.

- Replacing procedure **P** by a procedure allowing helix planning in the parallelepipeds. This new procedure will provide more general trajectories than polygonal tracings and will allow to deal with the lower bounded gyration radius constraint which appears in most mobile robots.

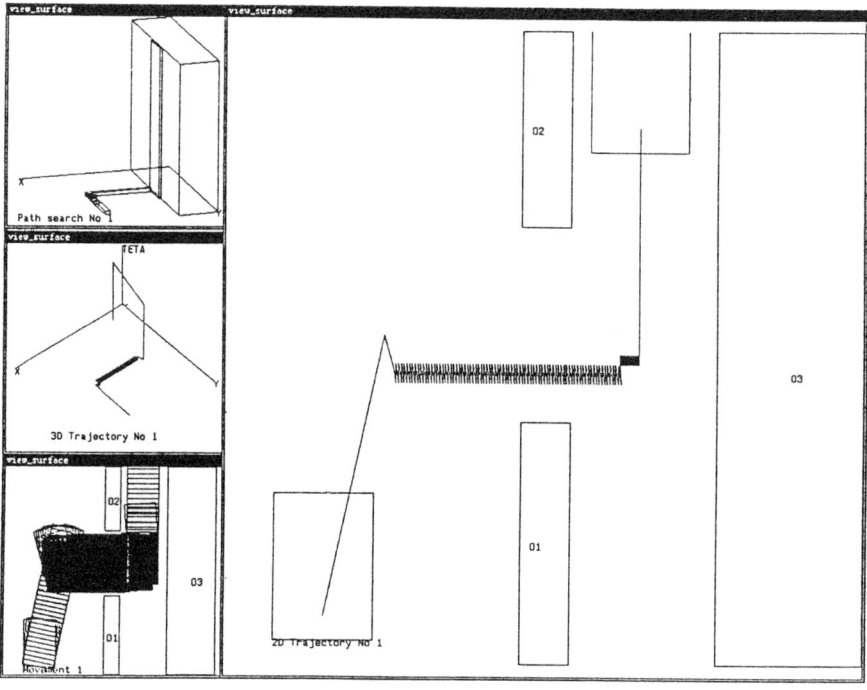
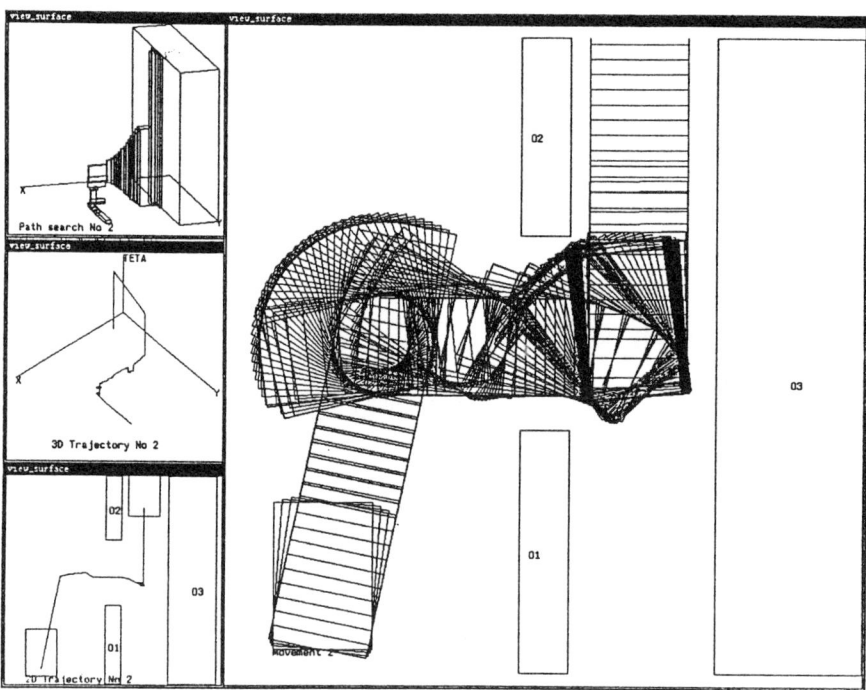

Figure 5: Trajectories produced by the algorithm

The approach can be applied whenever a $\mathcal{R}^2 * \mathcal{S}^1$ configuration space representation and exploration system is available. These systems are often complex and highly sophisticated. The choice made in our implementation is certainly not optimal since the tool described in section 3.3.1 is a general purpose tool (valid for spaces of any dimension k, efficient for k= 2,3,4). It has been used for obtaining rapidly the initial feasible trajectories for the mobile robot.

4 Trajectory Execution

As we already mentioned before, while the road-following type of methods rely on *physical* features of the local structured environment (*e.g.*, road boundaries) to guide robot motion, in our case such features are not always available. We will then replace this information by the precomputed trajectory.

After producing the trajectory, the problem is to control robot motion so that it stays on this trajectory. Due to inaccuracies in the measurement of robot position (by odometrical dead-reckoning for example), the movement will not follow the computed trajectory in general. Furthermore, the precomputed trajectory is based on a model of the environment that is also inaccurate.

On the other hand, some trajectories produced by a search in the configuration space, while guaranteeing collision-free motion, are not "easily" feasible, or require very slow movements because of their shape (*e.g.*, saw-like trajectories).

In order to take into account the mentioned two kinds of errors on the one hand, and to smoothen the considered constrained trajectory so that the movement is more continuous on the other hand, we propose to consider the local model of the environment together with the precomputed trajectory as inputs to the control system.

Let us recall that the trajectory planning method produces a polygonal tracing, *i.e.*, a trajectory constituted by straight line segments on which the robot can move either forward (if the motion agrees with its orientation) or backward (if both are opposite), and turns.

If there are turns along the trajectory between two consecutive segments, they correspond to corners or cusps according as the movements on the segments are identical (both forward or backward), or opposite. In each case the velocity vector (derivative vector of the position) must be zero. Cusps impose a mandatory stop.

We show in this section that corners can be smoothed. More precisely we show how to link the two segments with a doubly differentiated curve (*i.e.*, without a zero velocity vector) which is as close as necessary to the corner.

In order to link two straight line segments with a curve C such that the union is a doubly differentiated curve, C (assumed to be parametrized by time) has to pass in finite time from an infinite curvature radius to a finite one.

From equations (1) the curvature radius ρ of robot trajectory is given by : $\rho = \frac{l(v_1+v_2)}{2(v_1-v_2)}$ where l designates the distance between the two wheels and v_1 and v_2 their respective linear velocities. A particular solution is given by *clothoids*.

Several papers investigate the use of clothoids in mobile robot motion planning [8] [14]. The major property of a clothoid is that its curvature is in inverse ratio to the curviline abscisse : $\rho(k,t) = k*t/V$ where k is the proportionnality ratio, V the constant norm of the velocity vector along the clothoid, and t the time parameter.

The advantage of this choice is an easy command of the two driving wheels [16]. Indeed the vehicle describes a clothoid when wheel accelerations are constant and opposite, which furthermore leads to an optimal command *w.r.t* energy consumption.

A clothoid thus permits to pass from an infinite curvature radius to a finite one in finite time (and *vice-versa*). In order to link two segments we must use two tangent clothoids arcs. A property established in [16] shows that it is possible to compute a doubly differentiable curve consisting of a pair of clothoid arcs that connect two intersecting segments such that this curve remains inside *any* given region bounded by the two segments and an arc of circle tangent with them. From this property we deduce easily that the clothoid arcs can be as close to the corner as the environment constraints may impose it.

Because the polygonal tracing we have planed is in an open component of the free-space, we have the guaranty that the corners can be smoothable whitout vehicule stops. The only points where the velocity vector has to be zero are the cusps.

The resulting final trajectory, smoothed when possible, will be executed using sensor data (*e.g.*, ultrasonic sensors). If the trajectory is already smooth (*e.g.*, straight lines, clothoids, *etc.*), then sensor data, matched to the local environment model, will help to localize the robot along its trajectory, thus correcting the dead-reckoning system's error. In the case of *small* variations of the environment with respect to the model, the use of sensor data enables to control the motion in order to avoid collisions, thus departing from the computed trajectory. We rely on a basic assumption: the computed trajectory is not unique but belongs to a family of very close trajectories such that we can actually replace it by the family's envelope. Indeed, we have produced a non-contact feasible trajectory by means of paving the free-space with open cylinders. Within an open cylinder, there exists at least one trajectory between any two configurations, and in general more than one. Therefore, while staying inside the same open cylinder, the robot can actually move within this family of trajectories, using for this purpose sensor readings. Notice that characterizing the amplitude of the small authorized variations is a difficult open problem linked with the precise study of the topological structure of equivalent trajectories.

References

[1] F. Avnaim and J-D. Boissonnat. A practical exact motion planning algorithm for polygonal objects admist polygonal obstacles. In *IEEE, International Conference on Robotics and Automation, Philadelphia (USA)*, 1988.

[2] Zhang Bo, Zhang Ling, and Zhang Jianwei. *An efficient algorithm for findpath with rotation*. Report Department of Computer Science, Beijing Univ., 1986.

[3] M. Brady, J.M. Hollerbach, T.L. Johnson, T. Lozano-Perez, and M.T. Mason. *Robot motion : planning and control*. MIT Press, 1982.

[4] R. A. Brooks. Solving the find-path problem by good representation of free space. *IEEE journal on Systems, Man and Cybernetics*, 2(13), 1983.

[5] R. A. Brooks and T. Lozano-Perez. A subdivision algorithm in configuration space for findpath with rotation. *IEEE journal on Systems, Man and Cybernetics*, 2(15), 1985.

[6] R. Chatila. Path planning and environment learning in a mobile robot system. In *ECAI, Orsay (France)*, Juillet 1982.

[7] R. Chatila and G. Giralt. Task and path planning for mobile robots. In *NATO ARW on Machine Intelligence and Knowledge Engineering, Maratea (Italy)*, Mai 1986.

[8] H. Chochon and Leconte B. *Etude d'un module de locomotion pour un robot mobile*. Rapport de fin d'étude ENSAE, Laboratoire d'Automatique et d'Analyse des Systèmes (C.N.R.S.), Toulouse (France), Juin 1983.

[9] B. Faverjon. Object level programming of industrial robots. In *IEEE, International Conference on Robotics and Automation, San Francisco (USA)*, 1986.

[10] B. Faverjon and P. Tournassoud. A local based approach for path planning of manipulators with a high number of degrees of freedom. In *IEEE, International Conference on Robotics and Automation, Raleigh (USA)*, 1987.

[11] S. Fortune and G.T. Wilfong. *Planning constrained motion*. Technical Report, ATT Bell Laboratories, Murray Hill, Mai 1988.

[12] L. Gouzènes. Strategies for solving collision-free trajectories problems for mobile or manipulator robot. *International Journal of Robotics Research*, 3(4), Winter 1984.

[13] L. Guibas, L. Ramshaw, and J. Stolfi. A kinetic framework for computational geometry. IEEE Symp. on FOCS, 1983.

[14] Y. Kanayama and N. Miyake. Trajectory generation for mobile robots. In G. Giralt O. Faugeras Eds., *Robotics Research 3*, MIT Press, 1986.

[15] O. Khatib. Real time obstacle avoidance for manipulators and mobile robots. *International Journal of Robotics Research*, 1(5), 1986.

[16] A. Khoumsi. *Pilotage, asservissement sensoriel et localisation d'un robot mobile autonome*. Thèse de l'Université Paul Sabatier, Toulouse (France), Laboratoire d'Automatique et d'Analyse des Systèmes (C.N.R.S.), Toulouse (France), Juin 1988.

[17] J. P. Laumond. Obstacle growing in a nonpolygonal world. *Information Processing Letters*, 25(1), Avril 1987.

[18] J.P. Laumond. Feasible trajectories for mobile robots with kinematic and environment constraints. In F. C. A. Groen L. O. Hertzberger, editor, *Intelligent Autonomous Systems*, North Holland, 1987.

[19] J.P. Laumond. Finding collision-free smooth trajectories for a non-holonomic mobile robot. In *10th IJCAI, Milan (Italy)*, 1987.

[20] T. Lozano-Perez. A simple motion planning algorithm for general robot manipulators. Robotics Research: The Third International Symposium, O. Faugeras and G. Giralt Eds., MIT Press, 1986.

[21] T. Lozano-Perez. Spatial planning : a configuration space approach. *IEEE Transaction Computer*, 32(2), 1983.

[22] B. Mysliwetz and E.D. Dickmanns. A vision system with active gaze control for real-time interpretation of well structured dynamic scenes. In F. C. A. Groen L. O. Hertzberger, editor, *Intelligent Autonomous Systems*, North Holland, 1987.

[23] O'Dunlaing and C. Yap. A retraction method for planning a motion of a disk. *J. of Algorithms*, 6:104–111, 1985.

[24] J. Reif. Complexity of mover's problem and generalizations. In *IEEE Symposium on FOCS* pages 421–427, 1979.

[25] J. T. Schwartz and M. Sharir. On the piano mover : the case of a two dimensional rigid polynomial body moving amidst polygonal barriers. *Communication on Pure and Applied Math*, (36), 83.

[26] J. T. Schwartz, M. Sharir, and J. Hopcroft. *Planning, Geometry and Complexity of Robot Motion. Artificial Intelligence*, Ablex Pub., 1987.

[27] T. Siméon. Collision-Free Trajectories Planning using a Configuration Space Approach. (In this book)

[28] P. Tournassoud. Motion planning for a mobile robot with a kinematic constraint. In *IEEE Int. Conf. on Robotics and Automation*, 1988.

[29] R. Wallace, K. Matsuzaki, Y. Goto, J. Crisman, J. Webb, and T. Kanade. Progress in robot road-following. In *IEEE Int. Conf. on Robotics and Automation*, 1986.

[30] E.T. Whittaker. *A treatise on the analytical dynamics of particles and rigid bodies.* Cambridge University Press, 4th Ed., 1965.

Motion planning for a mobile robot with a kinematic constraint

Pierre Tournassoud

INRIA, Domaine de Voluceau
BP 105, 78153 Le Chesnay Cedex, France

Abstract

This paper explores the problem of model-based motion planning for a mobile robot. The kinematic model we use is that of a mobile platform with two degrees of freedom. The mobile robot is subject to a kinematic constraint, the number of its degrees of freedom being less than the dimension of its Configuration Space.

The problem posed here is then essentially to control the number of backing-up maneuvers along a path. Addressing first the "local" problem of turning in a corridor, we show that sliding continuously on the outer wall in a turn constitutes a canonical contact trajectory, in the sense that if it is not safe, there cannot be another safe trajectory with no backing-up maneuver. In the case no smooth trajectory exists, we propose some simple maneuvers.

A heuristic path planning algorithm based on a decomposition of Free Space into cones connected by turns is then presented. Finally we discuss how this can apply within the more realistic framework of a real mobile robot at work, by implementing these results as a set of rules of behaviour for the robot.

1 Introduction

Very little attention has been devoted so far to one basic issue posed by motion planning for mobile robots : as opposed to manipulator robots, whose bindings between bodies effectively decrease the dimension of their Configuration Space, those are usually subject to kinematic constraints that turn them into non-holonomic systems. Indeed, a mobile robot with two driving wheels mounted on the same axis can only move in the direction normal to the axis, a car can only turn around a point located along the rear axle... Such systems, though their position is described by three-dimensional configuration vectors, have only two degrees of freedom.

We recall that path planning for a manipulator robot moving among obstacles is now well understood. This is a difficult problem [SS83], but some practical algorithms have been proposed following the pioneering work of Lozano-Pérez and Wesley [LW79]. The basic idea behind most algorithms is to build a representation of Free Space in Configuration Space as a connectivity graph of free regions, that can then be searched for a path. The practical limitation due to the complexity of the representation can be alleviated by limiting the number of degrees of freedom taken into account exhaustively [Fav84,Loz86]. We have proposed another line of work, characterized by a combined use of local and global models, the global model being updated from the results of execution of the local planner [FT87a,FT87b].

In the plane, motion planning for one object is easy if it performs only translations, the boundary of Free Space being simply obtained by growing obstacles to represent sliding motions of the robot on their contours. When rotations are allowed, the explicit computation of configuration obstacle surfaces, if more complex, remains possible [ABF88]. Most algorithms nevertheless rely on a discretization of orientation ranges [LW79,BL85].

Much of our interest in model-based motion planning for a mobile robot must be credited to Laumond. [Lau86] demonstrates that the existence of a safe trajectory for the robot is unaffected by the kinematic constraint. [Lau87] addresses more the problem of a lower bounded turning radius than that of the kinematic constraint stated above. Indeed, as it uses a mobile robot of circular symmetry, raising the constraint on the turning radius will make the problem equivalent to planning a path for a disc without any specific constraint. We recall this 2D problem can be solved in a straightforward manner by using a generalized visibility graph (composed of arcs of circles and line segments) built from the obstacles grown by the radius of the disc.

The decomposition used by Brooks [Bro82] is of interest in our context. Free Space is directly represented in 2D Cartesian Space by a union of generalized cones, that capture freeways between pairs of facing walls. The motion of the robot is decomposed into pure translations along the spines of the cones, and pure rotations at points where spines intersect. Planning is performed by propagating constraints on the range of legal orientations along a given sequence of cones. No specific orientation is imposed along a translation, so this does not answer the question of path planning with a kinematic constraint. It could be done simply by imposing that the robot is oriented along the axis of the spine during a translation. Yet this would be very restrictive, as not much would be left to be propagated for a search.

In this paper we begin by addressing the more restricted problem of turning in a corridor. We introduce contact trajectories, that precise the notion of C-surface in this context : they are defined by the contact of the robot with an obstacle at one point, plus the kinematic constraint. We show that sliding continuously on the outer wall in a turn constitutes a canonical contact trajectory : it is such that if it is not safe, there cannot be another safe trajectory with no backing-up maneuver.

We then describe a heuristic motion planning algorithm, which makes it possible to control the number of maneuvers necessary along a path. It is based on a decomposition of Free Space into cones connected by turns : a connectivity graph is built whose arcs represent contact trajectories in turns, and possible links in corridors between them.

Finally we show how these ideas can be incorporated in the more realistic framework of a real mobile robot at work. We choose to implement these results as a set of rules of behaviour for the robot : using a combination of information from on-board sensors and higher level knowledge on its environment, the robot reproduces canonical trajectories without pre-computing them.

2 Kinematics of the Mobile Robot

The mobile robot is a vehicle with two driving wheels mounted on the same axis, controlled independently, plus one free wheel. In the sequel, we assume for simplicity that its body is rectangular, though any polygonal shape would lead to similar results. A configuration vector **q** consists in the 3-tuple (x, y, θ), with x and y the coordinates in the plane of point O of the robot at mid-distance from the driving wheels, and θ its orientation. Nevertheless, the robot has only two degrees of freedom : we write **v** the vector of components v and v', products of angular velocities of the wheels by their radius (see fig. 1a).

We get the equations of motion :

$$\dot{x} = \cos\theta \, \frac{v + v'}{2} \quad (1)$$

$$\dot{y} = \sin\theta \, \frac{v + v'}{2} \quad (2)$$

$$\dot{\theta} = \frac{v - v'}{2d} \quad (3)$$

with d equal to half the distance between the driving wheels. We also write this block of equations

$$\dot{\mathbf{q}} = \mathbf{K}(\theta)\mathbf{v} \quad (4)$$

as **K** depends only on angle $\theta = \theta_0 + \int_{t_0}^{t} (v - v') \, 2d \, dt$.

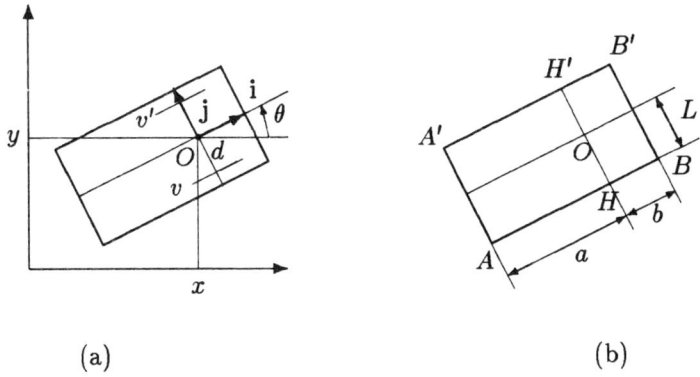

Figure 1: Kinematics (a) and geometry (b) of the mobile robot.

Equations 4 implicitely contain the non-holonomicity constraint of the vehicle :

$$dy = \tan\theta \; dx \tag{5}$$

It simply states that the robot can only move in the direction normal to the axis of the driving wheels.

For a point **m** bound to the robot, with coordinates (p,q) in its relative frame $(O, \mathbf{i}, \mathbf{j})$ (as represented on fig. 1a), we get :

$$\dot{\mathbf{m}} = \mathbf{L}(\theta)\dot{\mathbf{q}} = \begin{pmatrix} 1 & 0 & -p\sin\theta - q\cos\theta \\ 0 & 1 & p\cos\theta - q\sin\theta \end{pmatrix} \dot{\mathbf{q}} \tag{6}$$

Writing the standard differential relation $\dot{\mathbf{m}} = \mathbf{J}\,\mathbf{v}$, where $\mathbf{J}$ is the jacobian matrix at point **m** for configuration **q**, we derive from equations 4 and 6 that :

$$\mathbf{J} = \mathbf{J}(\theta) = \frac{1}{2}\begin{pmatrix} \cos\theta & \cos\theta \\ \sin\theta & \sin\theta \end{pmatrix} + \frac{1}{2d}\begin{pmatrix} -p\sin\theta - q\cos\theta & p\sin\theta + q\cos\theta \\ p\cos\theta - q\sin\theta & -p\cos\theta + q\sin\theta \end{pmatrix} \tag{7}$$

A trajectory of the robot is a curve in 3D-space (see figure 2a) :

$$t \rightsquigarrow \Gamma(t) = \mathbf{q}_0 + \int_{t_0}^{t} \mathbf{K}(\theta)\mathbf{v}\;dt. \tag{8}$$

For a point bound to the robot it yields a trajectory in the plane :

$$t \rightsquigarrow \gamma(t) = \mathbf{m}_0 + \int_{t_0}^{t} \mathbf{J}(\theta)\mathbf{v}\;dt. \tag{9}$$

In the sequel γ will denote the trajectory traced by point O of the robot. The radius of curvature of γ is equal to :

$$\rho = d\,\frac{v+v'}{v-v'} \tag{10}$$

Hence a cusp point of γ either corresponds to a turn on the spot ($v = -v' \neq 0$), or a backing-up maneuver ($v = v' = 0$). As far as path planning is concerned, the only significant cusp points of the trajectory are backing-up maneuvers. In this paragraph, let us define Free Space as the set of configurations such that the robot does not come into contact with any obstacle. Free Space is then an open set in Configuration Space, and a turn on the spot can always be regularized so as to yield a smooth trajectory for which the surface swept by the robot is as close from the original than we wish. Without any loss of generality, we can impose that the trajectory γ of point O be smooth unless at finitely many cusp points that we know correspond to backing-up maneuvers.

Proposition 1 *From the trajectory γ of point O, we can derive the trajectory Γ of the robot in its Configuration Space.*

Proof: Consider the parametrization $t \rightsquigarrow s(t) = \int_{t_0}^{t} \sqrt{\dot{x}^2 + \dot{y}^2} \, dt$ of trajectory γ. Combining equations 1 and 2 we get $v + v' = 2\,\dot{s}(t)$. From the radius of curvature ρ of γ we derive the ratio of the velocities :

$$\frac{v'}{v} = \frac{\rho - d}{\rho + d} \tag{11}$$

We thus compute the values of the command parameters v and v'. □

The proposition below, from Laumond [Lau86], states that the *existence* of a safe trajectory for the mobile robot between an initial and a goal configuration is unaffected by the kinematic constraint.

Proposition 2 (Laumond) *Between two safe configurations that belong to the same connected component of Free Space, there exists a trajectory that respects the kinematic constraint.*

We give the skeleton of the proof that appears in [Lau86]. Let Γ be a trajectory of finite length in Configuration Space, for which the robot does not come into contact with any obstacle. By compactness, Γ is contained in a finite sequence of open cells $C_k, k = 1, \ldots, n$ verifying the following properties. Each cell C_k is a small cylinder of configurations (x, y, θ), cartesian product of a disc by a range of orientations, such that no configuration in the cell intersects an obstacle. We build a sequence of configurations $\mathbf{q}_k$ such that $\mathbf{q}_0 \in C_1$ is the origin, $\mathbf{q}_n \in C_n$ is the goal, and $\mathbf{q}_k \in C_k \cap C_{k+1}$ for $k \in \{1, \ldots, n-1\}$. We then replace the part of the trajectory inside C_k between $\mathbf{q}_{k-1}$ and $\mathbf{q}_k$ by a trajectory with the same tangent lines at end-points $\mathbf{q}_{k-1}$ and $\mathbf{q}_k$, which respects the kinematic constraint, and remains within cell C_k (figure 2b,c). By pasting the new pieces together we obtain a trajectory in Free Space that respects the kinematic constraint. Yet this can only be done at the cost of adding for each new piece a number of backing-up maneuvers. Hence this property, even if it is contructive, is of limited practical interest for the computation of a trajectory that we wish has the smallest number of maneuvers possible.

3 Turning in a Corridor

The problem is essentially that of controlling the number of maneuvers needed on a path between two given configurations. We first concentrate on a more restricted problem, that of making a turn in a corridor, described by two walls, each one with an angular point. These are the "outer" obstacle S_e and "inner" obstacle S_i of figure 3. We assume S_i and S_e^c are convex (S^c denotes the complementary of a set S). Say the robot comes from the left. In the sequel, edge e^- will stand for the first edge of obstacle S_e, e^+ for the latter one.

This can be considered as a "local" motion planning problem, that we will in section 6 incorporate in a global planning scheme (for instance, the algorithm of Brooks [Bro82] uses a similar decomposition). In the absence of any kinematic constraint, that problem is simple. In the case we want to move a rectangular object through the corridor, we would simply check (with notations of fig. 1b) whether when moving with point A sliding on edge e^- of external wall S_e, point B on e^+, the surface swept by the robot intersects the inner wall S_i or not. Recall that for such trajectories defined by a double vertex-edge contact between the moving object and the obstacles, points bound to the robot follow ellipses in the plane, and the envelope of segment $A'B'$ is an astroid. As we now deal with a system incorporating a kinematic constraint, it is natural to study trajectories that remain *at one point* in contact with an obstacle, the remaining degree of freedom being determined by the kinematic constraint. This refines the notion of C-surface in this context.

In the case of a rectangular shaped body and the given kinematics, contact motions can be of the following types. The case when a vertex of the robot remains in contact with an edge of some obstacle (see fig. 4a) will be examined in detail in the next section. In the case of a contact between edges AB or $A'B'$ of the robot and a vertex of an obstacle, the robot can only move in a straight line, unless contact occurs at points H or H' (points on the axis of rotation of the driving wheels), in which case it can turn on the spot around this point (fig. 4b,c). A last case is the contact between edges AA' or BB' of the robot and a vertex of the obstacles, as discussed in section 5 (fig. 4d).

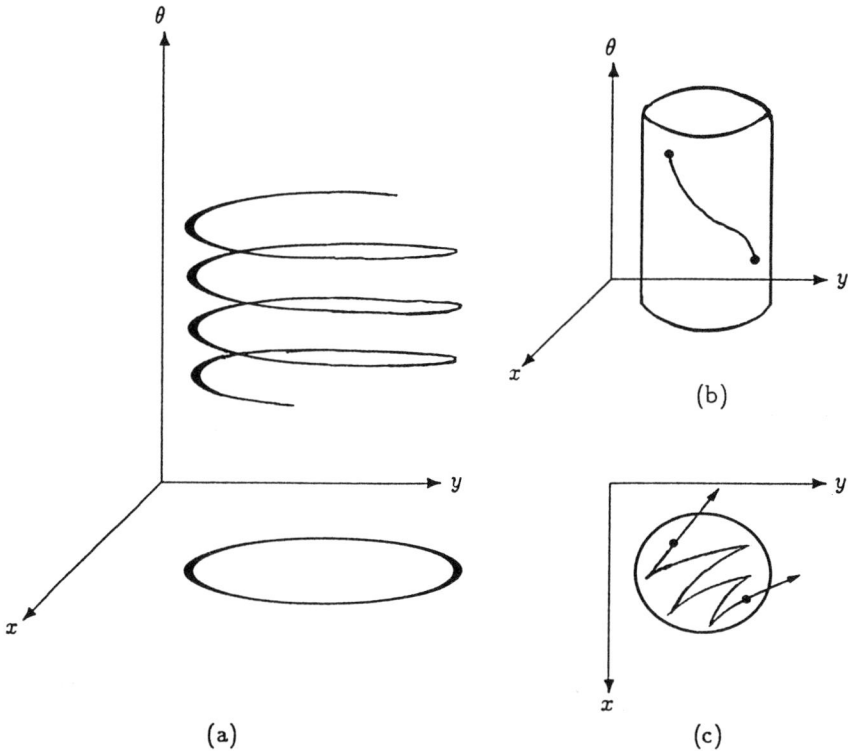

Figure 2: a) A trajectory of the mobile robot in its Configuration Space. b,c) Building a trajectory that respects the kinematic constraint (c) from an arbitrary trajectory in Configuration Space (b).

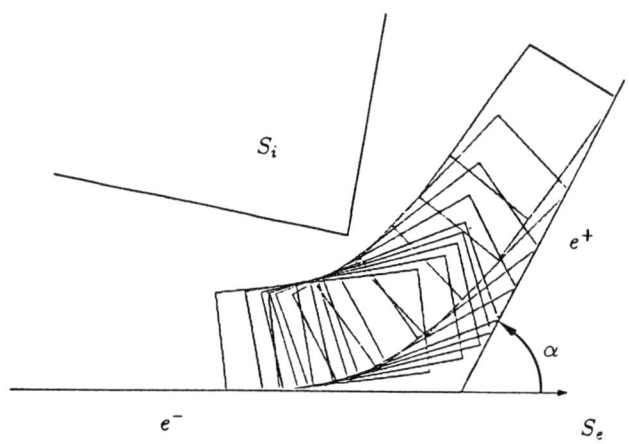

Figure 3: Description of the corridor and canonical contact trajectory.

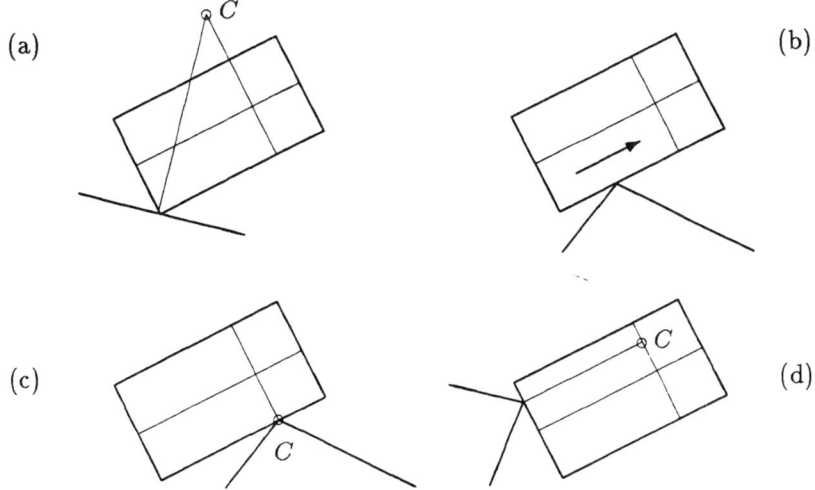

Figure 4: Possible contact motions (point C is the center of rotation).

4 Sliding along the outer wall

We now explicitely write the equations of motion for the robot sliding on the outer wall. The parameters for the geometry of the robot are those of figure 1b : $a = \|HA\|$, $b = \|HB\|$, L is the half-width of the robot. To make the equations simpler, we align the x-axis with edge e^-.

Let us assume the robot slides on the outer wall before the turn (horizontal edge e^- of figure 3), remaining in contact with point A. We get easily :

$$y(\theta) = a \sin \theta + L \cos \theta \qquad (12)$$

From that equation and the kinematic constraint (eq. 5), we derive :

$$x = x_0 + a(\cos \theta + \ln |\tan \frac{\theta}{2}|) - L \sin \theta \qquad (13)$$

The radius of curvature for the trajectory γ of point O equals :

$$\rho = \frac{a}{\tan \theta} - L \qquad (14)$$

Note that there exists a cusp point on γ for $\tilde{\theta} = \arctan(a/L)$.

Let us call α the angle between edges e^- and e^+ of outer wall S_e (fig. 3). In this section, we will suppose that $\alpha \leq \pi/2$. If we want the mobile robot to slide continuously

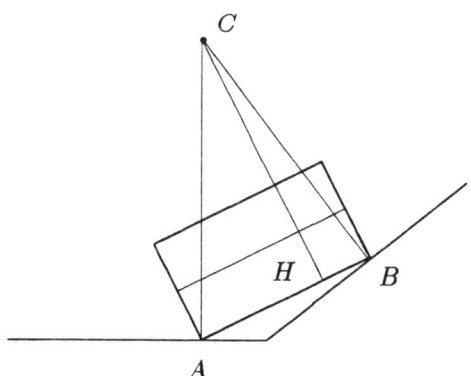

Figure 5: The double contact configuration.

along edge e^-, then e^+, the double contact configuration must be such that normals to the edges at contact points intersect along the axis of the driving wheels (see fig. 5).

The corresponding orientation θ^* thus verifies :

$$\rho = \frac{a}{\tan \theta^*} - L = \frac{b}{\tan(\alpha - \theta^*)} - L \qquad (15)$$

This is, we get $\tan(\alpha - \theta^*) = b/a \ \tan \theta^*$, or :

$$\theta^* = \frac{1}{2}(\alpha + \arcsin(\frac{a-b}{a+b} \sin \alpha)) \qquad (16)$$

We also derive the corresponding value for x_0 in equation 13, with the origin of the x-axis at the vertex of S_e :

$$x_0 = \frac{a+b}{\tan \alpha} \sin \theta^* - b \cos \theta^* - a \ln |\tan \frac{\theta^*}{2}| + L \sin \theta^* \qquad (17)$$

If we slightly translate that trajectory to the right, then the robot necessarily collides with edge e^+ of obstacle S_e. If we translate it to the left, the robot never touches that edge. Note that a nice consequence of equation 14 is that the trajectory of point H, and a fortiori that of B, has no cusp point and is convex. We call γ^* the trajectory we obtain by pasting the two pieces together, as illustrated on figure 3.

In the case $L < \sqrt{ab}$, that is, a robot that is not "larger than long", trajectory γ^* remains smooth, as $\theta^* < \tilde{\theta}$ and the converse for the second part of the trajectory is equally true. To see this, check that $\tan \theta^*$ increases with α between 0 and $\pi/2$, and for $\alpha = \pi/2$ equals $\sqrt{a/b}$, which is less than $\tan \tilde{\theta} = a/L$. In the sequel, we will assume for clarity that the condition $L < \sqrt{ab}$ is verified.

We will denote $R(\gamma)$ the surface swept by the body of the robot when it describes trajectory γ. We say a trajectory γ for which $R(\gamma)$ does not intersect the obstacles is *safe*. The following proposition characterizes the corridors in which the robot does not need to maneuver.

Proposition 3 *There exists a safe trajectory with no backing-up maneuver only if γ^* is safe.*

Proof: It proceeds in two steps. First, from a smooth and safe trajectory γ, we will derive another smooth admissible trajectory that is convex (θ keeps increasing along the trajectory) (1). Then we will show that γ' can only be "above" γ^*, so that γ^* does not intersect the inner wall (2), which will complete the proof.

(1) Let γ be a smooth safe trajectory as illustrated by figure 6. Then let γ' be the boundary of the union of half-planes below γ that do not intersect γ, and whose bounding lines have an orientation lying in between 0 and α. γ' is composed of a finite alterning sequence of arcs of γ and of line segments, with equal tangents at common end-points.

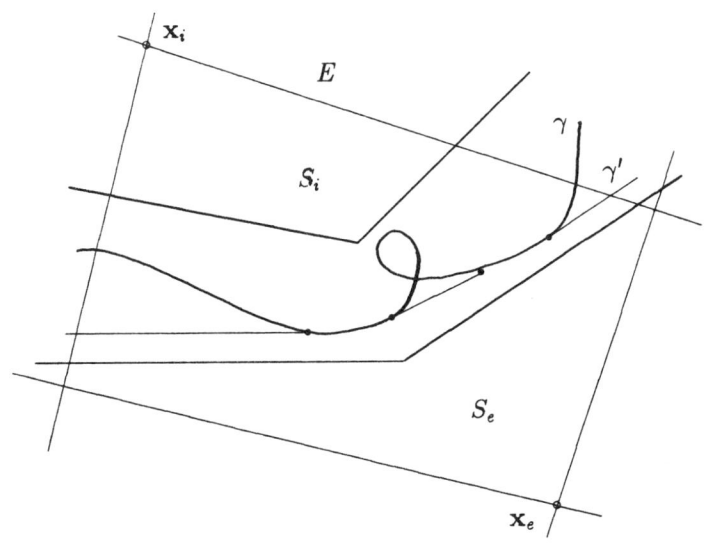

Figure 6: Construction òf a convex trajectory.

It is easily seen that $R(\gamma')$ does not intersect the outer wall. This is obvious at points **m** of $\gamma \cap \gamma'$. Other points of γ' lie on segments $[\mathbf{m}^-, \mathbf{m}^+]$, with both $\mathbf{m}^-$ and $\mathbf{m}^+$ on $\gamma \cap \gamma'$. The conclusion follows by convexity of S_e^c.

We now show that $R(\gamma')$ does not intersect the inner wall. First, for a set S that cuts Free Space F inside the corridor in two connected components, we define $Sup(S)$ as the connected component of $F \setminus S$ "above" S. For this to be done rigorously, we consider two points $\mathbf{x}_i$ and $\mathbf{x}_e$ far inside obstacles S_i and S_e respectively. We call E the bounded portion of the plane as delimited on figure 6. For any set S such that $\mathbf{x}_i$ and $\mathbf{x}_e$ belong to two distinct connected components of $E \setminus S$, we define $Sup(S)$ as being the connected component of $E \setminus S$ containing $\mathbf{x}_i$.

Let $\oplus$ denote the Minkowski sum, and D_L a disc of radius L. We will need the following lemma.

Lemma 1 *For any convex trajectory γ' :*

$$Sup(R(\gamma')) = Sup(\gamma' \oplus D_L)$$

This follows from the fact that the trajectory of point H' of the robot is tangent to its body at that point. Note that even in the case the trajectory of point H' has cusp points, the lemma holds.

Lemma 2 *For any trajectory γ :*

$$Sup(\gamma \oplus D_L) = (Sup(\gamma)^c \oplus D_L)^c$$

Lemma 3 *Let γ designate any smooth trajectory. We denote $\gamma \oplus^\perp S_L$ the union for $\mathbf{m} \in \gamma$ of points $\mathbf{m} + \xi \mathbf{n}$, with $\mathbf{n}$ the unit vector normal to γ at point $\mathbf{m}$ and $\xi \in [-L, L]$. Then :*

$$Sup(\gamma \oplus D_L) = Sup(\gamma \oplus^\perp S_L)$$

We derive that :

$$\begin{aligned}
Sup R(\gamma') &= Sup(\gamma' \oplus D_L) && \text{by lemma 1} \\
&= (Sup(\gamma')^c \oplus D_L)^c && \text{by lemma 2} \\
&\supset (Sup(\gamma)^c \oplus D_L)^c && \text{as } Sup(\gamma') \supset \gamma \\
&= Sup(\gamma \oplus D_L) \\
&= Sup(\gamma \oplus^\perp S_L) && \text{by lemma 3} \\
&\supset Sup(R(\gamma)) && \text{as } R(\gamma) \supset \gamma \oplus^\perp S_L \\
&\supset S_i
\end{aligned}$$

This ends the first part of the proof.

(2) It remains to be shown that γ' necessarily lies "above" γ^*, this is $Sup(\gamma^*) \supset \gamma'$. Then equations very similar to those above show that γ^* does not intersect the inner wall, which completes the proof.

Let us assume the contrary. We call $\mathbf{m}^-$ and $\mathbf{m}^+$ two consecutive intersection points of γ and γ' such that in between γ' lies below γ^*.

- First suppose that at $\mathbf{m}^-$ and $\mathbf{m}^+$, the contact of the robot for trajectory γ^* occurs with the same edge of S_e, say edge e^-. Then at $\mathbf{m}^+$, γ' being convex, $\theta^+_{|\gamma^*} < \theta^+_{|\gamma'} < \alpha$, and the robot intersects the outer wall at $\mathbf{m}^+$ for γ', which is wrong.
- If for γ^* the robot is at $\mathbf{m}^-$ in contact with edge e^- of S_e, at $\mathbf{m}^+$ with edge e^+, then :
  - at $\mathbf{m}^-$ $\theta^-_{|\gamma'} < \theta^-_{|\gamma^*}$,
  - at $\mathbf{m}^+$ $\theta^+_{|\gamma'} > \theta^+_{|\gamma^*}$.

The point $\mathbf{m}^*$ of double contact with the outer wall for γ^* lies in between $\mathbf{m}^-$ and $\mathbf{m}^+$. Then, as γ' is convex, there exists on γ' a point $\mathbf{m}'$ located between $\mathbf{m}^-$ and $\mathbf{m}^+$, such that $\theta' = \theta^*$. As $\mathbf{m}'$ lies below the line tangent to γ^* at $\mathbf{m}^*$, it contradicts the fact that at $\mathbf{m}'$ the robot does not intersect the outer wall. □

We have thus proved the existence of a canonical motion that remains in contact with the outer wall : it is such that if we want to verify the existence of a trajectory without maneuver, we only have to check that for this particular motion the robot does not intersect the inner wall. This is done by simply testing whether the trajectory traced by point H' of the robot stays below that wall.

Still in the case γ^* is safe, we introduce another class of canonical trajectories of interest, defined by the simultaneous contact of the robot with the inner and outer walls. γ^- is the trajectory obtained by sliding on edge e^- of S_e while coming in contact with inner obstacle S_i at point H' of the robot (fig. 7a). In fact, another eventuality is that for all possible sliding motions of the robot on edge e^-, no contact with S_i ever occurs (in which case, easy strategies exist for turning in the corridor). Symmetrically, we define γ^+ as the trajectory obtained by sliding on edge e^+ of S_e with point B and coming in contact with S_i (fig. 7b).

We define θ^- and θ^+ to be the orientations of the robot for the two point contact configurations when they exist, on γ^- and γ^+ respectively. Necessarily, $\theta^+ < \theta^-$.

5 More information when the contact trajectory γ^* is not safe

We now study the case when γ^* is not safe, but there exist safe paths with maneuvers. Recall this last point is checked by verifying that the envelope of segment $A'B'$ for the double contact sliding motion on S_e stays below S_i (section 3). We again write θ^- and θ^+ the orientations of the robot for the two point contact configurations with the inner and outer wall, when they exist. Note that for those configurations there cannot be collision with a wall at a third point, for then, there would not exist any safe trajectory for the robot, regardless of the number of maneuvers.

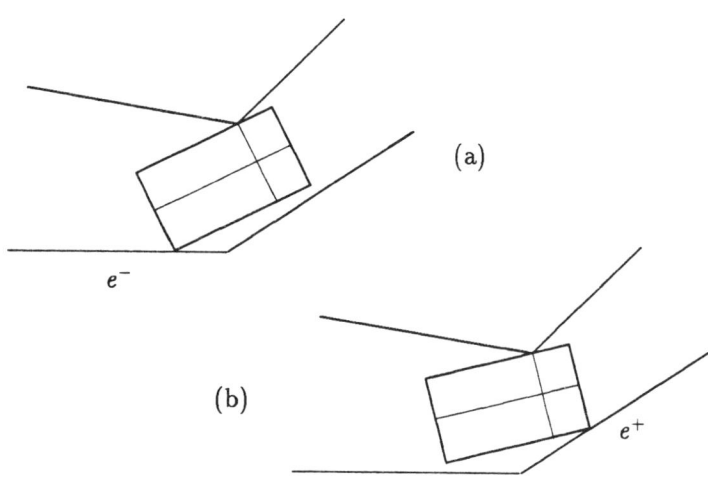

Figure 7: Simultaneous contacts with the inner and outer walls.

Proposition 4 *Suppose there exist some safe paths, but γ^* is not safe (or equivalently, there exists no safe path without maneuver). Then necessarily one of the two point contact configurations with S_i and S_e does exist. The corresponding inequality $\theta^* < \theta^+$ or $\theta^- < \theta^*$ holds. Contact with the inner wall occurs at point H'. The two point contact configuration is reached by the corresponding contact trajectory, γ^- or γ^+.*

Proof: $R(\gamma^*)$ intersects S_i. Translate γ^* along edge e^- to the right until $R(\gamma^*)$ is tangent with S_i at point H'. Indeed, this can always be done, as the minimum distance between S_i and S_e must be more than the width of the robot $2L$. Otherwise, there clearly exists no safe path through the corridor, even including maneuvers. Let θ_x be the orientation of the robot at the double contact configuration. Symmetrically, translate γ^* along edge e^+ towards the bottom until $R(\gamma^*)$ is tangent with S_i. Call θ_y the orientation at the double contact configuration. It is easy to check that $\theta_x < \theta_y$. Then at least one of the following inequalities is true, $\theta_x < \theta^*$ or $\theta^* < \theta_y$. If, for instance, $\theta_x < \theta^*$ holds, then necessarily the configuration corresponding to θ_x is on γ^-, θ_x equals θ^-, and this is a double contact configuration. Contact with S_i occurs at point H'. □

For the rest of this section, we suppose that $\theta^- < \theta^*$ and concentrate on trajectory γ^-. If this is not the case, similar results can be derived using γ^+.

Proposition 5 *Again suppose there exist some safe paths, but only with maneuvers. Call θ' the orientation of the robot when it collides with edge e^+ of S_e on trajectory γ^-. Then with $\theta^- < \theta^*$, necessarily $\theta^- < \theta' < \theta^*$.*

Proof: Suppose this is not true, then $\theta' < \theta^- < \theta^*$. Now imagine we widen the corridor by translating edge e^+ to the right in the direction of e^-. While we push the wall θ' keeps increasing (it would ultimately reach θ^*). On the way we necessarily find a position of wall e^+ for which $\theta' = \theta^-$, this is, a three point contact. For this enlarged corridor, there thus exists no safe trajectory for the robot. Moreover, this is true of the original problem. □

We can then assume that the robot touches the inner wall before colliding with edge e^+. In that case, we propose the following sequence of simple maneuvers, starting at the two point contact configuration on γ^-. An element of the sequence is composed of :
- *a move in straight line until point B touches edge e^+*,
- *then a rotation around H anti-clockwise until point A comes into contact with edge e^- (this is a backing-up maneuver)*.

We can check that no other contact occurs during those moves. As we repeat them, either the orientation of the robot converges towards some angle less than α, and it is blocked, or we reach an orientation superior to α and the turn can be completed. In practice, we would limit ourselves to testing a small number of iterations and declare we found no path if the orientation of the robot is still less than α.

Remark: *What to do in a corridor with a sharp turn ?*

Even in the case $\alpha > \pi/2$, we can exhibit a trajectory for which the robot slides continuously on edge e^-, then e^+. In case the turn is very sharp, this can necessitate sliding with point A' of the robot in contact with e^- after sliding with point A. What has been stated above holds with little modification, the main difference being that the contact trajectory can incorporate two backing-up maneuvers.

In a sharp turn, another class of interesting trajectories is obtained by sliding on the vertex of inner obstacle S_i with edge AA' in contact, as illustrated on figure 8a. This implies one backing-up maneuver, and the robot keeps moving backwards after the turn is completed.

We recall ρ is the radius of curvature for point O of the robot. The differential equation of this particular contact motion writes :

$$d\rho = a\, d\theta \qquad (18)$$

We derive the equations of motion, with ξ and χ the coordinates of the vertex of obstacle S_i in an absolute frame :

$$x = \xi + a\cos\theta + a(\theta - \theta_0)\sin\theta \qquad (19)$$

$$y = \chi + a\sin\theta - a(\theta - \theta_0)\cos\theta \qquad (20)$$

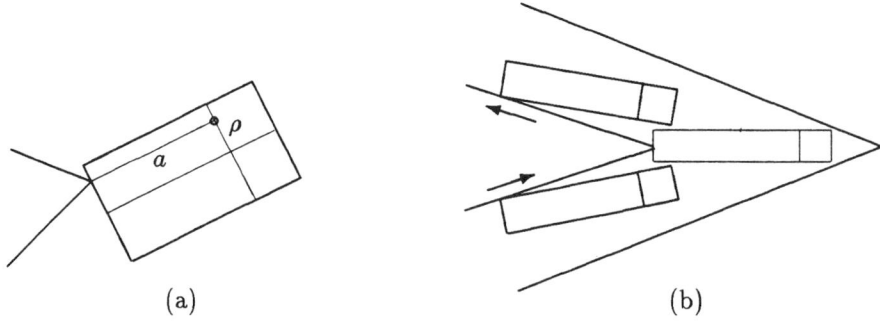

Figure 8: Backing-up maneuver in a corridor with a sharp turn.

Equation 18 implies that the change of orientation after the contact motion is completed equals $\Delta\theta = 2L/a$. Indeed, if the angle between the two edges of S_i is less than $\Delta\theta$, then a complete motion would include :
- *sliding on the first edge of S_i with point A' in contact,*
- *maneuvering keeping edge AA' in contact with the vertex of S_i,*
- *and finally sliding on the latter edge of S_i with point A in contact.*

This is illustrated by the example of figure 8b. A furniture remover has to move a long and heavy sofa in a narrow corridor with a sharp turn. Then he would place wheels at the very front of the sofa, stay at the back and slide along the inner wall, moving backwards after the turn.

6 A planner for a mobile robot with a kinematic constraint

We build a model of the corridors in which the robot can navigate from a planar map of the environment in the following manner. We first construct the Voronoi diagram of the obstacles, represented by polygons, and call *cone* the two linear constraints representing two facing walls at mid-distance from a line segment of the Voronoi diagram. The adjacency of two cones can be easily determined by checking that at least two of the corresponding segments share a common end-point, or that one segment is common to both cones. Our model of Free Space is thus simply composed of pairs of adjacent cones, each pair defining a turn as studied above.

We are aware of difficulties arising when using this simple decomposition technique. Essentially, we can miss legal paths at star-shaped intersections, because we constrain the envelope of the robot to remain within a given sequence of cones. Nevertheless the decomposition we obtain is quite straightforward for most indoor environments. Some very simple pre-processing can be done, such as eliminating cones whose minimum clearance (computed on the edges of the Voronoi diagram) is less than the width $2L$ of the robot.

We then build a connectivity graph representing paths of the robot along the corridors. Nodes of the graph stand for cuts perpendicular to the spines of cones, located just before and after a turn (see fig. 9). The cut before the turn, for instance, must be placed such that when rotating with point O on the cut, the robot can only come into contact with edges of S_i and S_e located before the turn. Recall the robot can turn on the spot, so that the height of point O on the cut provides all relevant information.

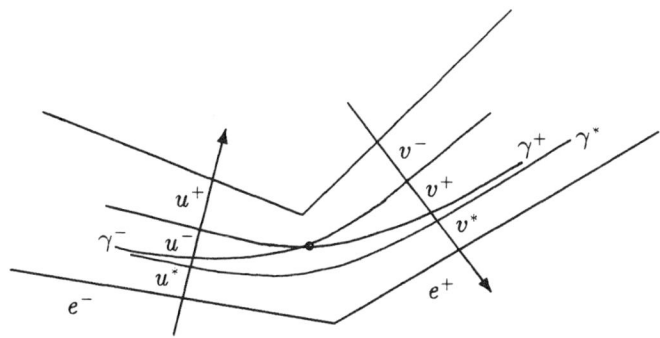

Figure 9: A turn, cuts and reference paths.

Transitions in the graph are of two types. The first class of transitions represent trajectories between the two cuts of a turn. The latter stands for paths of the robot within a cone, connecting a cut at the exit of a turn with the entry cut of the next turn.

The graph can embody different levels of sophistication. Say first that we are only interested in paths with no backing-up maneuver. Then there will be no transition in turns whose γ^* trajectories intersect the inner wall. Other cuts will be typically split into three nodes, representing entry or exit points of γ^*, γ^- and γ^+ trajectories. As illustrated on figure 9, let u^*, u^-, u^+ represent the corresponding points on the entry cut of the turn, and v^-, v^+, v^* points on the exit cut. We make use of the four transitions :

$$u^* \mapsto v^*,$$
$$u^- \mapsto v^-,$$
$$u^+ \to v^+,$$
$$u^+ \mapsto v^-.$$

The first three transitions are executed by staying on the corresponding canonical trajectories. The last one decomposes into a sliding motion on e^+, a turn on the spot around the vertex of S_i, and finally a sliding motion on e^-. The asumption behind this limited choice of transition trajectories is that they represent well the propagation of constraints in a sequence of turns. For instance, it is better to stay on the right in a sequence of left turns, it is preferable to be on the leftmost trajectory in a left turn preceding a turn to the right, etc. If the γ^- and/or γ^+ trajectories do not exist because there is no corresponding double contact configuration (the corridor is too wide), transition trajectories with the same characteristics can nevertheless easily be computed.

There will exist a transition between two nodes at each end of a cone if we can connect them using a trajectory with no maneuver. This is always true if the corridor is long enough. Otherwise, this is easy to check. The only difficulty arises when the robot has got to slide on one wall at an end of the cone, on the opposite one at the other. Then we simply verify that the two curves we obtain (of the type of equation 13) can be smoothly connected (fig. 10a). Otherwise (fig. 10b), some maneuvers are needed.

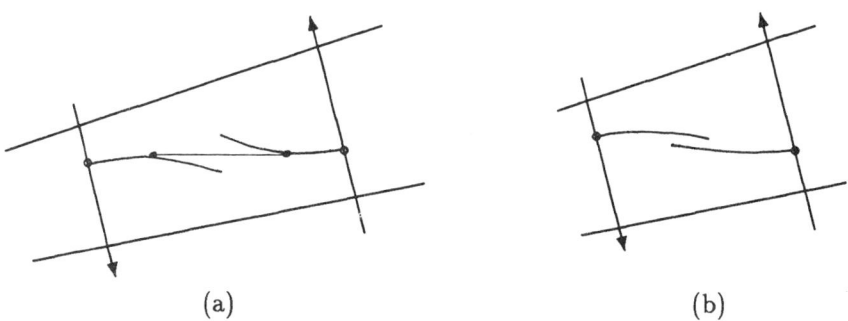

Figure 10: Checking a transition inside a cone.

A more elaborate version could incorporate sequences of maneuvers as described in section 5. Note that in the case of a sharp turn, the robot can be moving tail first after the turn. We must be aware that a turn is indeed described by two sets of nodes as described above, depending whether the robot comes head or tail first. Corridors that are large enough can then be used for intra-cone connections that necessitate U-turns. Planning is performed using a cost function combining the length of a path and a special penalty when maneuvering is necessary.

7 Implementing these results as rules of behaviour

We now describe our practical implementation of these results for the navigation of a mobile robot. Its essential features is that it relies more on information from sensors than on pre-defined planning, though it incorporates some features of global planning as described in the previous section. The robot can operate under two distinct modes.

In the first mode, the robot has no knowledge of its environment, except what it learns by reading the measures of its ultrasonic sensors. When a goal configuration is given to it, a free mode trajectory is first computed among those linking the origin and goal configurations with two tangent circular arcs. It is chosen so that it minimizes a compromise between the length of trajectory γ of point O, and the discontinuity of its radius of curvature when we switch from one circular arc to the other. The robot will eventually be deviated by obstacles on the way as explained below, so the whole computation of a free mode trajectory and constrained trajectory needs to be performed at each time increment. Deviating the robot is performed by applying very simple rules such that :

• If the robot sees a wall in front of him, he tries to re-orient himself parallel to it, temporarily forgetting his goal if necessary.

• If the back of the robot is very close to a wall and he wants to move away, then the best he can do is slide along the wall with nearest point in contact. For example, if point A of the robot is to remain in contact, this is done by respecting the following ratio for velocities (combine equations 10 and 14) :

$$\frac{v'}{v} = \frac{a - (L+d)\tan\theta}{a - (L-d)\tan\theta} \tag{21}$$

with θ the orientation of the robot with respect to the wall.

In the second mode, the sequence of corridors in which the robot has to navigate between the origin and the goal is given to it. The robot does not follow some pre-defined trajectory, but again uses sensor information and rules of behaviour.

We describe in detail the rule used to perform a turn with no maneuver, when the robot is coming with driving wheels at the front. Assume this is a left turn. Before the turn, the robot moves in straight line so that it finally comes to slide with edge $A'B'$ in contact with the vertex of the inner obstacle. The sliding motion goes on until the axis of the wheels gets aligned with the vertex (as wheels are at the front, there is little chance the robot will first bump into edge e^+ of the outer wall). Now the robot starts turning around point H'. Either the turn can be completed this way, or the robot will come into contact with edge e^- at the rear. This means we are now following the γ^- trajectory, and point A must slide on the outer wall. Again this is done by applying equation 21.

The trajectory the robot executes is finally made safe by imposing a velocity-damper type constraint to control the minimum distance with obstacles (as we introduced in [FT87a]), so that it will not hit them.

The following figures illustrate results of the program based on a simulation of the kinematics of the robot and of ultra-sonic sensors. Figure 11 shows how a free mode trajectory is modified to avoid obstacles. The last figures illustrate the planning algorithm. Figures 12a shows a sequence of cones in which the robot has to navigate, and figures 12b the resulting trajectories.

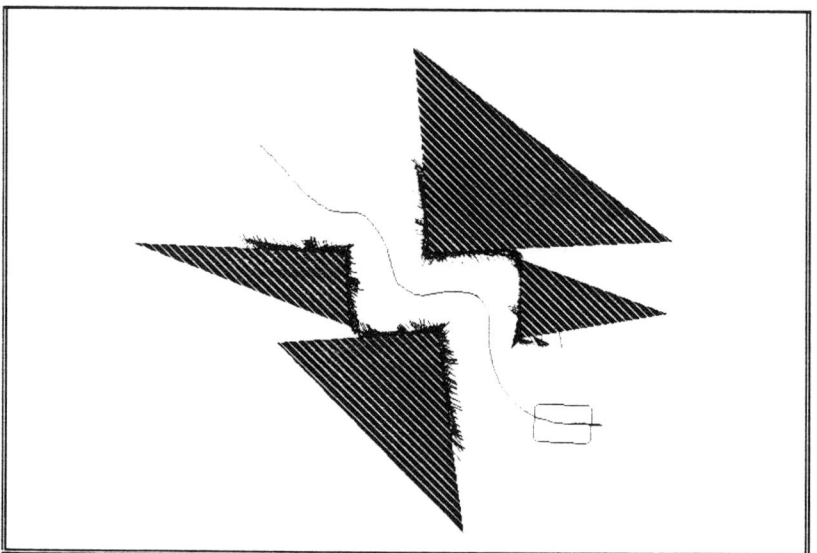

Figure 11: Safe trajectory between two configurations. Thin lines represent echos of the ultra-sonic sensors.

8 Conclusion

We have given in this paper results on the existence of trajectories with no maneuver, or a limited number of maneuvers, for motions of a mobile robot in a corridor with a single turn. Based on these results and a decomposition of Free Space into cones, a heuristic path planning algorithm has been proposed.

Our approach of implementing those results as a set rules of behaviour for the robot seems promising in many ways. It is realistic as it basically relies on sensor information, yet it permits to incorporate complex maneuvers whose result we know exactly.

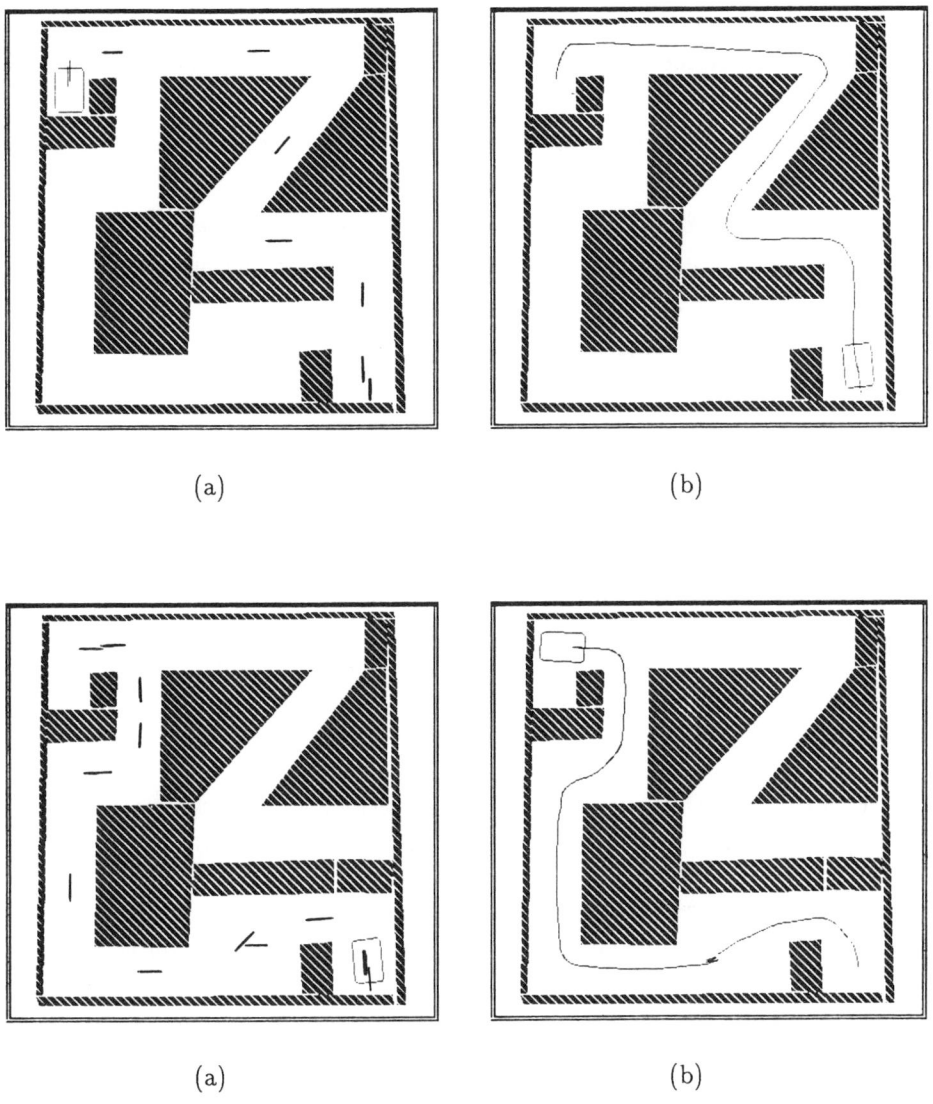

Figure 12: a) Sequences of cones in which the robot must navigate (center configurations of the cones only are drawn). b) Execution of the corresponding trajectories.

Interesting questions remain open :
- What can be said of the existence of a trajectory without maneuver in a complex environment ?
- For a given homotopy class of trajectories, does shortest length implies minimum number of maneuvers ?
- What is the computational complexity of finding a shortest path ?
- What does it become when we impose a higher bound on the number of maneuvers, or when no maneuver at all is allowed ?

Our goal, of course, should be to build a reasonably complex exact path-planning algorithm such that we can control the number of maneuvers along the path.

Acknowledgments

The author thanks Olivier Jehl for his contribution to the ideas presented in this paper, and Pierre Janière, Marc Boullé and Eric Iooss for sharing the work of implementing the planning and navigation algorithms.

References

[ABF88] F. Avnaim, J.D. Boissonnat, and B. Faverjon. A practical exact motion planning algorithm for polygonal objects amidst polygonal obstacles. In *Proceedings of IEEE Int. Conference on Robotics and Automation*, 1988.

[BL85]   R.A. Brooks and T. Lozano-Pérez. A subdivision algorithm in configuration space for findpath with rotation. In *IEEE Transactions on Systems, Man and Cybernetics, SMC-15(2)*, pages 224–233, 1985.

[Bro82] R.A. Brooks. Solving the find-path problem by representing free space as generalized cones. In *IEEE Transactions on Systems, Man and Cybernetics, SMC-13(2)*, pages 190–197, 1982.

[Fav84] B. Faverjon. Obstacle avoidance using an octree in the configuration space of a manipulator. In *Proceedings of IEEE Int. Conference on Robotics and Automation*, Atlanta, March 1984.

[FT87a] B. Faverjon and P. Tournassoud. A local based approach for path planning of manipulators with a high number of degrees of freedom. In *Proceedings of IEEE Int. Conference on Robotics and Automation*, pages 1152–1159, Raleigh, April 1987. Also INRIA Research Report N 621, February 1987.

[FT87b] B. Faverjon and P. Tournassoud. The mixed approach for motion planning: learning global strategies from a local planner. In *Proceedings of the Int. Joint Conference on Artificial Intelligence*, pages 1131–1137, Milan, August 1987.

[Lau86] J.P. Laumond. Feasible trajectories for mobile robots with kinematic and environment contraints. In *Proceedings of the Int. Conference on Autonomous Systems*, Amsterdam, December 1986.

[Lau87] J.P. Laumond. Finding collision-free smooth trajectories for a non-holonomic mobile robot. In *Proceedings of the Int. Joint Conference on Artificial Intelligence*, pages 1120–1123, Milan, August 1987.

[Loz86] T. Lozano-Pérez. A simple motion planning algorithm for general robot manipulators. In *Proceedings of the 5th AAAI*, Philadelphia 1986.

[LW79] T. Lozano-Pérez and M. Wesley. An algorithm for planning collision-free paths among polyhedral obstacles. In *Communications of ACM*, pages 560–570, 1979.

[SS83] J.T. Schwartz and M. Sharir. On the piano movers' problem ii: general techniques for computing topological properties of real algebraic manifolds. *Advances in Applied Mathematics*, (4):298–351, 1983.

Motion from point matches: multiplicity of solutions

Olivier D. Faugeras

INRIA - Domaine de Voluceau - Rocquencourt

78153 Le Chesnay Cedex - FRANCE

Steve Maybank

GEC - MARCONI - Marconi Command and Control System Limited

Surrey GU16 5PE - ENGLAND

Abstract

In this article, we study the multiplicity of solutions of the motion problem. Given n point matches between two frames, how many solutions are there to the motion problem ? we show that the maximum number of solutions is 10 when five point matches are available. This settles a question which has been around in the Computer Vision community for a while. We present two approaches:

- the first one is based on algebraic geometry has been developed by Demazure [Dem88]. We show that it provides a very simple answer to the multiplicity of solutions when more than 5 point matches are available, namely not more than 3.

- the second one is based on projective geometry and is based on the work of Kruppa [Kru13]. We correct Kruppa's result and show that it is compatible with Demazure's.

We then describe a computer implementation of the second approach that uses MAPLE, a language for symbolic computation. The program allows us to exactly compute the solutions for any configurations of 5 points. Some preliminary experiments are described.

1 Introduction

Given two images taken at different times by a camera moving in a static environment, we want to estimate the position and orientation of the camera with respect to its first position. Many approaches are possible and we consider in this paper only the approach consisting in identifying within the two images corresponding points, i.e. pairs of points which are images of the same physical points viewed from the two different positions.

Given a number of such pairs, the following questions may be asked:

1. can the displacement of the camera be recovered ?

2. if yes, what is the minimal number of pairs necessary ?

3. is there a unique solution to the motion ?

Partial answers to these questions have been known for a while in the Computer Vision community. For example, it is known that given a sufficient number of pairs (greater than or equal to 5), the displacement of the camera can be recovered, if eight pairs or more are given it can even be recovered by linear techniques [Longuet-Higgins] unless the 3D points and the optical centers of the cameras fall on some special second degree surfaces. When they do fall on such surfaces, it is known [Longuet-Higgins, Maybank, Horn] that the solution is not unique and that there may not be more than 3 different solutions to the motion problem.

Not much is known so far on what happens when the number of pairs is between 5 and 7. We shed some light on these problems and show in particular that when 5 pairs are given, the maximal number of solutions is 10. When more than 5 pairs are given, then the number of solutions is at most three. It might be thought that studying these problems is a bit remote from the harsh reality of applied Computer Vision but we believe that their solution is essential to understand the behaviour (stability or unstability) of algorithms for recovering motion from point matches. It is the mathematical structure of the set of solutions that governs this behaviour and only if we understand it thoroughly will we be able to design reliable systems for motion analysis.

The paper is then organised as follows; in the next Section we briefly introduce notations; we then derive some properties of the so-called essential matrix, and show that the solutions have to lie on a specific real algebraic manifold; it turns out that in order to really understand the problem, we have to abandon for a while the set of real numbers and deal with complex numbers; we present some results recently obtained by Demazure [Dem88] which completely characterize the structure of this manifold; the main result is that its degree is 10; we also relate some of his results to those of Longuet-Higgins ([Lon87,Lon88], Maybank ([May85,May87]), and Horn ([Hor87b]) on 3 being the maximum number of solutions; we then introduce another approach based on projective geometry which allows

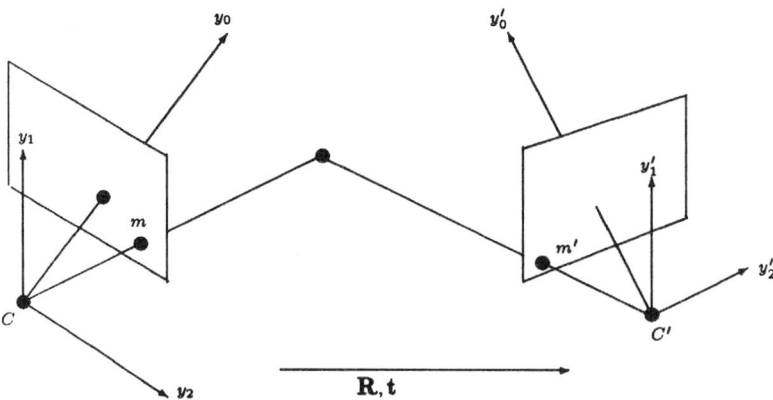

Figure 1: Geometry of the motion problem

us to explicitely compute the 10 solutions and describe a computer implementation of the method which makes heavy use of symbolic computation. This method was originally developped by Joseph Kruppa [Kru13], an austrian mathematician who published his result in 1913; unfortunately, since computers were not available at the time, Kruppa had found 11 solutions; having implemented his technique allowed us to discover the error and then give a formal proof.

2 Statement of the problem

We model our camera as a pinhole and choose the origin of the coordinate system at the optical center which we assume, for simplicity, to be behind the retina plane. The y_0-axis is the optical axis, the retina plane is at a unit distance from the optical center, and parallel to the (y_1, y_2) plane.

The rotation matrix from the first position to the second is $\mathbf{R}$, the translation is the vector $\mathbf{t} = \mathbf{CC'}$. A point M in 3D space has an image m in the first camera, and m' in the second. The problem is, given a number of pairs (m, m'), to recover $\mathbf{R}$ and the direction of $\mathbf{t}$.

Referring to Figure 1, the 3D point M has images m and m' at times t_1 and t_2. Knowing that m and m' are matches, what constraint can we derive on $\mathbf{R}$, and $\mathbf{t}$?

From Figure 1, it is clear that this is true if and only if the three vectors $\mathbf{Cm}$, $\mathbf{C'm'}$, and $\mathbf{CC'}$ are coplanar which we can write in the first coordinate system as:

$$\mathbf{Cm}.(\mathbf{t} \wedge \mathbf{RC'm'}) = 0 \qquad (1)$$

Introducing the antisymmetric matrix **T**:

$$\mathbf{T} = \begin{bmatrix} 0 & -t_z & t_y \\ t_z & 0 & -t_x \\ -t_y & t_x & 0 \end{bmatrix}$$

Matrix **T** is such that $\mathbf{Tx} = \mathbf{t} \wedge \mathbf{x}$ for all vectors **x**. Letting $\mathbf{E} = \mathbf{TR}$, equation 1 can be rewritten as:

$$\mathbf{Cm.EC'm'} = 0 \qquad (2)$$

Matrix **E** is called the essential matrix and has a number of important properties.

2.1 The essential matrix

Let us start with a simple property of matrix **E**.

Proposition 1 *Matrix* **E** *satisfies* $\mathbf{E}^T\mathbf{t} = 0$.

Proof : $\mathbf{E}^T = \mathbf{R}^T\mathbf{T}^T = -\mathbf{R}^T\mathbf{T}$. Therefore $\mathbf{E}^T\mathbf{t} = -\mathbf{R}^T(\mathbf{Tt}) = -\mathbf{R}^T(\mathbf{t} \wedge \mathbf{t}) = 0$. One consequence is that the rank of matrix **E** is less than or equal to 2, and therefore, $det(\mathbf{E}) = 0$. □

Proposition 2 *Matrix* **E** *is such that* $\mathbf{EE}^T$ *depends only of the translation vector* **t**.

Proof : Clearly, we have:

$$\mathbf{EE}^T = \mathbf{TRR}^T\mathbf{T}^T = -\mathbf{T}^2 = \begin{bmatrix} t_z^2 + t_y^2 & -t_x t_y & -t_x t_z \\ -t_y t_x & t_x^2 + t_z^2 & -t_y t_z \\ -t_z t_x & -t_z t_y & t_y^2 + t_z^2 \end{bmatrix}$$

□

An important consequence of this property is that the essential matrix **E** has two equal singular values. Indeed, we can choose a coordinate system such that the translation vector **t** can be written $[1, 0, 0]^T$ (**t** can be assumed of unit length since we know that it can be recovered only up to a scale factor). Therefore, if **E**' denotes **E** expressed in this coordinate system:

$$\mathbf{E'E'}^T = \begin{bmatrix} 0 & 0 & 0 \\ 0 & 1 & 0 \\ 0 & 0 & 1 \end{bmatrix} = \mathbf{A}$$

But there exists an orthogonal matrix $\mathbf{U}$ such that $\mathbf{E'} = \mathbf{U}\mathbf{E}\mathbf{U}^T$. Therefore:

$$\mathbf{EE}^T = \mathbf{U}^T \begin{bmatrix} 0 & 0 & 0 \\ 0 & 1 & 0 \\ 0 & 0 & 1 \end{bmatrix} \mathbf{U}$$

which completes the proof.

It is interesting to note that the reverse is true. More precisely we have the theorem:

Theorem 1 *A 3×3 matrix $\mathbf{E}$ can be decomposed as the product $\mathbf{TR}$ of an antisymmetric matrix $\mathbf{T}$ and a rotation matrix $\mathbf{R}$ if and only if it has one singular value equal to 0 and the other two are equal.*

Proof : [1] We have just proved the if part. Let us prove the only if part. If $\mathbf{E}$ has one singular value equal to zero and two singular values equal (and non zero), then there exists a rotation matrix $\mathbf{U}$ such that:

$$\mathbf{EE}^T = \mathbf{U}^T \begin{bmatrix} 0 & 0 & 0 \\ 0 & \Phi^2 & 0 \\ 0 & 0 & \Phi^2 \end{bmatrix} \mathbf{U}$$

Or:

$$(\frac{1}{\Phi}\mathbf{UE})(\frac{1}{\Phi}\mathbf{UE}^T) = \mathbf{A} \qquad (3)$$

Let us call $\mathbf{F}$ the matrix $\frac{1}{\Phi}\mathbf{UE}$, and $\mathbf{f}_1, \mathbf{f}_2$, and $\mathbf{f}_3$ its row vectors, from equation 3 we have:

$$\mathbf{f}_1\mathbf{f}_1^T = 0 \text{ and therefore } \mathbf{f}_1 = \mathbf{0}$$

$$\mathbf{f}_2\mathbf{f}_2^T = \mathbf{f}_3\mathbf{f}_3^T = 1 \; \mathbf{f}_2\mathbf{f}_3^T = 0$$

(4)

Matrix $\mathbf{F}$ is therefore "almost" orthogonal. It is easy to construct an orthogonal matrix $\mathbf{F'}$ such that $\mathbf{FF'}^T = \mathbf{A}$. Indeed, we can choose its row vectors $\mathbf{f}'_1, \mathbf{f}'_2$, and $\mathbf{f}'_3$ such that:

$$\begin{aligned} \mathbf{f}'^T_1 &= \mathbf{f}_2^T \wedge \mathbf{f}_3^T \\ \mathbf{f}'_2 &= \mathbf{f}_2 \\ \mathbf{f}'_3 &= \mathbf{f}_3 \end{aligned}$$

We can therefore write:

$$\mathbf{F} = \mathbf{AF'}$$

---

[1] The proof of this theorem was found by Huang and Faugeras, and independently by Maybank. The motivation for the theorem came from a conversation with Braccini (University of Genoa).

We notice that matrix $\mathbf{A}$ can be written:

$$\mathbf{A} = \begin{bmatrix} 0 & 0 & 0 \\ 0 & 0 & -1 \\ 0 & 1 & 0 \end{bmatrix} \begin{bmatrix} 1 & 0 & 0 \\ 0 & 0 & 1 \\ 0 & -1 & 0 \end{bmatrix} = \mathbf{T}'\mathbf{R}'$$

where $\mathbf{T}'$ is an antisymmetric matrix and $\mathbf{R}'$ a rotation matrix. We have now:

$$\frac{1}{\Phi}\mathbf{UE} = \mathbf{T}'\mathbf{R}'\mathbf{F}'$$

Which can be rewritten as:

$$\mathbf{E} = \mathbf{U}^T(\Phi\mathbf{T}')\mathbf{R}'\mathbf{F}' = (\mathbf{U}^T(\Phi\mathbf{T}')\mathbf{U})(\mathbf{U}^T\mathbf{R}'\mathbf{F}') = \mathbf{TR}$$

Indeed, $\mathbf{T} = \mathbf{U}^T(\Phi\mathbf{T}')\mathbf{U}$ is antisymmetric, and $\mathbf{R} = \mathbf{U}^T\mathbf{R}'\mathbf{F}'$ is the product of three rotation matrixes and therefore a rotation matrix. □

As a final property of the essential matrix $\mathbf{E}$, we derive the conditions on its coefficients that imply (and are implied by) the equality of its two nonzero singular values.

The characteristic equation $det(\mathbf{EE}^T - \lambda\mathbf{I})$ of matrix $\mathbf{EE}^T$ can be written:

$$-\lambda^3 + a\lambda^2 + b\lambda + c$$

since $c = det(\mathbf{EE}^T) = det^2(\mathbf{E})$, we have $c = 0$. Therefore the condition we seek is equivalent to writing that the quadratic equation $-\lambda^2 + a\lambda + b$ has two equal roots:

$$a^2 + 4b = 0$$

If we write:

$$\mathbf{EE}^T = \begin{bmatrix} a_1 & b_3 & b_2 \\ b_3 & a_2 & b_1 \\ b_2 & b_1 & a_3 \end{bmatrix}$$

it is easy to show that:

$$\begin{aligned} a &= a_1 + a_2 + a_3 \\ b &= b_1^2 + b_2^2 + b_3^2 - a_1 a_2 - a_2 a_3 - a_3 a_1 \end{aligned}$$

Therefore:

$$a^2 + 4b = (a_1 + a_2 + a_3)^2 + 4(b_1^2 + b_2^2 + b_3^2 - a_1 a_2 - a_2 a_3 - a_3 a_1)$$

Expressing the a_i's and the b_i's as functions of the row vectors e_i of matrix E yields:
$$(e_1^2 + e_2^2 + e_3^2)^2 + 4((e_2.e_1)^2 + (e_3.e_1)^2 + (e_1.e_2)^2 - e_1^2 e_2^2 - e_2^2 e_3^2 - e_3^2 e_1^2) = 0$$
Or:
$$(e_1^2 + e_2^2 + e_3^2)^2 = 4(\|e_1 \wedge e_1\|^2 + \|e_2 \wedge e_3\|^2 + \|e_3 \wedge e_1\|^2)$$
We have proved the following property.

Proposition 3 *Theorem 1 is equivalent to the following two relations between the row vectors of $\mathbf{E}$:*
$$e_1.(e_2 \wedge e_3) = 0$$
$$(e_1^2 + e_2^2 + e_3^2)^2 = 4(\|e_1 \wedge e_2\|^2 + \|e_2 \wedge e_3\|^2 + \|e_3 \wedge e_1\|^2)$$

Even though Proposition 3 has a nice interpretation in terms of the row vectors of matrix $\mathbf{E}$, it can also be expressed differently:

Proposition 4 *Theorem 1 is equivalent to the following two relations:*
$$det(\mathbf{E}) = 0$$
$$f(\mathbf{E}) = \frac{1}{2} Tr(\mathbf{E}\mathbf{E}^T)^2 - Tr((\mathbf{E}\mathbf{E}^T)^2) = 0$$

Proof : The proof is simple; one notices that $Tr(\mathbf{E}\mathbf{E}^T)^2 - 2Tr((\mathbf{E}\mathbf{E}^T)^2) = -(a^2 + 4b)$
□

3 Demazure's approach [Dem88]

Demazure's idea is to study the application which, to each pair $(\mathbf{R}, \mathbf{t})$ associates the essential matrix $\mathbf{E}$. In his approach, he does not restrict $\mathbf{R}$ and $\mathbf{t}$ to be real ($\mathbf{R}$ must be orthogonal though).

Using Proposition 2 we can write:
$$\frac{1}{2} Tr(\mathbf{E}\mathbf{E}^t)\mathbf{I} - \mathbf{E}\mathbf{E}^t = \mathbf{t}\mathbf{t}^T$$
multiplying right by $\mathbf{E}$ and using Proposition 1:
$$\frac{1}{2} Tr(\mathbf{E}\mathbf{E}^T)\mathbf{E} - \mathbf{E}\mathbf{E}^T\mathbf{E} = 0$$

Since we also have $det(\mathbf{E}) = 0$ (Proposition 1), we obtain 10 homogeneous polynomial equations of degree 3 which must be satisfied by the coefficients of $\mathbf{E}$. Demazure shows that they are linearly independent and define therefore an algebraic manifold M whose structure characterises the motion problem.

3.1 Resolving an apparent contradiction

There is an apparent contradiction since in the previous section we characterized the manifold with 2 real equations and here we have 10. This contradiction can be eliminated as follows. Let $\mathbf{F}(\mathbf{E}) = \frac{1}{2}tr(\mathbf{E}\mathbf{E}^t)\mathbf{E} - \mathbf{E}\mathbf{E}^t\mathbf{E}$, then it can be shown that:

$$tr(\mathbf{F}(\mathbf{E})\mathbf{F}(\mathbf{E})^t) = \frac{1}{2}tr(\mathbf{E}\mathbf{E}^t)tr(\mathbf{F}(\mathbf{E})\mathbf{E}^t) + 3det(\mathbf{E})^2$$

using proposition 4, we have:

$$tr(\mathbf{F}(\mathbf{E})\mathbf{F}(\mathbf{E})^t) = \frac{1}{2}tr(\mathbf{E}\mathbf{E}^t)f(\mathbf{E}) + 3det(\mathbf{E})^2$$

if $\mathbf{E}$ is a real essential matrix, $\mathbf{F}(\mathbf{E}) = 0$, $det(\mathbf{E}) = 0$, and therefore $f(\mathbf{E}) = 0$. Reciprocally, if $f(\mathbf{E}) = det(\mathbf{E}) = 0$ and $\mathbf{E}$ is real, we have $\sum F_{ij}(\mathbf{E})^2 = 0$. Since the F_{ij}'s are real, each one of them is equal to 0 and $\mathbf{F}(\mathbf{E}) = 0$. this resolves the contradiction.

3.2 The degree of M

It can be shown that the dimension of M is 5. Its degree is by definition the number of its intersections with a linear manifold of codimension 5. Such a manifold can be considered as the intersection of 5 hyperplanes whose equation can be written as:

$$Cm^T \mathbf{E} C'm' = 0$$

Therefore, the degree of M is precisely the answer to the question of how many solutions we obtain from 5 point matches, Demazure shows that this number is 10. We prove the same result with a different approach later.

3.3 Degenerate cases

Another very interesting result that can be obtained from this approach is that of Longuet-Higgins, Maybank, and Horn. The question they solved is that of determining how many different motions (essential matrixes) were possible given an arbitrary number of point matches.

this is the same as asking how many essential matrixes $\mathbf{E}_i$ can satisfy:

$$cm^T \mathbf{E}_i c'm' = 0 \qquad (5)$$

for an arbitrary number of pairs (m, m'). If we interpret 5 in the 9-dimensional space of the coefficient of $\mathbf{E}_i$, it is equivalent to saying that the "points" $\mathbf{e}_i$ are on a line; since they also must on the manifold M the question is equivalent to how many points can have the intersection of a line with M. since we saw that the equations of M are of degree 3, the maximum number of intersection is 3 which corresponds to the result showed by the previous authors.

4 Projective geometry approach

We now present a proof that the number of solutions is ten in general which is based on the use of projective and algebraic geometry. the proof is built on the work of Erwin Kruppa [Kru13]. We correct Kruppa's result who had found 11 solutions.

4.1 Epipolar geometry

Given the camera geometry of figure 1, we consider the epipoles ep and ep', intersections of the line joining the optical centers with the retina planes; these points play an important role in Computer Vision since they form the basis of the so-called epipolar constraint, heavily used in stereo, for example. The epipolar constraint simply states that given a point m in 3D space, it has an image m in the retina plane 1 and m' in retina plane 2. The epipolar plane $CC'm$ intersects these planes along the lines $ep\, m$ and $ep'\, m'$. Therefore, given a point m (m') in the first (second) image, its image m' (m) in the second (first) image lies on the corresponding epipolar line.

4.1.1 Fundamental property of epipolar lines

For a given epipolar plane (Π) intersecting the retina planes along l and (l'), when m varies in (Π), its image m and m' vary along l and l'. When (Π) rotates about CC', it generates a pencil of planes and l and l' rotate about ep and ep', generating two pencils of lines, (l) and (l'). The fundamental property of these two pencils is that they are homographically related, we note $(l)\overline{\wedge}(l')$.

This is a direct consequence of a number of so-called incidence theorem and the proof can be found in [SK52, chapters 4 and 11]. This is equivalent to saying that if we consider four epipolar planes $\Pi_1, \Pi_2, \Pi_3, \Pi_4$ and the corresponding epipolar lines $l_1, l'_1, l_2, l'_2, l_3, l'_3$, and l_4, l'_4, then the cross-ratios (l_1, l_2, l_3, l_4) and (l'_1, l'_2, l'_3, l'_4) are equal. For the definition of the cross-ratio of four lines see [SK52, chapter 4].

4.1.2 Chasles problem

Probably the oldest instanciation of the motion problem is the version given by the french mathematician Chasles [Cha55]. We found a statement of this problem in a paper by Hesse [Hes63] who also produced an analytical solution:

"On donne dans le même plan deux systèmes de sept points chacun et qui se correspondent. faire passer par chacun de ces systèmes un faisceau de sept rayons, de telle sorte que les deux faisceaux soient homographiques."

A detailed analysis and solution of this problem can be found in [Stu69]. The final word on this is that there exist three solutions in general, real or complex which can be found as the intersection of two planar cubics. As we show next, this does not solve our problem since there is no guarantee that the corresponding transformation from the first camera to the second is a rigid displacement.

4.2 The absolute conic, and its relationship with respect to rigid transformations

Linear non degenerate transformations in real projective spaces of any dimension (but we are interested here in dimensions 2 and 3 only) form a group $PGL(i, R)$ ($i = 2, 3$ in our case). There are two subgroups of $PGL(i, R)$ which are of considerable practical interest, namely the affine subgroup corresponding to affine transformations, and the euclidean subgroup corresponding to euclidean transforamtions, the rigid displacements.

4.2.1 Affine and euclidean geometries in dimension 2

Affine geometry involves an invariant line; if we single out a line of the projective plane, call it the line at infinity, and consider the subgroup of $PGL(2, R)$ leaving this line globally invariant, we obtain the group $AGL(2, R)$ of affine transformations of the plane.

If we now choose two conjugate complex points I and J on the line at infinity we can specialize things even further and consider the subgroup of $AGL(2, R)$ that leaves I and J globally invariant. This subgroup $EGL(2, R)$ is the group of euclidean transformations of the plane, which contains as a subgroup the group of rigid displacements. To make things a bit more precise, let us suppose that the projective coordinates are (y_0, y_1, y_2), then the equation of the line at infinity can be chosen as $y_0 = 0$ and the absolute points I and J can be taken as $(0, 1, i)$ and $(0, 1, -i)$.

4.2.2 Affine and euclidean geometries in dimension 3

The situation in dimension 3 is very similar. Affine geometry involves an invariant plane; if we single out a plane in the projective space, call it the plane at infinity, and consider the subgroup of $PGL(3, R)$ leaving this plane globally invariant, we obtain the group $AGL(3, R)$ of affine transformations of the three dimensional space.

If we now choose in the plane at infinity a virtual conic Ω i.e. a conic containing no real points, we can consider the subgroup of $AGL(3, R)$ that leaves Ω globally invariant. This subroup, $EGL(3, R)$ is the group of euclidean transformations of space, which contains as a subgroup the group of rigid displacements.

To make things more precise, let us suppose that the projective coordinates are (y_0, y_1, y_2, y_3), then the equation of the plane at infinity can be chosen as $y_0 = 0$ and the equation of the absolute conic is given by:

$$\begin{cases} y_0 = 0 \\ y_1^2 + y_2^2 + y_3^2 = 0 \end{cases}$$

Notice that if $y_1 = 0$ is the equation of the line at infinity in this plane, then Ω goes through the points $(0, 1, i)$ and $(0, 1, -i)$.

4.3 A restricted version of Chasles problem

We are now ready to rephrase Chasles problem to suit our needs. We saw that in the statement of Chasles problem we were given seven point correspondences in two planes from which we could deduce a number of epipolar points. From each of these epipolar points we can infer a projective transformation of 3d space that brings the pair (C, Π') onto (C', Π') but there is no guarantee that this projective transformation is a rigid displacement.

To enforce this, we must introduce the absolute conic and make sure it is left invariant. A very simple way to do this is to consider the images ω and ω' by the two cameras of Ω. ω and ω' are virtual conics in Π and Π' and we can consider the two tangents t_1, t_2 from ep to ω and the two tangents t'_1, t'_2 from ep' to ω'. If we want that the transformation from (C, Π) to (C', Π') be a rigid displacement, a necessary conditoin is that these tangents correspond to each other and therefore be parts of the epipolar pencils.

The problem is then as follows:

given five point correspondences in Π and Π' and the images ω and ω' of the absolute conic Ω, find two points o and o' in Π and Π' such that if we denote by t (resp. t') the pairs of tangents from o (resp. o') to ω (resp. ω'), the seven lines (oa, ob, oc, od, oe, t) and $(o'a', o'b', o'c', o'd', o'e', t')$ are homographically related. We denote this correspondence by:

$$o(a, b, c, d, e, \omega) \overline{\wedge} o'(a', b', c', d', e', \omega')$$

4.4 Writing the equations

This problem was solved in 1913 by Erwin Kruppa [Kru13] who found that there were in general 11 solutions, real or complex. This contradicts Demazure's result given in section 3 who proved that the number of solutions is 10. In what follows we derive Kruppa's equations and show that in fact there are in general 10 solutions. The interest of Kruppa's method as compared to Demazure's is that it allows us to explicitely construct the solutions.

We will not give here all the details of the computation and only provide the main ideas and results. The details can be found either in the original paper of Kruppa or in [FM88].

After a change of projective coordinates, we can always consider that the points (a, b, c, d) have coordinates:

$$a : \begin{Vmatrix} 1 \\ 0 \\ 0 \end{Vmatrix} \quad b : \begin{Vmatrix} 0 \\ 1 \\ 0 \end{Vmatrix} \quad c : \begin{Vmatrix} 0 \\ 0 \\ 1 \end{Vmatrix} \quad d : \begin{Vmatrix} 1 \\ 1 \\ 1 \end{Vmatrix}$$

we will denote the coordinates of e by (e_0, e_1, e_2), those of o by (x_0, x_1, x_2) and the absolute conic in plane Π is given by its equation:

$$\sum_{i=0}^{2} a_{ik} y_i y_k \quad \text{with} \quad a_{ik} = a_{ki}$$

We use the same notations for plane Π' by adding ' whenever required.

The idea is to consider the intersections of the rays with the lines ab and $a'b'$ and to write that these intersections are homographically related.

4.4.1 Intersections of the rays with ab

We need the following result.

Suppose we are given a conic c of equation:

$$S_{yy} \equiv \sum_{i=0}^{2} a_{ik} y_i y_k$$

and a point x of coordinates (x_0, x_1, x_2). Then the equations of the two tangents drawn from x to c are given by ([SK52, chapter 5]):

$$S_{xy}^2 - S_{yy} S_{xx} = 0$$

where:

$$S_{xy} = x^T A y = y^T A x$$

¿From this we can deduce that the coordinates of the intersections of the rays oa, ob, oc, od, oe with the line ab are given by:

$$(1,0); (0,1); \tilde{c}:(x_0, x_1); \tilde{d}:(x_0 - x_2, x_1 - x_2); \tilde{e}:(x_0 e_2 - x_2 e_0, x_1 e_2 - x_2 e_1)$$

Using the results of the previous section and after some algebra, it can be shown that the coordinates on ab of the intersections u and v of the tangents from O to ω are the roots of the following equation:

$$A_{00} y_0^2 + 2 A_{01} y_0 y_1 + A_{11} y_1^2$$

in which the coefficients A_{ij} are quadratic functions of x:

$$\begin{aligned} A_{00} &= \delta_{01} x_1^2 + \delta_{20} x_2^2 + 2\delta_0 x_1 x_2 \\ A_{11} &= \delta_{01} x_0^2 + \delta_{12} x_2^2 + 2\delta_1 x_0 x_2 \\ A_{01} &= \delta_2 x_2^2 - \delta_{01} x_1^0 - \delta_0 x_0 x_2 - \delta_1 x_1 x_2 \end{aligned}$$

where the coefficients δ_i and δ_j are quadratic functions of the coefficients of the equation of ω:

$$\delta_0 = a_{01} a_{02} - a_{00} a_{12}$$

δ_1 and δ_2 are obtained by circular permutations.

$$\delta_{01} = a_{01}^2 - a_{00} a_{11}$$

δ_{12} and δ_{20} are obtained by circular permutations.

We now have to express that there exists a homographic transformation from line ab to line $a'b'$ mapping a onto a', b onto b', $\tilde{c}$ onto $\tilde{c}'$, $\tilde{d}$ onto $\tilde{d}'$, $\tilde{e}$ onto $\tilde{e}'$, u onto u', and v onto v'.

Since such a transformation is determined by three point correspondences and we have seven, we obtain four equations. Since we have four unknowns, two for the coordinates of o and two for those of o' we should be able to solve the problem.

4.4.2 Writing the equations

The homographic correspondence between ab and $a'b'$ exchanges (a,b) and (a',b') therefore it has a very simple form:

$$\begin{cases} \rho y_0' = \alpha y_0 \\ \rho y_1' = \beta y_1 \end{cases}$$

Applying this to point $\tilde{c}, \tilde{d}$, and $\tilde{e}$ we have:

$$\begin{cases} \lambda x_0' = \alpha x_0 \\ \lambda x_1' = \beta x_1 \end{cases} \qquad \begin{cases} \mu(x_0' - x_2') = \alpha(x_0 - x_2) \\ \mu(x_1' - x_2') = \beta(x_1 - x_2) \end{cases}$$

$$\begin{cases} \nu(x_0' e_2' - x_2' e_0') = \alpha(x_0 e_2 - x_2 e_0) \\ \nu(x_1' e_2' - x_2' e_1') = \beta(x_1 e_2 - x_2 e_1) \end{cases}$$

Taking ratios we eliminate $\alpha, \beta, \lambda, \mu$ and ν:

$$\frac{x_0' - x_2'}{x_1' - x_2'} \times \frac{x_1'}{x_0'} = \frac{x_0 - x_2}{x_1 - x_2} \times \frac{x_1}{x_0} \tag{6}$$

$$\frac{x_0' e_2' - x_2' e_0'}{x_1' e_2' - x_2' e_1'} \times \frac{x_1'}{x_0'} = \frac{x_0 e_2 - x_2 e_0}{x_1 e_2 - x_2 e_1} \times \frac{x_1}{x_0} \tag{7}$$

Applying the homographic transformation to the pairs (u,v) and (u',v') we obtain:

$$\begin{cases} \sigma A_{00} = \alpha^2 A_{00}' \\ \sigma A_{11} = \beta^2 A_{11}' \\ \sigma A_{01} = \alpha\beta A_{01}' \end{cases}$$

Again, we can eliminate σ, α, and β:

$$\frac{A_{00}}{A_{01}} \times \frac{x_0}{x_1} = \frac{A_{00}'}{A_{01}'} \times \frac{x_0'}{x_1'} \tag{8}$$

$$\frac{A_{11}}{A_{01}} \times \frac{x_1}{x_0} = \frac{A_{11}'}{A_{01}'} \times \frac{x_1'}{x_0'} \tag{9}$$

We have thus obtained our four equations.

4.4.3 Simplifying the equations

We can simplify equation 6 to 9 by applying the change of variables Φ_1 defined by:

$$x_0 = k u_1 u_2$$
$$x_1 = k u_2 u_0$$
$$x_2 = k u_0 u_1$$

We denote this transformation by:

$$x_0 : x_1 : x_2 = u_1 u_2 : u_2 u_0 : u_0 u_1$$

Φ_2 is defined in the same way by adding ':

$$x'_0 : x'_1 : x'_2 = u'_1 u'_2 : u'_2 u'_0 : u'_0 u'_1$$

Applying Φ_1 and Φ_2 to equations 6 to 9, we obtain:

$$\frac{u_2 - u_0}{u_2 - u_1} = \frac{u'_2 - u'_0}{u'_2 - u'_1} \tag{10}$$

$$\frac{u_2 e_2 - u_0 e_0}{u_2 e_2 - u_1 e_1} = \frac{u'_2 e'_2 - u'_0 e'_0}{u'_2 e'_2 - u'_1 e'_1} \tag{11}$$

This is after simplification by $u_0 u_1 u_2$ and $u'_0 u'_1 u'_2$ and, after simplification by $u_0^2 u_1 u_2$ and $u'^2_0 u'_1 u'_2$:

$$\frac{\delta_{01} u_2^2 + \delta_{02} u_1^2 + 2\delta_0 u_2 u_1}{\delta_2 u_0 u_1 - \delta_{01} u_2^2 - \delta_0 u_1 u_2 - \delta_1 u_0 u_2} = \frac{\delta'_{01} u'^2_2 + \delta'_{02} u'^2_1 + 2\delta'_0 u'_2 u'_1}{\delta'_2 u'_0 u'_1 - \delta'_{01} u'^2_2 - \delta'_0 u'_1 u'_2 - \delta'_1 u'_0 u'_2} \tag{12}$$

and, after simplification by $u_0 u_1^2 u_2$ and $u'_0 u'^2_1 u'_2$:

$$\frac{\delta_{01} u_2^2 + \delta_{12} u_1^2 + 2\delta_2 u_0 u_2}{\delta_2 u_0 u_1 - \delta_{01} u_2^2 - \delta_0 u_1 u_2 - \delta_1 u_0 u_2} = \frac{\delta'_{01} u'^2_2 + \delta'_{12} u'^2_1 + 2\delta'_2 u'_0 u'_2}{\delta_2 u_0 u_1 - \delta_{01} u_2^2 - \delta_0 u_1 u_2 - \delta_1 u_0 u_2} \tag{13}$$

4.5 Finding the solutions in one plane

We now notice that the two equations 10 and 11 define a tranformation Σ from the plane (u_0, u_1, u_2) to the plane (u'_0, u'_1, u'_2). After some algebra, it can be shown that Σ is defined by:

$$u'_0 : u'_1 : u'_2 = \{e'_1[eu]_1[1u]_2 - e'_2[eu]_2[1u]_1\} :$$
$$= \{e'_2[eu]_2[1u]_0 - e'_0[eu]_0[1u]_2\} :$$
$$= \{e'_0[eu]_0[1u]_1 - e'_1[eu]_1[1u]_0\}$$

in which we have used:

$$[eu]_0 = e_1u_1 - e_2u_2 \qquad [1u]_0 = u_1 - u_2$$

$[eu]_1, [1u]_1$ and $[eu]_2, [1u]_2$ being obtained by circular permutations.

Σ (as well as Φ_1 and Φ_2) are quadratic transformations which are the simplest generalization of the linear projective transformations. We will return to this kind of transformation in a later section.

If we now replace u'_0, u'_1, and u'_2 by their values in equations 12 and 13, we obtain two polynomial equations of degree 6 in u_0, u_1, u_2. Each equation represents a curve in the plane (u_0, u_1, u_2), a sextic, and the solutions for $\Phi_1(o)$ are among the points of intersection of these two sextics, which we denote by A and B.

Algebraic geometry [SK52] tells us that the number of intersections of two algebraic curves of degrees m and n is mn, therefore we may have here as many as 36 solutions, real or complex. We are now going to see that not all these solutions are possible and that only 10 remain.

4.5.1 Throwing away the impossible solutions

There are three sources of impossible solutions:

1. those which make the products $u_0u_1u_2$ and $u'_0u'_1u'_2$ by which we simplified our equations to obtain equations 12 and 13 equal to 0.

2. the points m such that $\Sigma(m) = (0,0,0)$. Such points are called the fundamental points of a quadratic transformation ([SK52, Chapter 9]).

3. if we denote by D and D' the denominators in equations 12 and 13, the points of intersection of the curves $D = 0$ and $\Sigma^{-1}(D') = 0$.

Let us consider them in turn.

Case 1:

In order for the product $u_0u_1u_2$ to be equal to 0, at least one of the factors has to be equal to 0. If we assume $u_2 = 0$, then we realize using for example a system for symbolic computation (in our case MAPLE) that in the equation of A, the coefficients of the terms in u_0^6, u_0^5, u_0^4, and u_2^6 are equal to zero. This indicates that the point $(1,0,0)$ is on A and has order 3 and the point $(0,1,0)$ is also on A and has order 1. Similarly, it can be shown that $(1,0,0)$ is of order 1 and $(0,1,0)$ is of order 3 on B. It can also be verified that the point $(0,0,1)$ is of order 1 on both curves.

We thus have the following table:

	(1,0,0)	(0,1,0)	(0,0,1)
A	3	1	1
B	1	3	1

These three points are included in the intersection of A and B and cannot be solutions. Counting the orders, we see that we have eliminated 7 solutions.

Case 2:

We have to find the fundamental points of the quadratic transformation Σ. There are three such points in general [SK52, chapter 9] which we can easily discover.

Let us define the two linear applications x and y by:

$$x(u) = \begin{Vmatrix} [1u]_0 \\ [1u]_1 \\ [1u]_2 \end{Vmatrix} \qquad y(u) = \begin{Vmatrix} e'_0[1u]_0 \\ e'_1[1u]_1 \\ e'_2[1u]_2 \end{Vmatrix}$$

They are defined by the two matrixes:

$$\mathbf{X} = \begin{bmatrix} 0 & 1 & -1 \\ -1 & 0 & 1 \\ -1 & -1 & 0 \end{bmatrix} \qquad \mathbf{Y} = \begin{bmatrix} 0 & e'_0 e_1 & e'_0 e_2 \\ -e'_1 e_0 & 0 & e'_1 e_2 \\ e'_2 e_0 & -e'_2 e_1 & 0 \end{bmatrix}$$

and we have:

$$\Sigma(u) = \rho x(u) \wedge y(u)$$

where ρ is a nonzero constant. The fundamental points are therefore those points u such that $y(u) = 0$ or $x(u) = 0$ or such that $x(u)$ and $y(u)$ are proportional.

The case $x(u) = 0$ corresponds to the point $\bar{d} = (1,1,1)$, the case $y(u) = 0$ corresponds to the point $\bar{e} = (\frac{1}{e_0}, \frac{1}{e_1}, \frac{1}{e_2})$ and the third case is solved as follows. Let $\bar{p}$ be a point such that:

$$x(\bar{p}) = \lambda y(\bar{p}) \quad \text{or} \quad \mathbf{X}\bar{p} = \lambda \mathbf{Y}\bar{p}$$

This is equivalent to:

$$(\mathbf{X} - \lambda \mathbf{Y})\bar{p} = 0$$

The values of λ for which the determinant of $\mathbf{X} - \lambda \mathbf{Y}$ is equal to 0 are $\lambda = 0$ (corresponding to matrix $\mathbf{X}$), $\lambda = \infty$ (corresponding to matrix $\mathbf{Y}$), and a third value, λ_0, which can be computed using MAPLE. For this value, it can be verified that the rank of $\mathbf{X} - \lambda_0 \mathbf{Y}$ is in general 2 and $\bar{p}$ is represented by any vector in the null space. Notice that the coordinates of $\bar{p}$ can be explicitely computed as functions of the coordinates of e and e'.

Using again the power of the symbolic processing, it is easy to show that the three points $\bar{d}, \bar{e}$, and $\bar{p}$ are points of order 2 on A and B.

	$\bar{d}$	$\bar{e}$	$\bar{p}$
A	2	2	2
B	2	2	2

These three points are included in the intersection of A and B and cannot be solutions. Counting the orders, we see that we have eliminated 12 solutions.

Case 3:

The curve of equation $D = 0$ is a conic; the curve of equation $\Sigma^{-1}(D') = 0$, obtained by replacing u'_0, u'_1, u'_2 by their values in D' is a quadric. These two curves have in general 8 points in common among which, as it can be verified, the points $(1,0,0)$ and $(0,1,0)$ which have already been removed in case 1). Since each of the remaining 6 points is of order 1 on A and B we have obtained six more solutions to eliminate, thus a total of 25, leaving 11 solutions to be considered.

4.5.2 Kruppa's forgotten solution

It turns out that in the equations of A and B, the coefficients of u_2^6 are equal to 0 since the curves contain the point $(0,0,1)$ but also that the coefficients of u_2^5 are equal. Again, this can be verified, in fact discovered would be the right word, using MAPLE.

But because the coefficients of u_2^5 in A and B are equal, A and B are tangent at the point of coordinates $(0,0,1)$; incidentally the equation of the tangent at this point is the coefficient of u_2^5, a linear polynomial in u_0 and u_1.

Therefore, this point must count twice in the intersection and not once; this eliminates 26 solutions, leaving only a maximum of 10, real or complex, in agreement with Demazure's result.

4.6 Computer implementation

We have already mentioned that we had implemented in MAPLE Kruppa's paper and that it had allowed us to correct his error.

We need to introduce some simple results of the theory of elimination.

Let $f(x)$ and $g(x)$ be two arbitrary polynomials of degree m and n:

$$f(x) \equiv a_0 x^n + a_1 x^{n-1} + \ldots + a_n$$
$$g(x) \equiv b_0 x^m + b_1 x^{m-1} + \ldots + b_m$$

The resultant $R(f,g)$ of f and g is the determinant of order $m+n$:

$$R(f,g) = \left. \begin{vmatrix} a_0 a_1 & \ldots & a_n 0 & \ldots & 0 \\ 0 a_0 & \ldots & a_{n-1} a_n & \ldots & 0 \\ \vdots & & & & \\ 0\ldots & ..a_0 a_1 & \ldots & \ldots & a_n \\ b_0 b_1 & \ldots b_m & 0\ldots & \ldots & 0 \\ 0 b_0 & \ldots b_{m-1} & b_m & \ldots & 0 \\ \vdots & & & & \\ 0\ldots & & \ldots & b_0 \ldots & b_m \end{vmatrix} \right\} \begin{matrix} m \\ \\ \\ n \\ \\ \end{matrix}$$

The fundamental property of this resultant is that it is equal to 0 iff $a_0 = b_0 = 0$ or $f(x)$ and $g(x)$ have a non-constant common factor.

Writing the equations of A and B as polynomials in u_2 with coefficients in $K[u_0, u_1]$, we obtain:

$$A \equiv a_0 u_2^5 + a_1 u_2^4 + \ldots + a_5$$
$$B \equiv b_0 u_2^5 + b_1 u_2^4 + \ldots + b_5$$

We saw in the previous Section that a_0 and b_0 are proportional. Also, according to the analysis of the previous Sections, the resultant R of A and B, considered as polynomials in u_2 whose coefficients are polynomials in u_0 and u_1, is dividable by:

$$u_0^3 u_1^3 (u_0 - u_1)^4 (e_1 u_0 - e_0 u_1)^4 (\overline{p}_1 u_0 - \overline{p}_0 u_1)^4$$

We also apply the same treatment to D and $\Sigma^{-1}(D')$:

$$D = c_0 u_2^2 + c_1 u_2 + c_2 \Sigma^{-1}(D') = c'_0 u_2^4 + c'_1 u_2^3 + \ldots + c'_4$$

We compute their resultant R', a polynomial of degree 8 in u_0, u_1 which is dividable by $u_0 u_1$. Letting Q be the quotient, Q is a homogeneous polynomial of degree 6 in u_0 and u_1 which also divides R.

Finally, we divide R by a_0, thus obtaining a homogeneous polynomial of degree 10 in u_0 and u_1.

Up to now, we have not made any approximation, and all the computations have been done either symbolically or in the field Q of rational numbers. In order to actually compute the displacements which are solutions of the problem, we have simulated a camera configuration such as the one represented in figure 1, the translation vector **t** is represented by three rational numbers, the rotation matrix **R** is represented by a quaternion $\mathbf{q} = [l, m, n, s]^T$ (for example see [FH86,Hor87a]), a four-dimensional vector with rational coefficients (it is not necessary to enforce the unit norm constraint since everything is projective, i.e defined up to a multiplicative constant. The relation between **R** and **q** is:

$$\mathbf{R} = \begin{bmatrix} s^2 + l^2 - m^2 - n^2 & 2(lm - sn) & 2(ln + sm) \\ 2(ln + sn) & s^2 - l^2 + m^2 - n^2 & 2(mn - sl) \\ 2(ln - sm) & 2(mn + sl) & s^2 - l^2 - m^2 + n^2 \end{bmatrix}$$

We consider five points A, B, C, D, E in 3D space, with rational coordinates. From these five points, **R**, and **t**, we can compute their image coordinates in the two retina planes. We then perform in each plane the change of projective coordinates such that relations 6 are satified. This provides the equations of ω and ω' in the two retina planes. We then compute the polynomial of degree 10 as explained in the previous Sections. Since the coordinates of the epipole ep are simply the coordinates of **t**, we know one root of this polynomial.

Dividing it by the corresponding linear factor yields a homogeneous polynomial of degree 9 in u_0 and u_1. We then solve for the real roots of this polynomial, replace u_0 and u_1 by the corresponding values in the equations of A and B and find, if any, the real common root in u_2. To each triplet $\tilde{ep} = (u_0, u_1, u_2)$ corresponds through the quadratic transformation Σ a unique triplet $\tilde{ep}' = (u'_0, u'_1, u'_2)$. Applying Φ_1^{-1} and Φ_2^{-1} to $\tilde{ep}$ and $\tilde{ep}'$, respectively yields two points to which we then have to apply the inverse changes of projective coordinates, yielding ep and ep'. ep yields the translation, ep' is equal to $-\mathbf{R}^T\mathbf{t}$. This fixes the displacement from the first camera to the second up to a rotation along CC', rotation which can be determined by writing that, for example Ca and $C'a'$ are coplanar. We can then reconstruct the points A, B, C, D, and E in 3D space.

Acknowledgements

We want to acknowledge the role of Thomas Buchanan who introduced us to the German literature on photogrametry and motion and opened our eyes to the beauty of projective geometry. We also want to acknowledge the help of the MAPLE language for symbolic computation without whom the experimental part of this work would have been totally impossible.

References

[Cha55] M. Chasles. Question No. 296. *Nouv. Ann. Math.*, 14:50, 1855.

[Dem88] M. Demazure. *Sur deux problèmes de reconstruction*. Technical Report, INRIA, 1988. To appear.

[FH86] O.D. Faugeras and M. Hebert. The Representation, Recognition, and Locating of 3D Shapes from Range Data. *The International Journal of Robotics Research*, 5(3):27–52, 1986.

[FM88] O.D. Faugeras and S.J. Maybank. *Multiplicity of solutions for motion problems*. Technical Report, INRIA, 1988. To appear.

[Hes63] O. Hesse. Die cubische Gleichung, von welcher die Lösung des Problems der Homographie von M. Chasles abhängt. *J. reine angew. Math.*, 62:188–192, 1863.

[Hor87a] B.K.P. Horn. Closed-form Solution of Absolute Orientation using Unit Quaternions. *Journal of the Optical Society A*, 4(4):629–642, April 1987.

[Hor87b] B.K.P. Horn. *Relative orientation*. A.I. Memo 994, MIT, October 1987.

[Kru13]   E. Kruppa. Zur Ermittlung eines Objektes aus zwei Perspektiven mit innerer Orientierung. *Sitz.-Ber. Akad. Wiss., Wien, math. naturw. Kl., Abt. IIa.*, 122:1939–1948, 1913.

[Lon87]   H.C. Longuet-Higgins. *Mental Processes: Studies in Cognitive Science*, chapter Configurations that defeat the eight-point algorithm, pages 395–397. MIT Press, Cambridge, MA-London, 1987.

[Lon88]   H.C. Longuet-Higgins. Multiple interpretations of a pair of images of a surface. *Proc. Roy. Soc. Lond. A.*, 1988.

[May85]   S.J. Maybank. The angular velocity asociated with the optical flow field due to a rigid moving body. *Proc. Roy. Soc. Lond. A*, 401:317–326, 1985.

[May87]   S.J. Maybank. *A theoretical study of optical flow*. PhD thesis, University of London, Birkbeck College, November 1987.

[SK52]   J.G. Semple and G.T. Kneebone. *Algebraic Projective Geometry*. Oxford: Clarendon Press, 1952. Reprinted 1979.

[Stu69]   R. Sturm. Das Problem der Projektivitat und seine Anwendung auf die Flächen zweiten Grades. *Math. Ann.*, 1:533–574, 1869.

Singular configurations of parallel manipulators and Grassmann geometry

Jean-Pierre MERLET

INRIA, Centre de Sophia Antipolis

2004 Route des Lucioles
06560 Valbonne, France

Abstract

Parallel manipulators have a specific mechanical architecture where all the links are connected both at the basis and at the gripper of the robot. By changing the lengths of these links we are able to control the positions and orientations of the gripper. In general, for a given set of links lengths there is only one position for the gripper. But in some cases more than one solution may be found for the position of the gripper : this is a singular configuration. To determine these singular configurations the classical method is to find the roots of the determinant of the jacobian matrix. In our case this matrix is complex and it seems to be impossible to find these roots. We propose here a new method based on Grassmann line-geometry. If we consider the set of lines of P^3, it constitutes a linear variety of rank 6. We show that a singular configuration is obtained when the variety spanned by the lines associated to the robot links has a rank less than 6. An important feature of the varieties of this geometry is that they can be described by simple **geometric** rules. Thus to find the singular configurations of parallel manipulators we have to find the configuration where the robot matches these rules. Such an analysis is performed on a special parallel manipulator and we show that we find all the well-known singular configurations but also new ones.

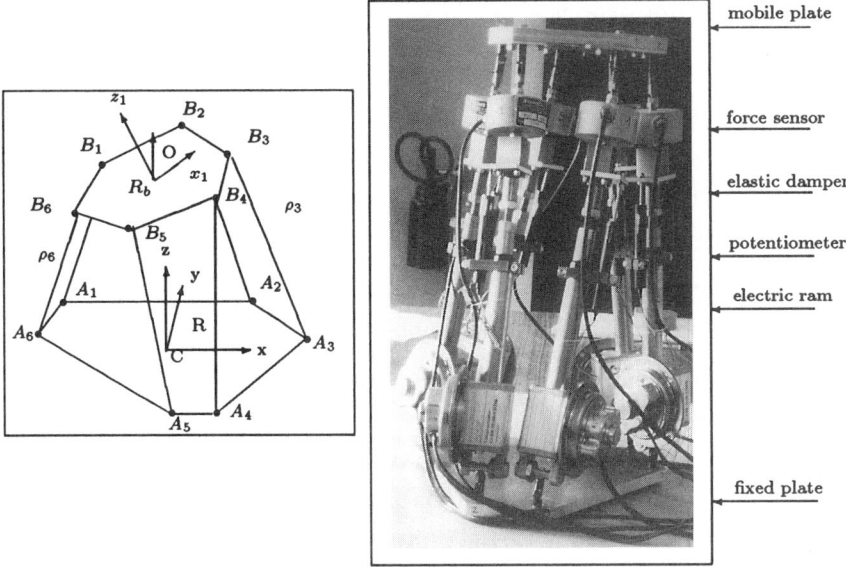

Figure 1: notation

Figure 2: INRIA parallel manipulator prototype

1 Parallel manipulator

1.1 Introduction

We deal here with the study of parallel manipulators like the model presented in Figure 2.

Basically it consists of two plates connected by 6 articulated links. In the following chapters the smaller plate will be called the *mobile* and the greater (which is in general fixed) will be called the *base*. In each link there is one linear actuator.

The main application of this mechanical architecture consists of the flight simulators (see for example Stewart [Stewart 1965]). The first design as a manipulator system has been done by Mac Callion in 1979 for an assembly workstation [Mac Callion 1979].

Some other researchers have also addressed this problem: Reboulet [Reboulet 1985], Inoue [Inoue 1985], Fichter [Fichter 1986],Yang [Yang 1984], Zamanov [Zamanov 1984]. This kind of manipulator has a great positionning ability and is very convenient for forcefeedback control. A prototype of parallel manipulators has been developed at INRIA (figure 2). The links articulations are universal joints on the fixed plate and ball-andsocket joint at the mobile plate. The linear actuator are electric rams and the lengths variations are measured through linear potentiometers.

The height of the prototype is about 51 cm, its weight about 11kg.

1.2 Notation

We introduce the absolute frame R with origin C and a relative frame R_b fixed to the mobile with origin O (see Figure 1). The rotation matrix relating a vector in R_b to the same vector in R will be denoted by M.

The centers of the articulations on the base for link i will be denoted A_i and those on the mobile B_i. The length of link i will be noted ρ_i, and the unit vector of this link n$_j$. The coordinates of A_i in frame R are (xa_i, ya_i, za_i), the coordinates of B_i in frame R_b are (x_i, y_i, z_i) and the coordinates of O, the origin of the relative frame, (x_o, y_o, z_o). We use the Euler's angles ψ, θ, ϕ to characterize the orientation of the mobile.

For the sake of simplicity the subscript i is omitted whenever it is possible and vectors will be noted in **bolt** character. A vector whose coordinates are expressed in the relative frame will be denoted by the subscript $_r$.

We will consider the case where all the articulation points of both the base and the mobile lie in a plane and are symmetric along one axe (see Figure 3). The articulation points on the mobile are located only in three different positions. The mobile is homothetic to the base and is rotated at 180 degrees for the connection of the links. In this case, without loss of generality, we will define R such that $za_i = 0$ and R_b such that $z_i = 0$. The symmetry axes will be used as an axe of each frame R, R_b. We exclude the case where three or more articulation points are collinear. We will call this architecture the triangular simplified symmetric manipulator (TSSM). We denote by C_{ij} the intersection point of lines $A_k A_i$ and $A_j A_l$ and by P_{12}, P_{34}, P_{56} the planes defined by $A_1 A_2 B_1, A_3 A_4 B_3, A_5 A_6 B_5$. We may remark that:

$$C_{23} \in P_{12} \qquad C_{23} \in P_{34}$$
$$C_{45} \in P_{56} \qquad C_{45} \in P_{34}$$
$$C_{61} \in P_{12} \qquad C_{61} \in P_{56}$$

1.3 Singular configurations and the Jacobian matrix

Let us calculate the fundamental relations relating the links lengths to the position of the mobile. We have :

$$\mathbf{AB} = \rho\mathbf{n} \quad \mathbf{AB} = \mathbf{AC} + \mathbf{CO} + \mathbf{OB} \quad \mathbf{OB} = M\mathbf{OB_r} \qquad (1)$$

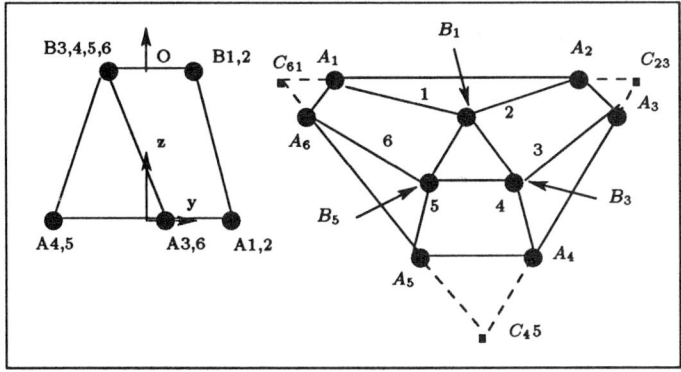

Figure 3: triangular simplified symmetric manipulator TSSM

where OB_r means the coordinates of the articulation points with respect to the frame R_b. n being a unit vector :

$$\rho = ||AC + CO + MOB_r|| = ||U|| \qquad (2)$$

If the position of the mobile is given we are able to calculate the components of U and thus the length of the segment. On the opposite we have to solve a system of 6 non-linear equations of type 2 to get the position of the mobile from the links lengths. At this time no theoretical solution of this system has been established. From the rank theorem we know that the solution is unique if the rank of the jacobian matrix J of this system is equal to 6 with:

$$J = ((\frac{\partial \rho}{\partial q})) \qquad (3)$$

where q is the position parameters vector. Note that this matrix is in fact the inverse jacobian (in a robotics sense) of the manipulator. The symbolic computation of the determinant of J is rather tedious. Mac Callion [Mac Callion 1979] used a numerical deflation method to find all the roots of the determinant. Mac Callion has found up to nine roots to this determinant, all outside the range of the links lengths, and Hunt has shown that there can be up to 16 roots [Hunt 1983]. Bricard [Bricard 1897] has shown that the resolution of the above system is equivalent to solve a complex trigonometric equation.

Hunt [Hunt 1978] describes a singular configuration (Figure 4). In this case all the segments intersect one line (line B_3B_5). We will see later that a simple mechanical analysis explains why this is a singular configuration.

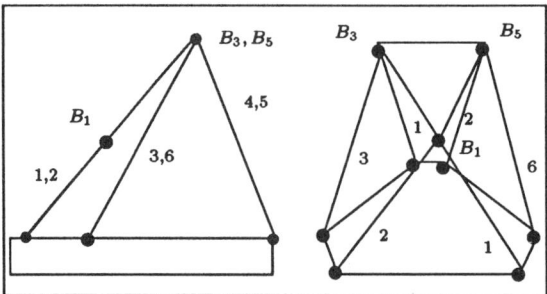
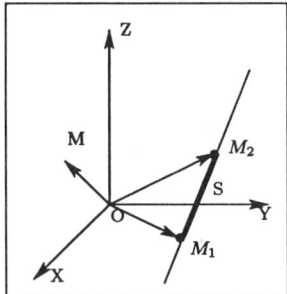

Figure 4: Hunt's singular configuration for the TSSM

Figure 5: Plücker coordinates

Fichter [Fichter 1986] describes another singular configuration which is obtained when the mobile plate is rotated around the z axis with an angle of $\pm\frac{\pi}{2}$ (see figure 13). This configuration was obtained by noticing that in this case two lines of the jacobian matrix were constant. But outside these two particular configurations no systematic method was proposed to find *all* the singular configurations of a parallel manipulator. Let us now investigate a geometrical approach.

2 Plücker coordinates of lines and rigidity

It is well known that a line can be described by its Plücker coordinates. Let us briefly introduce these coordinates. We consider two points on a line, say M_1 and M_2, and a reference frame R_0 which origin is O (see Figure 5). Let us now consider the two three dimensional vectors **S** and **M** defined by :

$$\mathbf{S} = \mathbf{M_1 M_2}$$

$$\mathbf{M} = \mathbf{OM_1} \wedge \mathbf{OM_2} = \mathbf{OM_2} \wedge \mathbf{S} = \mathbf{OM_1} \wedge \mathbf{S}$$

If we assemble these vectors to form a 6-dimensional vector we get the vector **U** of the Plücker-coordinates of this line.

$$\mathbf{U} = [S_x, S_y, S_z, M_x, M_y, M_z]$$

It is useful to introduce the normalized vector $\mathbf{U}'$ defined by:

$$\mathbf{U}' = \frac{\mathbf{U}}{\|\mathbf{S}\|} = [S'_x, S'_y, S'_z, M'_x, M'_y, M'_z]$$

It may be seen that the first three components of this vector are the components of the unit vector n_i of the line. The last three components are given by :

$$OM \wedge n_i$$

M being any point of the line. Let us consider now the matrix P defined by:

$$P = ((U'_1, U'_2, ... U'_6))$$

where U'_i is the coordinate vector of line i. If we denote by $\mathcal{T}$ the generalized force vector it is easy to show that the equilibrium state of the manipulator may be written as:

$$\mathcal{T} = P\mathbf{f} \qquad (4)$$

From this relation it is easy to show that we have:

$$J^T = P \qquad (5)$$

Equation 4 is a linear system of equations in term of the articular forces. If the system is rigid this means that whatever the generalized forces are these exists one set of articular forces such that the system is in an equilibrium state (and in our case the solution will be unique). This will be true if the matrix P is of full rank which is equivalent to say that the Plücker vectors are linearly independent. Thus *a singular configuration of a parallel manipulator corresponds to a configuration where it is not rigid.* One main result for the rigidity of polyhedron was obtained by Cauchy [Cauchy 1813]. He shows that a *convex* polyhedron with invariable faces is always rigid. But in our case a parallel manipulator may be not convex. However from this result we may say that all the singular configurations we will find must be such that the parallel manipulator is not convex.

As a matter of example let us consider Hunt's singular configuration. We notice that the torque around the axis B_3B_5 exerted by the segments on the mobile is always equal to zero (remember that every lines intersect B_3B_5). Thus if we apply an external force on the mobile such that the resulting external torque around the axis B_3B_5 is not equal to zero, the mobile cannot be in an equilibrium state: it is not rigid.

Let us assume now that the Plücker vectors belong to a vector space V_6 and we consider the one-dimensional subspaces of V_6 as points of a projective P_5. Then every line g in P_3 corresponds to exactly one point G in P_5.

It is well known that point G belongs to a quadric Q_p (see [Crapo 1973], [Veblen 1910], [Behnke 1986]). Indeed we have for every line of P_3 :

$$S_x M_x + S_y M_y + S_z M_z = 0$$

This equation defines the quadric Q_p which is called the *Grassmannian* or the *Plücker quadric*. At this point we have defined a one-to-one relation between the set of lines in the real P_3 and the quadric Q_p in P_5. The rank of this mapping is 6 (there are at most 6 independent Plücker vectors).

Let us consider now the various sub-spaces of P_5 (or more precisely their intersection with Q_p). We get various varieties which rank ranges from 0 to 6. As a matter of example a point in P_5 (rank=1) corresponds to a line in P_3. As for Q_p (which represents the set of line of P_3) it is defined through 6 linearly independent Plücker vectors and is therefore of rank 6.

Let us come back to the rigidity of parallel manipulators. We have seen that this manipulator is rigid (and therefore not in a singular configuration) if and only if the 6 lines are linearly independent. Therefore **a parallel manipulator will be in a singular configuration if, and only if, there is a subset spanned by n of its lines which has a rank less than n.** At this point the problem is far to be solved because we are not able to find the generalized coordinates of the mobile for which there is a linear dependency between the n Plücker vectors. But these dependencies can be described by geometric rules.

3 Grassmann Geometry

The varieties of lines have been studied by H. Grassmann (1809-1877). The purpose of this study was to find a geometric characterization of each variety. We will introduce now the various results which can be found in [Dandurand 1984] or ,with more mathematical justifications, in [Veblen 1910].

Let us begin with the varieties of rank 0 through 3 (Figure 6). We have first the empty set of rank 0. Then the *point* (rank=1), which is a line in the 3D space. The *lines* (rank=2) are either a pair of skew lines in R^3 or a flat pencil of lines: those lying in a plane and passing through some point on that plane.

The *planes* (rank=3) are of four types:

- all lines in a plane (3d)

- all lines through a point (3c)

- the union of two flat pencils having a line in common but lying in distinct planes and with distinct centers (3b)

- a regulus (3a)

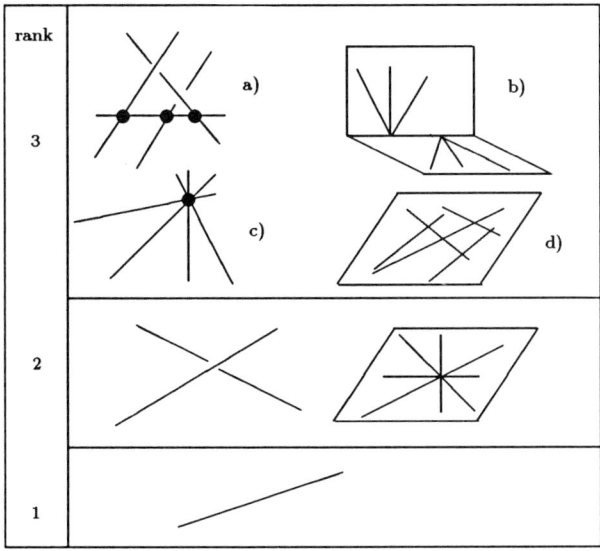

Figure 6: Grassmann varieties of rank 1,2,3

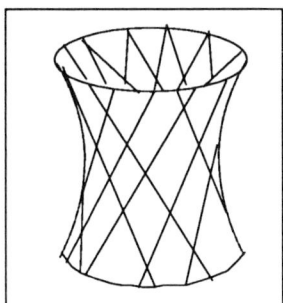

Figure 7: hyperboloid of one sheet

Let us define the regulus. Let three skew lines be in space and consider the set of lines which intersect these three lines : it builds a surface which is an hyperboloïd of one sheet (a quadric surface, Figure 7) and the set of the three skew lines is called a **regulus**. Each line belonging to the regulus is called a *generator* of the regulus.

It is shown in [Hilbert 1952],[Veblen 1910] that this surface is *doubly ruled*. This means that there exist two reguli (a regulus and its "complementary" regulus) which generate the same surface or that each point on the surface is on more than one line.

Therefore there are two families of straight lines on the hyperboloïd and each family covers the surface completely. A line on this surface is dependent on the lines of either the regulus or the complementary regulus. An interesting property is that a line of one family intersects all the lines of the other family and that any two lines of the same family are mutually skew (see [Tyrrell 1971] for the hairy details).

Let us describe now the varieties of higher rank of the Grassmann geometry (Figure 8). Varieties of dimension 4 are called *congruences* and are of four types:

- a linear spread generated by 4 skew lines i.e. no line meets the regulus generated by the three others lines in a proper point (*elliptic congruence*, 4a)

- all the lines concurrent with two skew lines (*hyperbolic congruence*, 4b)

- a one-parameter family of flat pencil, having one line in common and forming a variety (*parabolic congruence*, 4c)

- all the lines in a plane or passing through one point in that plane (*degenerate congruence*, 4d)

Varieties of dimension 5 are called *complexes* and are of two types:

- *non singular* (or *general*): generated by 5 independent skew lines (5a)

- *singular* (or *special*): all the lines meeting one given line (5b)

The geometric characterization of a general complex is that through any point of the space there is one and only one flat pencil of line such that all the lines which belong to the pencil belong also to the complex. In other words all the lines of a complex which are coplanar intersect one point.

4 Study of the TSSM

We will deal now with the case of the TSSM (Figure 3).

Let us make a preliminary remark: for the TSSM we may have at most 2 coplanar lines. Indeed we notice that there are at most two segments with collinear articulation points on the base. Therefore we have not to consider the degeneracy of subset where more than two lines must be coplanar. This will be the case for

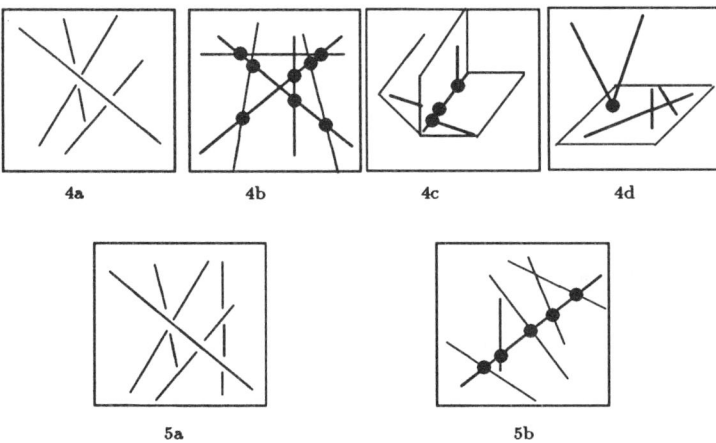

Figure 8: Grassmann varieties of rank 4,5

- the subset of two lines
- the subset of three lines,
- the subset of four lines in configuration 3d, 3b
- the subset of five lines in configuration 4c

4.1 Subset of 4 bars

4.1.1 Type 3c

(Four lines cross the same point).

Among a set of four bars two have a common articulation point on the mobile. Thus this common point must be the common point to the four lines. We will assume that the two lines with a common point are 1,2. Lines 3,4 and 5,6 have a common point different from B_1 and thus cannot have another one. Thus the only sets to be considered are (1,2,3,5),(1,2,3,6), (1,2,4,5) and (1,2,4,6). The demonstration of the following result will be the same in each case and we will study only the case of the set (1,2,3,5).

If 3 crosses B_1 then 3 is collinear to the edge B_1B_3. In the same manner if 5 crosses B_1 then 5 is collinear to the edge B_1B_5 and thus 3 and 5 are coplanar. Thus we get a singular configuration if lines 3,5 are coplanar and intersect the articulation point B_1.

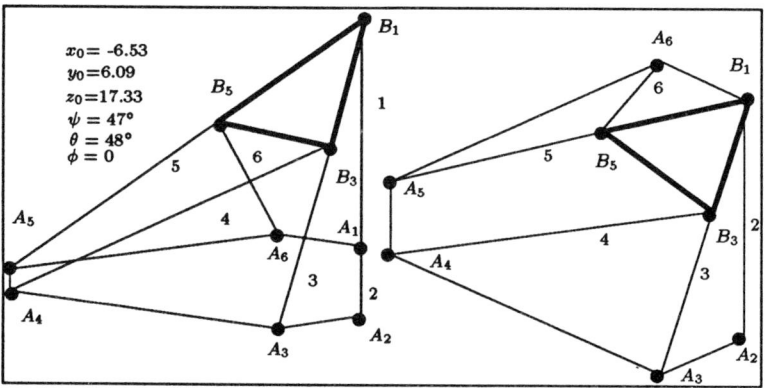

Figure 9: Perspective and top view of a singular configuration of type 3c

This condition enables to calculate the Euler's angle ψ by:

$$\tan \psi = \frac{ya_3 - ya_4}{xa_3 + xa_4}$$

and the position of the center of the mobile plate may be defined as a function of the Euler's angles. Figure 9 shows such a case.

4.1.2 Type 3a

The problem is to find 4 lines which are on the same regulus. A hyperboloïd of one sheet has two regulus $\Re_1$ and $\Re_2$ and we denote by (1) the family of lines which are spanned by $\Re_1$ and (2) the family of lines spanned by $\Re_2$. Remember that each line of (1) has an intersection point with every lines of (2) and none with the other lines of (1).

Let us suppose that line 1 belongs to the family (1) spanned by the regulus. Line 2 intersects line 1 and thus belongs to the family (2) spanned by the complementary regulus. For the same reason lines 3,4 and 5,6 cannot belong to the same family. Therefore 4 lines cannot belong to the same regulus.

4.2 Set of five bars

4.2.1 Configuration 4d

(five lines in a plane or crossing a point of this plane)

Let us remember that we have at most 2 coplanar lines. Among a set of 5 lines two pairs are coplanar and have a common point which is their articulation point on the

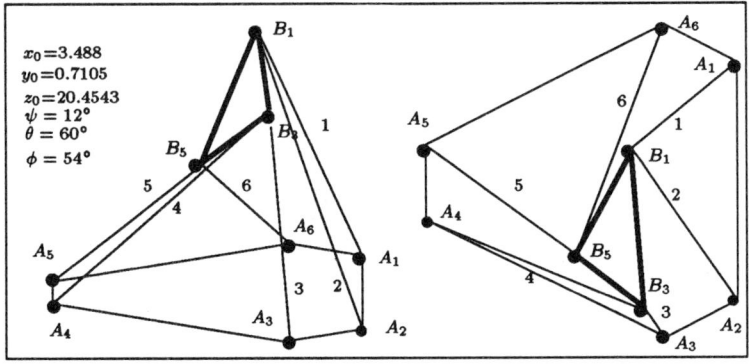

Figure 10: Perspective and top view of a singular configuration of type 4d

mobile. These two points being different this implies that the plane to be considered is spanned by a pair of lines and that the common point to the three other lines is the articulation point of the second pair. For example if we consider lines 1,2,3,4,5 we will consider the plane spanned by 1,2 , put the articulation point B_3 common to 3,4 in this plane and look if 5 can intersect B_3. It may be shown that this condition is fulfilled if the coordinates of the center of the mobile plate are expressed as functions of the Euler's angle. Figure 10 shows such a case.

4.2.2 Configuration 4b

(five lines intersect two skew lines)

Let us consider first lines 1,2,3,4. We have to find two skew lines D_1, D_2 which intersect these 4 lines. We have 4 possibilities for a line D which intersects lines 1,2,3,4:

-$D \in P_{12}$ and D intersects B_3

-$D \in P_{34}$ and D intersects B_1

-$D = P_{12} \cap P_{34}$

-D intersects both B_1 and B_3

4.2.3 $D_1 \in P_{12}$ and D_1 intersects B_3

If D_2 is skew to D_1 then $D_2 \notin P_{12}$ and D_2 does not intersect B_3. Therefore D_2 cannot be neither $P_{12} \cap P_{34}$ nor $B_1 B_3$. Thus the only remaining case is $D_2 \in P_{34}$ and D_2 intersects B_1. In this case we have :

$$B_1 \in P_{34} \quad B_1 \in P_{12} \Rightarrow B_1 \in P_{12} \cap P_{34}$$

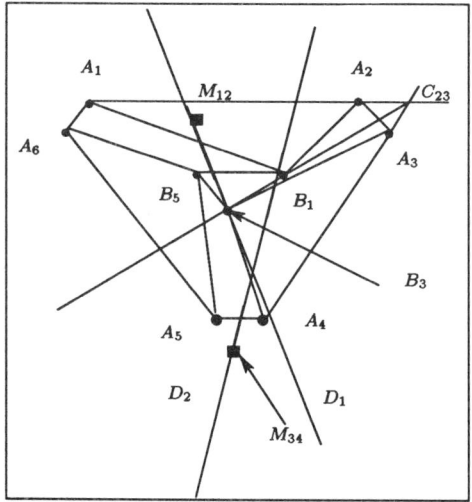

Figure 11: 2 skew lines intersecting 1,2,3,4,5, configuration 4b, first case

$$B_3 \in P_{34} \quad B_3 \in P_{12} \Rightarrow B_3 \in P_{12} \cap P_{34}$$

Thus C_{23}, B_1, B_3 belong to the same line. Let M_{12} be the intersection point of line 5 with P_{12} and M_{34} the intersection point of line 5 with P_{34}. If M_{12} is different from M_{34} then the lines $B_3 M_{12}$ and $B_1 M_{45}$ are skew and intersect the lines 1,2,3,4,5. This is then a singular configuration (Figure 11). Here some tedious calculations show that the parameters x_0, y_0 are defined as functions of z_0, ψ, θ, ϕ.

4.2.4 $D_1 \in P_{34}$ and $B_1 \in D_1$

As in the previous part we get:

$$D_2 \in P_{12} \quad B_3 \in D_2$$

which is the case we investigated above.

4.2.5 $D_1 = P_{12} \cap P_{34}$

If D_2 is not coplanar with D_1 then we must have:

$$D_2 = B_1 B_3$$

But if line 5 intersects D_2 then line 5 and $B_1 B_3$ are coplanar and thus the mobile and line 5 are coplanar. We must then investigate if line 5 intersects D_1. If we write that line 5

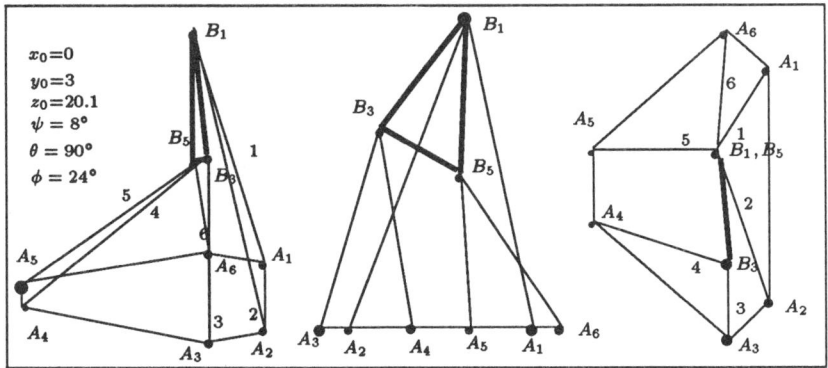

Figure 12: Perspective, side and top view of a singular configuration of type 5a ($\theta = \pm\frac{\pi}{2}$)

intersects D_1 we get four linear equations in term of the intersection point. A numerical resolution of the three first equations gives the position of the intersection point and then we have to verify if the fourth equation is satisfied. This will be the second case of degeneracy of type 4b.

4.3 Set of six bars

4.3.1 Configuration 5a

In this case the variety spanned by the 6 lines is a general complex. We consider the lines D_i belonging to the flat pencils spanned by lines 1-2, 3-4, and 5-6 and lying in the mobile plane. We get a general complex if and only if these 3 lines intersect the same point.

We will consider first the case where we have only rotation around the vertical axis. We have shown [Merlet 1988], with the help of MACSYMA, that we get then a singular configuration if, and only if, we have $\psi = \pm\frac{\pi}{2}$, $\theta = \phi = 0$ whatever the position of the center of the mobile is. This is Fichter's singular configuration.

In a second part we consider the general case. We have shown [Merlet 1988] that we may find two more different solutions:

$$\theta = \pm\frac{\pi}{2} \quad ou \quad \psi = \phi$$

and in both cases z_0 is solution of polynomial of degree 3 or x_0, y_0 are solutions of a linear equation. Figure 12 and Figure 13 show examples of each configuration.

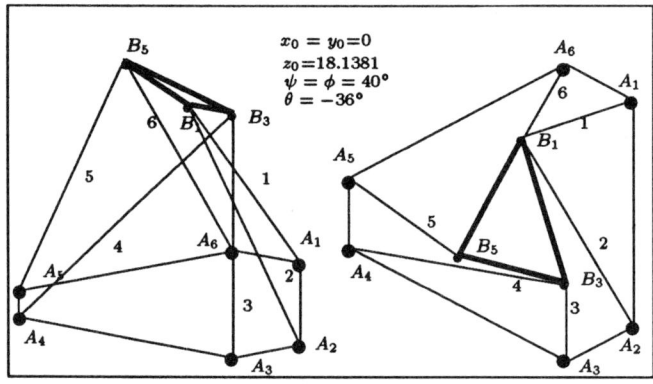

Figure 13: Perspective and top view of a singular configuration of type 5a ($\psi = \phi$)

4.3.2 Configuration 5b

We have to consider the case where the 6 segments cross the same line. Let us consider lines 1,2,3,4.

We have 4 possibilities for line D to intersect 1,2,3,4:

- $D = P_{12} \cap P_{34}$
- D intersects both B_1 and B_3
- $D \in P_{12}$ and D intersects B_3
- $D \in P_{34}$ and D intersects B_1

Let us now consider lines 5,6 in each of these cases.

4.3.3 $D = P_{12} \cap P_{34}$

We may have:

- $B_5 \in P_{12} \cap P_{34}$
- $D = P_{56} \cap P_{12} \cap P_{34}$

In the first case we may deduce from the preliminary remark that line D intersects both B_5 and C_{23}. In this case it is possible to show that x_0, y_0 are defined as functions of z_0, ψ, θ, ϕ. Figure 14 shows such a singular configuration.

Let us consider now the second case. The three planes must have a line in common. Let us consider the intersection line of plane P_{34}, P_{56}. We know that C_{45} belongs to this line. If the intersection line also lie in the plane P_{12} then C_{45} must also lie in this plane. This is impossible under our assumption and therefore the three planes cannot have a line

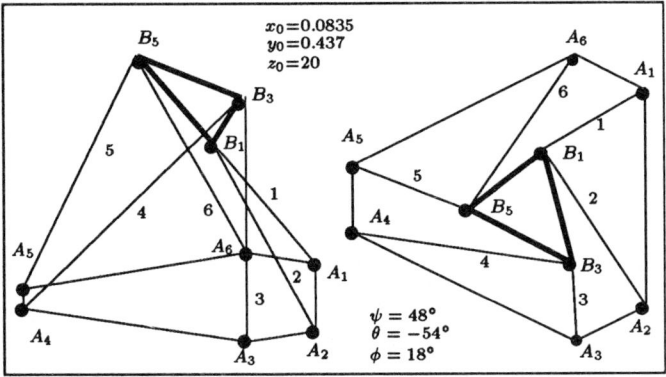

Figure 14: Perspective and top view of a singular configuration of type 5b

in common.

4.3.4 D intersects both B_1 and B_3

Thus the line common to the 6 segments is the edge B_1B_3 of the mobile. If lines 5,6 intersect both this edge this means that the edge is coplanar to P_{56}. This is Hunt's singular configuration.

4.3.5 $D \in P_{12}$ and D intersects B_3

Thus B_3 belongs to P_{12}. If D intersects also 5,6 we may have two possibilities.
- $B_5 \in P_{12}$
- the intersection line $P_{12} \cap P_{56}$ intersects B_3

In the first case D is the edge B_3B_5 of the mobile and two of the segments are coplanar to the mobile. This is Hunt's singular configuration.

In the second case we may deduce that the intersection line must be the line joining C_{16} and B_3. We have dealt with a similar problem in a previous part ($D = P_{12} \cap P_{34}$ and $B_5 \in D$).

4.3.6 $D \in P_{34}$ and D intersects B_1

This case is similar to the previous one.

5 Conclusion

The study of the singular configurations by the use of Grassmann geometry yields to interesting results and new singular configurations. With this method we have found *every* singular configurations of the TSSM. We have established the constraints on the position parameters which must be satisfied to obtain the various singular configurations. This work has been extended in [Merlet 1988] for various architectures of parallel manipulators. What appears in this work is that in the most general case of parallel manipulators it may be difficult to find out the conditions on the mobile parameters so that the geometric rules which indicate a singular configuration are fulfilled. But we are at the beginning of this approach and we hope that by the use of the powerful theorem which has been established in this part of the mathematics we may be able to obtain new results.

Another extension of this work will be to deduce from the constraints on the *position parameters* the constraints on the *links lengths*. From this point it will be possible to determine, for a given architecture, if the singular configurations are in the working area.

References

[Behnke 1986] BEHNKE H. and all "Fundamentals of mathematics, Geometry", Vol II, The MIT Press, third edition 1986

[Bricard 1897] BRICARD R.

"Mémoire sur la théorie de l'octaèdre articulé", Journal de Mathématiques pures et appliquées, Liouville, cinquième série, tome 3, 1897

[Cauchy 1813] CAUCHY A.

"Deuxième mémoire sur les polygones et les polyèdres" Journal de l'Ecole Polytechnique, XVIeme cahier, 1813

[Crapo 1973] CRAPO H.

"A combinatorial perspective on algebraic geometry", Colloquio Int. sulle Teorie Combinatorie, Roma, september 3-15, 1973

[Dandurand 1984] DANDURAND A.

"The rigidity of compound spatial grid", Structural Topology 10, 1984

[Fichter 1986] FICHTER E.F. "A Stewart platform based manipulator: general theory and practical construction" , The Int. J. of Robotics Research, Vol.5, n 2, Summer 1986, pp. 157-181

[Hilbert 1952] HILBERT D., COHN-VOSSEN S.

" Geometry and the imagination" , Chelsea Publ. Company, 1952

[Hunt 1978] HUNT K.H. "Kinematics geometry of mechanisms", Clarendon Press, Oxford, 1978

[Hunt 1983] HUNT K.H.

"Structural kinematics of in Parallel Actuated Robot Arms" ,Trans. of the ASME, J. of Mechanisms,Transmissions, and Automation in design, Vol. 105, December 1983, pp.705-712

[Inoue 1985] INOUE H., TSUSAKA Y., FUKUIZUMI T.

"Parallel manipulator" , 3th ISRR, Gouvieux, France,7-11 Oct.1985

[Mac Callion 1979] Mac CALLION H., PHAM D.T.

"The analysis of a six degree of freedom work station for mechanized assembly", 5th World Congress on Theory of Machines and Mechanisms, Montreal, July 1979

[Merlet 1988] MERLET J-P. "Parallel Manipulator, Part 2: Singular configurations and Grassmann geometry", INRIA research Report,791, February 1988

[Reboulet 1985] REBOULET C., ROBERT A.

"Hybrid control of a manipulator with an active compliant wrist", 3th

ISRR, Gouvieux, France, 7-11 Oct.1985, pp.76-80

[Stewart 1965] STEWART D.

"A platform with 6 degrees of freedom" Proc. of the institution of mechanical engineers 1965-66, Vol 180, part 1, number 15, pp.371-386

[Tyrrell 1971] TYRRELL J.A., SEMPLE J.G.

"Generalized Clifford Parallelism", Cambridge Univ. Press, 1971

[Yang 1984] YANG D.C.H., LEE T.W.

"Feasibility study of a platform type of robotic manipulator from a kinematic viewpoint", Trans. of the ASME, J. of Mechanisms, Transmissions, and Automation in design, Vol 106, June 1984, pp.191-198

[Zamanov 1984] ZAMANOV V.B, SOTIROV Z.M.

"Structures and kinematics of parallel topology manipulating systems", Int. Symp. on Design and Synthesis, July 11-13 1984, Tokyo, pp.453-458

[Veblen 1910] VEBLEN O.,Young J.W. "Projective geometry", The Athenaeum Press, 1910

Applications of Geometric Homology

Henry Crapo
Bât 10, INRIA, B.P. 105, 78153 Le Chesnay Cedex, France

Introduction

We present a brief introduction to the calculus of arithmetic invariants of geometric objects. This theory, an extension of combinatorial topology to the domain of projective geometry, makes use of the concept of a sheaf, and of its cohomology. The invariants in question are the Betti numbers and the characteristic of this cohomology theory. The theory has applications to scene analysis, to polynomial approximation, to solid modeling, to mechanics, and to the study of the rigidity of structures.

Arithmetic invariants of geometric objects

Let P be a finite set of points, and let $\mathbb{F}$ be a vector space of real-valued functions defined on the set P, that is, a subspace of the vector space $\mathbb{R}^P$. (We will also consider the case of vector-valued functions, where $\mathbb{F}$ is a subspace of the vector space $(\mathbb{R}^d)^P$, but for the moment we adhere to the simpler notation.) Let k be the dimension of the vector space $\mathbb{F}$. The *action* â of a point $a \in P$ on the vector space $\mathbb{F}$ is a linear functional

$$â: \mathbb{F} \to \mathbb{R},$$

that is, *evaluation* at the point a:

$$â(f) = f(a).$$

If B, a subset of cardinality k, is a base for the vector space $\mathbb{F}$, then $M(\mathbb{F})$, the matrix in which the row "f", for each function $f \in B$, lists the values f(a) at the points $a \in P$, arranged in a fixed, but arbitrary, order. Then the columns of the matrix $M(\mathbb{F})$ *represent* the points of P by giving their *actions* on the space $\mathbb{F}$.

Some examples of such vector spaces $\mathbb{F}$ are:

$\mathbb{F} = \mathbb{C}$, the space of constant functions,

$\mathbb{F} = \mathbb{L}$, the space of linear functions,

$\mathbb{F} = \mathbb{Q}$, the space of quadratic functions, and

$\mathbb{F} = \mathbb{I}$, the space of velocity vector fields of rigid motions, that is, of isometries. (This final example uses vector-valued functions.)

For example if $\mathbb{F} = \mathbb{L}$, the space of linear functions defined on a finite set P of points in a Euclidean space of dimension k-1, then the base B can be the set of projections on the principal axes x, y, ... , together with the constant function "1". The points of P are then represented as points in $\mathbb{R}^k$. These vectors are the *standard projective coordinates* of the points. In the real plane, where k=3, every point (x, y) has a standard representation

$$(x, y) \rightarrow (x, y, 1)$$

and equivalent projective representations

$$(x, y) \rightarrow (\alpha x, \alpha y, \alpha) \quad \forall \, \alpha \in R$$

which differ only by an overall scalar multiple. Any such vector gives *homogeneous coordinates* of the point (x, y). We show the matrix M($\mathbb{L}$) for a set of six points in the plane, in **Figure 1**.

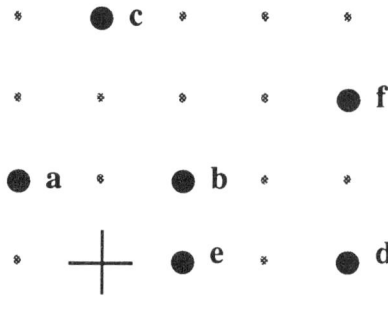

Figure 1

M:

	a	b	c	d	e	f	
	-1	1	0	3	1	3	x
	1	1	3	0	0	2	y
	1	1	1	1	1	1	1

Every subset A of P, as a set of linear functionals, spans a vector subspace $<A>_\mathbb{F}$ of $\mathbb{R}^k$, its *closure* with respect to the space $\mathbb{F}$. We write simply $<A>$ for $<A>_\mathbb{L}$, the projective space spanned by the set A. The dimension of this space $<A>_\mathbb{F}$ is the *rank* $r_\mathbb{F}(A)$ of the set A. It is equal to the cardinality of a minimal subset $B \subseteq A$ such that $<B>_\mathbb{F} = <A>_\mathbb{F}$. For example, if A is a set of collinear points, of which at least two are distinct, $r_\mathbb{L}(A) = 2$. We write simply $r(A)$ for $r_\mathbb{L}(A)$, the projective rank of the set A.

Let **E** be a family of subsets of P, $\mathbf{E} \subseteq 2^P$. We say that **E** is *open* if it is hereditary, that is if

$$A \subseteq B \in \mathbf{E} \Rightarrow A \in \mathbf{E},$$

We associate, with every open family **E**, the sub-family of those maximal elements A, B, ... , the *nodes* of **E**, and write

$$\mathbf{E} = (A, B, \ldots)$$

to indicate that **E** is the hereditary family *spanned* by those nodes. For example, if

$$\mathbf{E} = \{\emptyset, a, b, d, e, ab, ad, ae, abd\},$$

then **E** is open, with two nodes, and $\mathbf{E} = (abd, ae)$.

For every subset $A \subseteq P$, we define

$$\mathbb{F}(A) = \{ f|_A ; f \in \mathbb{F}\},$$

the vector space of functions obtained by *restriction* to the set A. By extension of this concept, for every open family **E** of subsets of P, we define

$$\mathbb{F}(\mathbf{E}) = \{ g; g: \cup\mathbf{E} \to R \text{ such that } g|_A \in \mathbb{F}(A), \forall A \in \mathbf{E}\},$$

the vector space of functions g defined on the union $\cup\mathbf{E} \subseteq P$ of the family **E**, such that the restriction $g|_A$ to every element $A \in \mathbf{E}$ is the restriction to A of a function in the space $\mathbb{F}$. We say that such a function is *locally in* $\mathbb{F}$, *on* **E**. This mapping $\mathbb{F}$, which associates a vector space $\mathbb{F}(\mathbf{E})$ with every open family **E**, is called a *sheaf*.

The dimension of the space $\mathbb{F}(E)$ is equal to β^0, the 0th Betti number of the open family **E** in the *cohomology* of the sheaf $\mathbb{F}$. The Betti number β^0 depends on *geometric* properties of the set P of points and on the choice of nodes. Fortunately, there is a *combinatorial* approximation of the value β^0. The *characteristic*

$$\chi = \beta^0 - \beta^1 + \beta^2 - \ldots,$$

the alternating sum of the Betti numbers, is expressible not only as the alternating sum of the Betti numbers of a dual *homology* theory,

$$\chi = \beta_0 - \beta_1 + \beta_2 - \ldots,$$

but also in a completely combinatorial fashion:

$$\chi = \alpha_0 - \alpha_1 + \alpha_2 - \ldots,$$

where α_i is the sum of the ranks $r_\mathbb{F}(B)$, the sum being over every expression of the form $B = A_1 \cap A_2 \cap \ldots$, the set-theoretic intersection of i+1 nodes of **E**. A further simplification of this latter formula is expressed in terms of the *Möbius function* μ of the semi-lattice L of intersections of nodes of the family **E**. The value of μ is given by the recursion

$$\Sigma_{B \in L; A \subseteq B} \, \mu(B) = 1,$$

and

$$\chi = \Sigma_{B \in L} \, \mu(B) \, r_\mathbb{F}(B).$$

To calculate β_0, it is sometimes possible to find an open family **E** which is *topologically trivial* ($\beta_i = 0$ for $i \geq 1$), and to replace β_0 by its combinatorial expression, $\beta_0(E) = \chi(E)$.

In order to calculate all the Betti numbers β_i of an open family **E** with respect to a sheaf $\mathbb{F}$, we need the notions of boundary operator, of vector spaces of chains, of cycles, of boundaries, and of homologies. An *i-chain* is a formal linear combination

$$p \, AB\ldots D + \ldots + q \, EF\ldots G$$

of (ordered) words AB…D of length i+1 in which the letters A, B, … are distinct nodes, and in which the *coefficients* are vectors in the space $<A \cap B \cap \ldots>_\mathbb{F}$, the space spanned by the points in the set-theoretic intersection of the nodes A, B, … . The *boundary operator* ∂ is first defined on words:

$$\partial(A) = 0,$$
$$\partial(AB) = B - A,$$
$$\partial(ABC) = BC - AC + AB,$$
$$\partial(ABCD) = BCD - ABD + ACD - ABC,$$
$$\ldots$$

Then, the operator ∂ is extended by linearity to linear combinations of words. For example:

$$\partial(pAB + qAC) = pB - (p+q)A + qC.$$

We see here that the coefficients of the boundary are correct:

$$\text{if } p \in <A \cap B>_\mathbb{F} \text{ and } q \in <A \cap C>_\mathbb{F},$$

$$\text{then } p \in <B>_\mathbb{F}, \quad p + q \in <A>_\mathbb{F}, \quad \text{and } q \in <C>_\mathbb{F}.$$

An i-*cycle* is an i-chaîne s pour laquelle $\partial s = 0$. An i-*boundary* is an i-chain for which there exists an (i+1)-chain t such that $\partial t = s$. The i-boundaries form a subspace of the vector space of i-cycles; the quotient of these spaces is the space of i-*homologies*, the dimension of which is the ith Betti number, β_i.

Combinatorial Topology ($\mathbb{F} = \mathbb{C}$)

Let $\mathbb{F} = \mathbb{C}$, the space of constant functions defined on P. Set

$$\mathbf{E} = (abc, de, df, ef)$$

in the example of six points, **Figure 1**. Let Q = abc, R = de, S = df, T = ef. The points are represented by the columns of the matrix $M(\mathbb{C})$,

a	b	c	d	e	f
1	1	1	1	1	1

So the coefficient of a word QR...T can be a non-zero scalar if and only if $Q \cap R \ldots \cap T \neq \emptyset$. There is no non-empty intersection of three nodes, so there is no non-zero 2-chain. The unique 1-cycle, up to a scalar multiple, is the chain

$$RS - RT + ST.$$

These coefficients 1, -1 are permissible because the intersections $R \cap S = d$, $R \cap T = e$, $S \cap T = f$ are non-empty. So $\beta_1 = 1$. The space $\mathbb{C}(\mathbf{E})$ is spanned by two locally constant functions, each representing a connected component of the complex E,

	a	b	c	d	e	f
	1	1	1	0	0	0
	0	0	0	1	1	1

so $\beta_0 = \beta^0 = 2$. If we calculate the characteristic, we find that

$$\chi = \alpha_0 - \alpha_1 = 4 - 3$$
$$= \beta_0 - \beta_1 = 2 - 1 = 1.$$

This calculation gives the expected result: there are two connected components, and one single cycle of edges that is not the boundary of a polygon "filled" with triangular simplices. From a geometric point of view, calculations in combinatorial topology concern only those geometric configurations of dimension 0!

Another way to see that $\beta_1 = 1$ is to look at those dependencies among the columns of the matrix $M(\mathbb{C})$ that are supported on the nodes of the open family E, and to find dependencies among those dependencies. Such a dependency is called a second order *syzygy*. In the given example, α = d-e, β = e-f, γ = f-d are dependencies supported on the nodes R, S, T, respectively, and $\alpha + \beta + \gamma = 0$ is a second order syzygy supported on the set {d, e, f} which is not in the open family. We will use this type of calculation in another context, for scene analysis.

Scene Analysis

For the set of six points shown in **Figure 1**, we choose the vector space $\mathbb{L}$ of linear functions, and set E = {abcd, abef, cdef} = {A, B, C}. The only non-empty intersections of nodes correspond to the words AB, AC, BC. A chain

$$\gamma = p\,AB + q\,AC + r\,BC$$

will have, as boundary, the chain

$$\partial \gamma = (-p - q)\,A + (p - r)\,B + (q + r)\,C.$$

In order for γ to be a cycle, it must be the case that $\partial \gamma = 0$, that is to say that

$$p = -q = r.$$

But in order for the chain p AB - p AC + p BC to be admissible, the point p must be simultaneously in the three linear extensions

$$\langle A \cap B \rangle_L = \langle ab \rangle,$$

$$\langle A \cap C \rangle_L = \langle cd \rangle,$$

$$\langle B \cap C \rangle_L = \langle ef \rangle.$$

In the present case, there is such a point, the point p = (2, 1, 1), because

$$2p = 3b - a, \quad 3p = c + 2d, \quad 2p = e + f.$$

So $\beta_1(\mathbb{L}, E) = 1$.

The characteristic $\chi(E)$ has value

$$\chi(E) = r(A) + r(B) + r(C) - r(A \cap B) - r(A \cap C) - r(B \cap C)$$

$$= 3 + 3 + 3 - 2 - 2 - 2 = 3.$$

Since $\beta_1 = 1$, $\beta_i = 0$ for $i \geq 2$, we know that $\beta_0 = 4$. Three independent functions, which are locally linear because they are globally linear, already form the rows of the matrix $M(\mathbb{L})$. So there is a function h, locally but not globally linear, definable on the set of six points. How can we find one?

In order for a function h to be locally linear, the vector formed from its six values h(a) ... h(f) must be orthogonal to the vectors of coefficients of the dependencies that are supported on those nodes. To find the dependence (unique up to one overall scalar multiple) among four coplanar points a, b, c, d, it suffices to calculate

$$[bcd] a - [acd] b + [abd] c - [abc] d,$$

the brackets indicating the determinants (twice the areas of triangles) of triples of points. These dependencies are as follows, the rows of the matrix D(E):

a	b	c	d	e	f	
3	-9	2	4	0	0	α, on A
2	-6	0	0	2	2	β, on B
0	0	-4	-8	6	6	γ, on C

For example, if we fix the values h(a) = h(b) = h(c) = h(d) = 0, we can still choose h(e) = 1, h(f) = -1. The fact that the restrictions of the function h to the subsets abcd, abef, cdef are linear implies that if we construct the *graph* of this function h in 3-dimensional space, the points a', ... f' which project vertically onto points a, ... f, respectively, with heights h(a), ... h(f), will be "locally coplanar" on the open family **E** = (abcd, abef, cdef). That is to say, the sets a'b'c'd', a'b'e'f', c'd'e'f' will be coplanar. See **Figure 2**.

If the three rows of the matrix D(**E**) had been independent, then the space of locally linear functions could not have had rank higher than 6-3 = 3, the number of points less the dimension of the space of dependencies. In the given example, this is not the case. There is a dependence $2\alpha - 3\beta + \gamma = 0$ among the rows of the matrix D(**E**), and the space orthogonal to the space of dependencies, that is, the space of locally linear functions, has dimension $\beta_0 = 4 = 6-2$.

a	b	c	d	e	f	
-1	1	0	3	1	3	x
1	1	3	0	0	2	y
1	1	1	1	1	1	1
0	0	0	0	1	-1	h

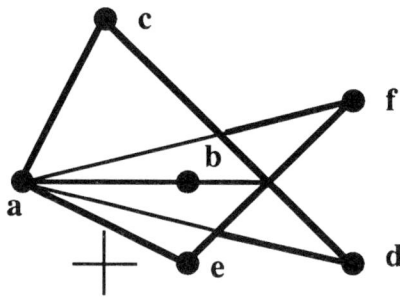

Figure 2

Recall: the dependence $2\alpha - 3\beta + \gamma$ is a second order syzygy.

This example illustrates an important principle. If the points had been in a slightly different position, or if the choice of nodes A, B, C had been different, it would not necessarily be the case that the intersection

$$<A \cap B>_L \cap <A \cap C>_L \cap <B \cap C>_L$$

would be non-empty. For example, if one were to take D = abcf, E = acde, F = bdef, the combinatorial structure is the same as before, but the intersection

$$<D \cap E>_L \cap <D \cap F>_L \cap <E \cap F>_L = <ac> \cap <bf> \cap <de> = \emptyset$$

is empty. So $\chi = \beta_0 = 3$, $\beta_i = 0$ for $i \geq 1$, and there is no locally linear function other than those in the 3-dimensional space of globally linear functions. There is no 3-dimensional model with the given projection. The condition that the three lines <ab>, <cd>, <ef> meet at a point is a typical *projective condition* permitting the construction of a model over a given geometric scene.

The calculation of the characteristic for polyhedral scenes is particularly simple. We always set $\mathbb{F} = \mathbb{L}$, and we choose, as nodes, the faces of the polyhedron. Every edge is at the intersection of two faces, and every vertex is surrounded by a cycle of faces and edges. The value of the Möbius function is 1 for a face, -1 for an edge, and 1 for a vertex surrounded by a cycle of faces and edges. So, for a polyhedron with f faces, a edges and s vertices,

$$\chi_{\mathbb{F}} = 3f - 2a + s.$$

Comparing this with the topological case ($\mathbb{F} = \mathbb{C}$) we see the difference between the "topological" homology and geometric homology. If $\mathbb{F} = \mathbb{C}$, the values of μ are the same as above, but the rank of every non-empty subset becomes 1. The characteristic

$$\chi_{\mathbb{C}} = f - a + s,$$

is the Euler characteristic, which is equal to 2 for every spherical polyhedron. The difference

$$\chi_{\mathbb{F}} - 2\chi_{\mathbb{C}} = f - s$$

is equal to

$$\beta_0 - \beta_1 - 4$$

because $\beta_i = 0$ for $i \geq 2$. The value $\beta_0 - 4$ is equal to the number of *projective choices* in the construction of a 3-dimensional model over a given polyhedral scene. The value β_1 is equal to the number of projective conditions required for a correct plane drawing of such a polyhedron. Whence the formula

$$\text{projective choices - projective conditions} = f - s.$$

Consider the projective cube, six plane faces meeting at 8 vertices. The caracteristic of the open family generated by the six faces has the value

$$\chi = 18 - 24 + 8 = 2,$$

In **Figure 3** we see the algorithm for calculating the number of choices and projective conditions. We start by choosing 3 heights, for the vertices shown in black. We can calculate one additional height, for the cross-hatched vertex, by making use of the coplanarity of the four vertices on the face indicated with a check-mark. We continue in this way, choosing one additional height and checking four faces in all by the time we have determined all eight heights. It remains to find out whether the two faces indicated by question marks are planar. This will be the case if and only if the drawing satisfies two independent projective conditions. In that case there will be

$$0 \text{ projective choices, 2 projective conditions, and } f - s = -2.$$

The Betti numbers have values $\beta_0 = 4$, $\beta_1 = 2$ for a correct plane drawing of a projective cube. In fact, there are $\beta_1 = 2$ independent relations among the first order syzygies (the dependences) supported on the six faces.

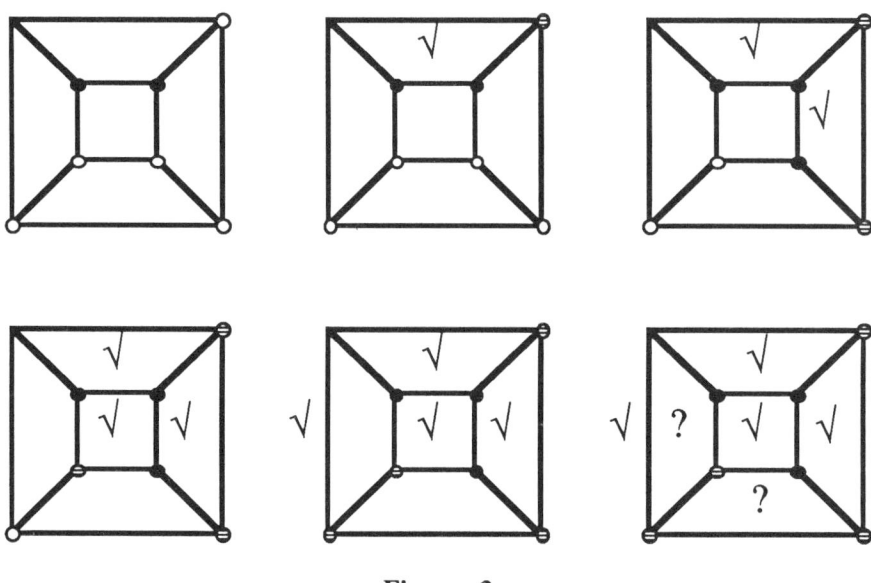

Figure 3

Mechanics and the rigidity of structures

For applications of geometric homology to mechanics, we take $\mathbb{F} = \mathbb{I}$, the space of vector fields of velocities of points moved by isometries of the space, that is, by rigid motions of the entire space. Passage to the concept of a *locally rigid* motion of a mechanical structure permits us to study the rigidity or mobility of such a structure by homological methods.

Let S be a structure in 3-space R^3, consisting of b rigid bars meeting at s vertices (universal joints). The vector space of isometries of a single bar **ab** is 5 dimensional: we can choose the velocity of the point **a** (3 choices), and then, the point **b** can only move relative to **a** in the directions tangent to a sphere with centre at the point **a** (2 choices). A point has 3 degrees of freedom. The Möbius function of a bar is equal to 1, that of a vertex incident to n bars is equal to -(n-1). Let s_n be the number of vertices incident to exactly n bars. Then the characteristic χ (𝕀, **E**) of the open family **E** spanned by those pairs of points that are linked by bars is equal to

$$\chi = 5b - 3 \,(0\, s_1 + s_2 + 2\, s_3 + 3\, s_4 + \ldots).$$

The sum of the s_i is equal to s:

$$s = s_1 + s_2 + s_3 + \ldots ,$$

and the total incidence of bars and joints yields

$$2b = s_1 + 2\, s_2 + 3\, s_3 + \ldots .$$

So

$$\chi - 3s = 5b - 3\, (s_1 - 2\, s_2 - 3\, s_3 - \ldots) = -b$$

and

$$\chi = 3s - b.$$

For structures consisting of bars and universal joints in d-space R^d, this calculation yields

$$\chi = ds - b.$$

For the structure based on a correct drawing of a projective cube, **Figure 3**, we find

$$\chi(𝕀, \mathbf{E}) = 16 - 12 = 4,$$

but the structure has 5 degrees of infinitesimal motion. Even if one fixes one bar (3 choices) and the position of an adjacent vertex (1 choice), a quadrilateral face will be rigid, but the other four vertices can still move with one infinitesimal degree of freedom. The difference

$$\beta_0 - \chi = 5 - 4$$

yields

$$\beta_1 = 1,$$

a value which counts the number of self-stresses (systems of tension and compression in the bars, the entire being in internal equilibrium). In the symmetric position indicated in **Figure 3**, the square on the outside can be in compression, the other 8 bars in tension, like a spiderweb in a window frame, the entire structure being in static equilibrium.

Imagine two rigid bodies in 3-space R^3, linked by six bars. The characteristic $\chi(\mathbb{I}, \mathbf{E})$ of the open family generated by the two bodies and six bars has the value

$$\chi = 2\, r_{\mathbb{I}}(\text{body}) + 6\, r_{\mathbb{I}}(\text{bar}) - 12\, r_{\mathbb{I}}(\text{point})$$

$$= 12 + 30 - 36 = 6.$$

The value of β_0 expected for such a structure in general position is 6, and β_1 will be zero. But it can happen that the structure is in special position, where the six constraints on the relative motion of the two bodies become dependent, the six bars lie in a single *line complex*, and $\beta_0 = 7$, $\beta_1 = 1$. A locally, but not globally, isometry will exist.

A still unsolved problem concerns the number of degrees of freedom of structures with bars and universal joints in general position in 3-space R^3. We conjecture that if we choose the open family $\mathbf{E}$ generated by those vertex sets which remain rigid relative to one another in every possible infinitesimal motion, then one will find that

$$\chi = \beta_0, \quad \beta_i = 0 \text{ for } i \geq 1,$$

so $\mathbf{E}$ is an open family that is homologically trivial with respect to the sheaf $\mathbb{I}$ of local isometries. The analogous theorem in R^4 is false: consider the bipartite structure $K(6, 6)$.

Bibliography

H. Crapo et J. Ryan, *Réalisations spatiales des scènes linéaires*, **Topologie structurale 13**, (1986), 33-68.

H. Crapo, *The Combinatorial Theory of Structures: Lectures on the application of combinatorial geometry in architecture and structural engineering*, **Matroid Theory, Colloq. Math. Soc. János Bolyai 40,** North-Holland, Amsterdam - New York (1985), 107-213.

Robin Hartshorne, **Algebraic Geometry,** Springer-Verlag, 1977.

Some Examples of Algorithms Analysis in Computational Geometry by Means of Mathematical Morphological Techniques

Michel Schmitt
INRIA Sophia-Antipolis
2004, route des Lucioles
06565 VALBONNE Cedex, FRANCE

Abstract

In this paper we show how the notion of convergence of a sequence of closed sets and that of random closed sets can be given a precise definition by means of mathematical morphological tools. Then we use these two notions on the one hand to analyze how the Delaunay triangulation enables us to get a good approximation of the skeleton of an object and on the other hand to estimate the performances of bucketing techniques in the average case.

1 Introduction

Mathematical morphology is essentially known in robotics by the erosion and the dilation transformations. For instance, if X represents the free space in which a robot K can move (translation only is allowed), the set of all its positions where it is inside free space is given by the erosion of X by K:

$$E_K(X) = X \ominus \check{K} = \{x : K_x \subset X\}$$

where K_x is the translation of K by vector $\vec{x}$ and $\check{K}$ the symmetrical set of X with respect to the origin of the space (The notations are recalled in section 7). In the same way, all the points reached by the robot will be given by the opening of X by K:

$$(X)_K = (X \ominus \check{K}) \oplus K = \bigcup\{K_x : K_x \subset X\}$$

where $\oplus$ is the Minkowski sum:

$$A \oplus B = \{a + b : a \in A \text{ and } b \in B\} = \{x : \check{B}_x \cap A \neq \emptyset\}$$

By means of tools developed in computational geometry, algorithms have been proposed to efficiently compute the erosion and opening of X in the case of a polygon in the plane (a polyhedron in the space) or a set whose boundary is composed of line segments and arcs of circle [5]. These problems widely escape traditional mathematical morphology, the aim of which is rather to investigate the properties of mappings based on Boolean relationships such as $K \cap X \neq \emptyset$ or $K \subset X$.

Two aspects of morphological theory have been studied:

The first one, operating on a complete lattice, is interested in the underlying algebraic properties of the transformations (see [6] and [12]). For example, it looks for the characterization of the class of openings (increasing, idempotent and antiextensive mappings) or of filters (increasing and idempotent mappings). This approach allows us to write compact formulae whose the proofs are purely algebraic: it is the case when one attempts to solve the problem of placing n forms in another by use of structuring functions ($n > 2$).

The second one, beginning in the 70's, studies the regularity (continuity, semicontinuity, etc.) of the transformations when we provide the set of closed sets of $\mathbb{R}^n$ with an appropriate topology. An associate σ-algebra has been defined in order to describe porous media as random closed sets. These notions led G. Matheron to write his famous book "Random Sets and Integral Geometry" [7]. In the sequel, we will be interested in this second aspect of mathematical morphology.

The paper will be organized in the following manner: we first expose the topology provided by mathematical morphology on the closed sets of $\mathbb{R}^n$ and illustrate it by studying the convergence towards the skeleton of the centers of the ball circumscribed about the simplexes of the Delaunay triangulation (Delaunay spheres). Then we describe the associated random frame and give as an application the complexity analysis of algorithms using bucketing techniques.

2 The hit or miss topology

One of the familiar tools when one wants to derive a distance between two shapes is the Hausdorff distance. This distance is defined as follows:

$$\varrho(K, K') = \max\left\{\sup_{x \in K} d(x, K'), \sup_{x \in K'} d(x, K)\right\}$$
$$= \inf\{\varepsilon : K \subset K' \oplus B_\varepsilon, K' \subset K \oplus B_\varepsilon\}$$

where B_ε is the closed ball of radius ε. However, this distance is defined only on the set of all non-empty compact sets of $\mathbb{R}^n$. This set itself is non compact and consequently do not allow to study cases such as "a sequence of compacts sets converging towards a half-plane" for instance.

Our goal in this section is to present a topology on the set of all closed sets of $\mathbb{R}^n$ which has not the drawbacks we already mentioned. A σ-algebra will be derived in section 4 in order to deal with random closed sets. In section 2 and 4, we follow the presentation of G. Matheron [7]. The proofs are omitted to focus on the tools developed. But, when we use these tools, proofs will be given (see section 3 and 5).

2.1 Definition of a topology on the sets of closed sets of $\mathbb{R}^n$

Let us denote $\mathcal{F}(\mathbb{R}^n)$ or simply $\mathcal{F}$ if there is no ambiguity, the set of the topologically closed sets of the Euclidean space $\mathbb{R}^n$ provided with its usual topology and $\mathcal{K}(\mathbb{R}^n)$ or simply $\mathcal{K}$ the one of compact sets. We have seen that mathematical morphology was found on relationships such as $K \cap X \neq \emptyset$ or $K \subset X$. From these relations, a topology $\mathcal{T}$ can be derived.

Definition 1

We call <u>hit or miss topology</u> on $\mathcal{F}$ the topology $\mathcal{T}$ generated by the neighborhoods

$$\mathcal{F}^K_{G_1,...,G_p} = \{F \in \mathcal{F} : F \cap K = \emptyset \text{ and } F \cap G_i \neq \emptyset\}$$

where $p \in \mathbb{N}$, $K \in \mathcal{K}$ is a compact set and $(G_i)_{i=1}^p$ is a finite family of open sets in $\mathbb{R}^n$.

Why only one compact set? Simply because

$$K_1 \cap X = \emptyset \text{ and } K_2 \cap X = \emptyset \iff (K_1 \cup K_2) \cap X = \emptyset$$

which shows that several compact sets need not to be simultaneously considered.

From these definitions, the following fundamental theorem can be proved:

Theorem 2

$(\mathcal{F},\mathcal{T})$ is compact, Hausdorff and separable.

Hausdorff means

$$F_1, F_2 \in \mathcal{F} \implies \exists O_1, O_2 \in \mathcal{T}, F_1 \in O_1, F_2 \in O_2, O_1 \cap O_2 = \emptyset$$

and separable means that $\mathcal{T}$ has a countable basis.

The property of compacity, used in the sequel, may be translated as follows: any sequence of closed sets has at least one cluster point in $\mathcal{F}$.

Let us illustrate this paradoxical point: $\mathbb{R}^n$ is not compact, but $\mathcal{F}$ is compact! Let be $(x_i)_{i \in \mathbb{N}}$ a sequence of points in $\mathbb{R}^n$ diverging towards infinity ($x_i \to \infty$). Then $(\{x_i\})_{i \in \mathbb{N}}$ is a sequence of closed sets <u>converging</u> towards $\emptyset$.

Because the topology $\mathcal{T}$ is separable, it is possible to restrict ourselves to sequences of closed sets. The following theorem gives a more physical interpretation of $\mathcal{T}$ by means of sequences of points in $\mathbb{R}^n$.

Theorem 3

A sequence $(F_i)_{i \in \mathbb{N}}$ of closed sets converges towards F if and only if we have simultaneously:

Criterion 1: $\forall x \in F$, $\exists x_i \in F_i$ for i great enough, $x_i \to x$.

Criterion 2: Let $(F_{i_k})_{i_k \in \mathbb{N}}$ be a subsequence of $(F_i)_{i \in \mathbb{N}}$ and $x_{i_k} \in F_{i_k}$. Then $(x_{i_k} \to x) \Rightarrow (x \in F)$.

The following properties can be derived:

1. The mapping $\cup: \mathcal{F} \times \mathcal{F} \to \mathcal{F}$ is continuous.

2. If $(F_i)_{i \in \mathbb{N}}$ is a decreasing sequence of closed sets, then $\lim_{i \to \infty} F_i = \cap F_i$.

3. If $(F_i)_{i \in \mathbb{N}}$ is a increasing sequence of closed sets, then $\lim_{i \to \infty} F_i = \overline{\cup F_i}$.

4. The Minkowski sum $\oplus$ of two closed sets is not always a closed set so that no property can be stated for the moment.

Unfortunately, the intersection $\cap: \mathcal{F} \times \mathcal{F} \to \mathcal{F}$ is not continuous so that the notions of upper and lower limits have to be examined with more care as well as the notion of semicontinuity.

The criteria 1 and 2 of theorem 3 allow us to define two closed sets which are respectively the intersection and the union of all the cluster points of the sequence $(F_i)_{i \in \mathbb{N}}$.

Definition 4

Let $(F_i)_{i \in \mathbb{N}}$ be a sequence of closed sets. We call

1. *lower limit* of $(F_i)_{i \in \mathbb{N}}$ the set of all limits of sequences of points $x_i \in F_i$. It is the greatest closed set verifying criterion 1.

2. *upper limit* of $(F_i)_{i \in \mathbb{N}}$ the set of all cluster points of sequences of points $x_i \in F_i$. It is the smallest closed set verifying criterion 2.

We have in particular:
$$F_i \to F \iff \limsup F_i = \liminf F_i.$$

The notion of semicontinuity can be deduced as in the case of real functions:

Definition 5

Let Ω be a topological separable space and $\psi: \Omega \to \mathcal{F}$.

1. ψ is lower semicontinuous (l.s.c.) at $\omega \in \Omega$ if and only if
   $\forall \omega_i \to \omega, \; \limsup \psi(\omega_i) \subset \psi(\omega)$.

2. ψ is upper semicontinuous (u.s.c.) at $\omega \in \Omega$ if and only if
   $\forall \omega_i \to \omega, \; \liminf \psi(\omega_i) \supset \psi(\omega)$.

With these definitions:

1. The mapping $\cap: \mathcal{F} \times \mathcal{F} \to \mathcal{F}$ is upper semicontinuous.

2. The mapping associating to F its boundary ∂F is lower semicontinuous.

3. The inclusion holds when we take the limit: $F_i \subset F'_j \Rightarrow \limsup F_i \subset \liminf F'_j$.

2.2 Relationships between the hit or miss topology and that induced by the Hausdorff metric

Let us now examine what happens when we provide the space $\mathcal{K}' = \mathcal{K} \setminus \{\emptyset\}$ with the topology induced by $\mathcal{T}$. It does not coincide with the one induced by the Hausdorff distance.

Definition 6

We call myope topology on $\mathcal{K}$ the topology generated by the neighborhoods

$$\mathcal{K}_{G_1,...,G_p}^{F} = \{K \in \mathcal{K} : K \cap F = \emptyset \text{ and } K \cap G_i \neq \emptyset\}$$

where $p \in \mathbb{N}$, $F \in \mathcal{F}$ is a closed set and $(G_i)_{i=1}^{p}$ is a finite family of open sets of $\mathbb{R}^n$. This topology coincides on $\mathcal{K}'$ with the one induced by the Hausdorff distance.

Proposition 7

The topology induced by the Hausdorff distance on $\mathcal{K}'$ is strictly thiner than the one induced by $\mathcal{T}$.

The links between the convergence according to $\mathcal{T}$ and according to the myope topology can be stated more precisely:

Proposition 8

Let $(K_i)_{i \in \mathbb{N}}$ be a sequence in $\mathcal{K}$. $(K_i)_{i \in \mathbb{N}}$ converges in $\mathcal{K}$ if and only if:

1. *$(K_i)_{i \in \mathbb{N}}$ converges in $\mathcal{F}$ according to $\mathcal{T}$.*

2. *$\exists K \in \mathcal{K}, \forall i\ K_i \subset K$.*

Remark: the expression "myope topology" is now self explanatory because a sequence of compact sets cannot converge if there are points going towards infinity. The myope topology is unable to take them into account. Going back to our example of a sequence of points $(x_i)_{i \in \mathbb{N}}$ diverging towards infinity, $(\{x_i\})_{i \in \mathbb{N}}$ converges towards $\emptyset$ in $\mathcal{F}$ but does not converge in $\mathcal{K}$.

Providing $\mathcal{F}$ with $\mathcal{T}$ and $\mathcal{K}$ with the myope topology, we can state:

1. The Minkowski sums, $\mathcal{F} \times \mathcal{K} \to \mathcal{F}$ and $\mathcal{K} \times \mathcal{K} \to \mathcal{K}$ are continuous.

2. The homotheties, $\mathbb{R}^{+*} \times \mathcal{F} \to \mathcal{F}$ and $\mathbb{R}^{+*} \times \mathcal{K} \to \mathcal{K}$ are continuous.

3. The erosions, $\mathcal{F} \times \mathcal{K} \to \mathcal{F}$ and $\mathcal{K} \times \mathcal{K} \to \mathcal{K}$ are lower semicontinuous.

4. The convex hull, $\mathcal{K} \to \mathcal{K}$ is continuous and $\mathcal{F} \to \mathcal{F}$ is lower semicontinuous.

5. The space of convex sets of $\mathcal{F}$ is closed in $\mathcal{F}$ and the space of convex sets of $\mathcal{K}$ is closed in $\mathcal{K}$.

3 Application: relationships between the skeleton and the Delaunay triangulation

Let X be an object in $\mathbb{R}^n$. We assume that X is a compact set whose boundary is a surface or an union of surfaces C^3 (X is neither necessarily simply connected nor even connected). This ideal object is only known to us by a set of points measured on its surface. When the points are dense enough on the surface, the centers of the Delaunay spheres circumscribed about the simplexes provided by the Delaunay triangulation of this set of points give a good approximation of the skeleton of X (figure 1). We will show how we can give a precise meaning to the word "approximation" and what is the type of convergence we can obtain when the density of points on the surface of the object increases indefinitely.

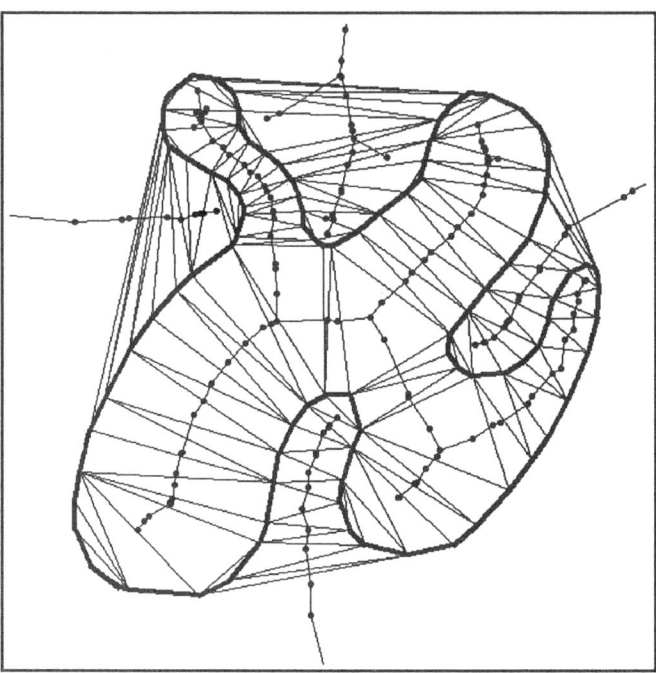

Figure 1: Approximation of the skeleton by the Delaunay triangulation.

Let us first define some basic notions which we will use throughout this section:

Definition 9
> *We call <u>set of points of density θ</u> on X any set $P = (p_i)_{i=1}^{q}$ of points on ∂X verifying:*
>
> $$\forall x \in \partial X,\ \exists i \in [1,q],\ p_i \in B(x, \theta^{-1})$$
>
> *where $B(x, \theta^{-1})$ is the closed ball of radius θ^{-1}.*

This definition shows how we want the boundary of X to be sampled as dense as possible when $\theta \to \infty$.

Definition 10

> The <u>interior skeleton</u> of X, which will be denoted by $Sk_I(X)$, is the topological closure of $Bmax_I(X)$, the locus of the centers of the maximal closed balls included in X. Formally,
>
> $$Bmax_I(X) = \{x : \exists r_x,\ B(x,r_x) \subset B(x',r') \subset X \Rightarrow x' = x,\ r' = r_x\}$$
> $$Sk_I(X) = \overline{Bmax_I(X)}$$
>
> The <u>exterior skeleton</u> $Sk_E(X)$ can be defined in the same way with the maximal balls $(Bmax_E(X))$ included in $\overline{X^c}$.
> The <u>skeleton</u> of X, denoted by $Sk(X)$, is the union of its interior skeleton and its exterior skeleton. In the same way, we will denote by $Bmax(X)$ the union of $Bmax_I(X)$ and $Bmax_E(X)$.

The Delaunay triangulation is a well-known notion and will not be defined here again (see for instance [9] [2]). It is sufficient to keep in mind that it yields a tessellation of the convex hull of a set of points with simplexes (triangles in the plane, tetrahedra in the space). One characteristic property of that triangulation is the following: the ball circumscribed about any simplex (called in the sequel Delaunay simplex) do not contain other points in its interior.

The study of the convergence of closed balls in $\mathbb{R}^n$ is based on the following proposition which is very intuitive. It is worth noting that this proposition does not remain true in $\mathcal{K}$.

Proposition 11

> Let $(B_i = B(x_i, r_i))_{i \in \mathbb{N}}$ be a convergent sequence of balls in $\mathcal{F}$.
>
> 1. $x_i \to x$ and $r_i \to r \iff B_i \to B(x,r)$ in $\mathcal{K}$ and consequently also in $\mathcal{F}$.
>
> 2. Otherwise, $(B_i)_{i \in \mathbb{N}}$ converges towards a generalized ball, i.e. either the empty set or a closed half space or $\mathbb{R}^n$ itself.

Notations:
In order to study convergence problems, we will give us a sequence of sets of points $(P_i)_{i \in \mathbb{N}}$ more and more dense on the surface of X. The points of P_i will be called sometimes "sampling points" on ∂X. We assume that the density of P_i is $\theta_i \to \infty$.
We denote by S_i the set of the centers of the Delaunay spheres corresponding to the triangulation of P_i.
Finally we assume that ∂X is an orientable manifold of smoothness C^3 at least.
With these assumptions, we are going to show that $S_i \to Sk(X) = Bmax(X)$.

3.1 Convergence of the centers of the Delaunay spheres towards the skeleton

In a heuristic way, let $B = B(x,r)$ be a ball circumscribed about a simplex of the triangulation of P_i. Then, by definition of a set of points of density θ_i, $B(x, r - \theta_i^{-1})$ is included into X or

$\overline{X^c}$. If θ_i is large, B is "almost maximal" in X or $\overline{X^c}$ and then approximates a maximal ball (see Figure 2).

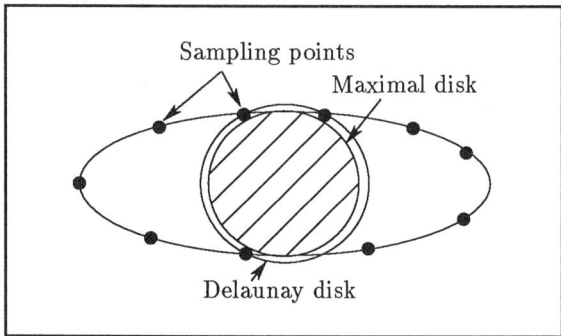

Figure 2: Delaunay ball approximating a maximal ball.

We first study the feasible limits of the centers of the Delaunay balls, then what kind of convergence happens for the set of the centers of these balls and we finally determine under what assumptions, all the skeleton is well reached.

Let us prove that each sequence of centers of Delaunay balls converges towards the center of a maximal ball, i.e. a point of the skeleton of X.

Proposition 12

> Let $(B_i)_{i \in \mathbb{N}}$ be a sequence of Delaunay balls where B_i is circumscribed about a simplex of the triangulation of P_i. If B is a cluster point of B_i, then B is maximal in X or in $\overline{X^c}$.
> In other words:
> $$\limsup_{i \to \infty} S_i \subset Bmax(X)$$

<u>Proof</u>: Let $B_{i_k} \to B$ be a subsequence. Because $B_i = B(x_i, r_i)$ is a Delaunay ball and P_i is a set of points of ∂X whose density is θ_i, the ball $B(x_i, r_i - \theta_i^{-1})$ does not intersect the boundary of X. Then it is included into X or $\overline{X^c}$. By extracting a subsequence if necessary, we can assume that $B(x_{i_k}, r_{i_k} - \theta_{i_k}^{-1})$ are all together in X or $\overline{X^c}$. Since $B(x_{i_k}, r_{i_k} - \theta_{i_k}^{-1}) \to B$ and since the inclusion is preserved by taking the limits, $B \subset X$ or $B \subset \overline{X^c}$. Let us assume, for example, $B \subset X$ (the other case can be proved in the same way), and let B' be such that $B \subset B' \subset X$. We are going to show that $B = B'$ which will prove that B is maximal in X.

Let us note $(a_{i_k}^0, \ldots, a_{i_k}^n)$ the vertices of the simplex which B_{i_k} is circumscribed about. Again by extracting a subsequence if necessary, we can assume that these $n+1$ sequences of points converge towards $a^0, \ldots, a^n$ respectively. All limits are points on the boundary of X and belong to B (see the criteria 1 and 2 of the theorem of convergence of the closed sets theorem 3).

- Let us assume that two of these limits are different. Because these points are on the boundary of B and B' (these balls are included in X) and because $B \subset B'$, we have: $B = B'$ and B is maximal into X.

- Let us now assume that all these limits are equal to a and let us first consider the planar case. B is the limit circle circumscribed about three points converging towards a single point a. B is then osculator to the boundary (we must here assume that ∂X is C^3). Therefore, $B \subset X$ is maximal into X. In $\mathbb{R}^n$, the proof is a little bit more tricky: let us orientate the surface according to the interior normal and let $(c_1, \ldots, c_n)$ be the principal curvature at point a, sorted in increasing order. The radius of B verifies $\frac{1}{c_n} \leq r \leq \frac{1}{c_1}$, since B is the limit ball of the ball circumscribed about $n+1$ points converging towards a. Because $B \subset X$ we have $r = \frac{1}{c_n}$. But the radius of the maximal ball included into X and containing a is at most $\frac{1}{c_n}$. We deduce the maximality of B.

This ends the proof ■

What happens if we assume the boundary of X less regular?

1. If X is limited by a piecewise C^∞ curve, some pathologies can appear at turn back points (see figure 3.1).

2. If X is a polyhedron, the proposition is still verified.

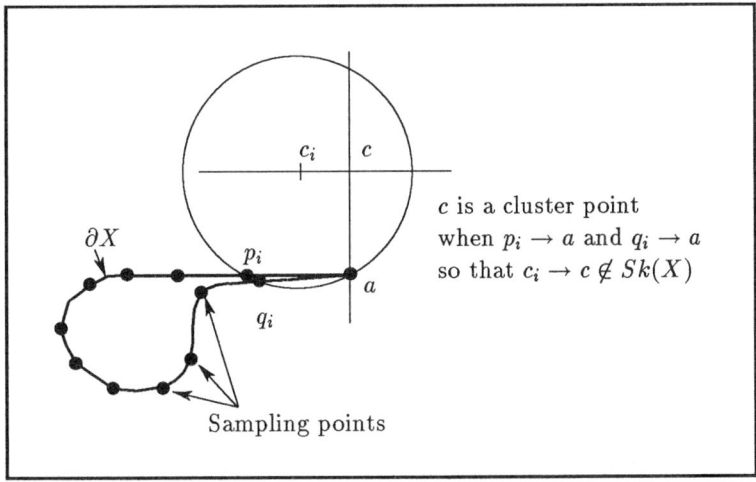

Figure 3: A sequence of centers of Delaunay disk does not converge towards a point of the skeleton.

After having shown a kind of simple convergence, we prove that in fact, on any compact set, this convergence is uniform.

Proposition 13

Let K be a compact set.

$$\forall \varepsilon \in \mathbb{R}^{+*}, \exists n_0 \in \mathbb{N}, \forall i > n_0, K \cap S_i \subset Bmax(X) \oplus B(0,\varepsilon)$$

i.e. the centers which are into K are at a distance less than ε from $Bmax(X)$ if the sampling points P_i are dense enough.

Proof: let us assume that the property is not true. Let ε be a strictly positive real number and $B_{i_k} = B(x_{i_k}, r_{i_k})$ be a sequence of Delaunay balls associated to the triangulation of P_{i_k} such that $d(x_{i_k}, Sk(X)) > \varepsilon$ for all k. Because $\mathcal{F}$ is a compact space, we can extract from $(B_{i_k})_{k \in \mathbb{N}}$ a converging subsequence, which will be denoted by $(B_{i_k})_{k \in \mathbb{N}}$ too $(B_{i_k} \to B = B(x,r))$. Only two situations can be considered:

- B is a true ball: then there is a convergence in $\mathcal{K}$ and x and r are not infinite. According to proposition 12, x is the center of a maximal ball, and then belongs to the skeleton. But $x_{i_k} \to x$, $x \in Bmax(X)$ and $d(x_{i_k}, Sk(X)) > \varepsilon > 0$. Contradiction.

- B is a generalized ball: B can be neither the whole space, because X is no-empty nor empty because B contains at least one point of ∂X (see the proof of proposition 12). Then B is a closed half space included into $\overline{X^c}$. Therefore, x_{i_k} diverges towards infinity, which contradicts the fact that $x_{i_k} \in K$ ∎

For the moment, we know nothing about $\liminf S_i$, i.e. if each point of the skeleton actually can be reached with the centers of the Delaunay spheres. The answer is still positive if we assume the surface ∂X to be C^2.

Proposition 14

$$\liminf_{i \to \infty} S_i \supset Bmax(X)$$

Proof: Let $B = B(x,r)$ be a maximal ball in X or $\overline{X^c}$. Let us note $A = B \cap \partial X$, the set of support points B on the boundary of X. $\overset{\circ}{B}$, the interior of B does not contain any point of the P_is. Then, for each i let $\overset{\circ}{B_i} = \overset{\circ}{B}(x_i, r_i)$ be a maximal ball in $\mathbb{R}^n \setminus P_i$ verifying $A \subset B \subset B_i$. We can find B_i', a Delaunay ball corresponding to the triangulation of P_i, lying on $B_i \cap P_i$ and containing the center of B_i. (We will not prove here this lemma. The idea is to increase the radius of B_i until at least $n+1$ points are on the surface as it is shown in figure 4).

$B_i \ominus B(0, \theta_i^{-1})$ does not contain any point of ∂X because P_i is a sampling set of points of density θ_i. Let B^0 be a cluster point of B_i. Then $A \subset B \subset B^0 \subset X$ or $\overline{X^c}$. Because B is maximal, $B = B^0$ is the unique cluster point of B_i. Therefore, $B_i \to B$. $(x_i)_{i \in \mathbb{N}}$, the sequence of the centers of B_i converges towards x, the center of B. Because $x_i \in B_i'$, $x \in B'$, a cluster point of (B_i') (which exists because $\mathcal{F}$ is compact). B' is a maximal ball since it is the limit of Delaunay spheres.

If the surface ∂X is C^1, there exist only two maximal balls which lie on $a \in A$, where $A = B \cap \partial X$, one of them D_{int} is interior to X, the other one is in $\overline{X^c}$, the radii of them is not equal to zero

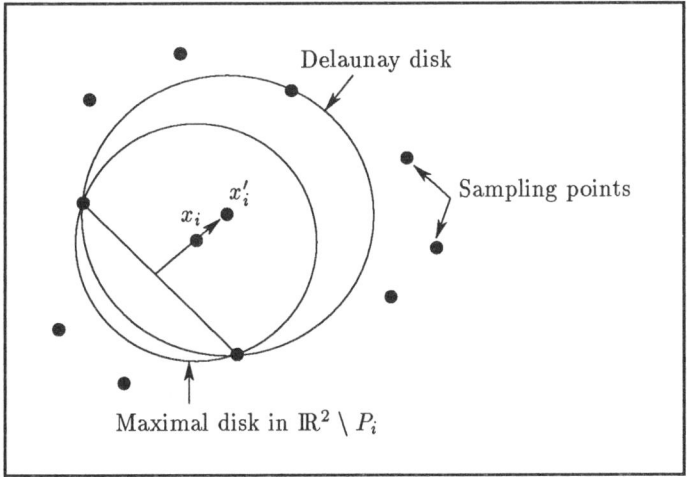

Figure 4: Transformation of a maximal circle in $\mathbb{R}^2 \setminus P_i$ into a Delaunay circle.

if the surface ∂X is C^2. Because $A \subset B'$, B' is one of the two balls B_{ext} and B_{int}. B' is the one which contains x, and therefore also B. Then, each cluster point of B'_i is B, i.e. $B'_i \to B$, which ends the proof ■

Remark: in the case of a C^∞ curve in the plane, so in the case of a polygon, the previous proposition does not hold, as it is shown in the example of figure 3.1. The line segment of the skeleton between a and b cannot be reached, if we assume that the points p_1 and p_2 are into P_i for all i.

The set of the proved propositions allows us to state the central theorem:

Theorem 15
> The set S_i of the centers of the Delaunay balls corresponding to the triangulation of P_i converges, in the sense of the topology of the closed sets $\mathcal{T}$ towards the skeleton of X.

Proof: We have seen that
$$\limsup S_i \subset Bmax(X) \subset \liminf S_i$$
Then $\lim S_i = Bmax(X)$. It also proves that $Bmax(X)$ is closed under the assumption that ∂X is C^3, i.e. $Bmax(X) = Sk(X)$ ■

3.2 Convergence speed

Although the convergence towards the skeleton is uniform on each compact, it does not prove that the convergence is very fast. We will partially answer this question in the plane by showing that only at the level of endpoints, this convergence may be very slow. We will not give the details of the computations.

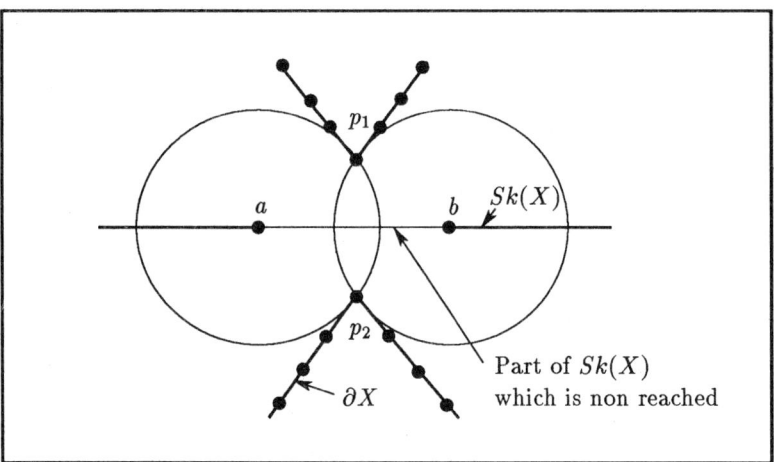

Figure 5: Some points of the skeleton are not reached by the limits of the Delaunay circles.

Proposition 16

Let X be the set intersection of the over-graph of the mapping $f(x) = x^2(1 - \sqrt{|x|})$ and of the square centered on the origin whose side is of length $1 \ll L$. Then the terminal point $(0,0)$ is reached in $O(\theta_i^{-1/2})$ (see figure 6).

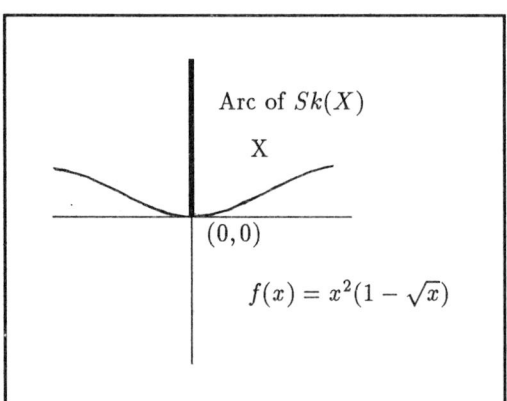

Figure 6: Example of an endpoint which is reached "slowly".

An endpoint of the skeleton may then be reached "slowly", but if f is C^3, the convergence is $k\frac{f^{(3)}}{f^{(2)}}\theta_i^{-1}$ where k is a constant independent of f.

Proposition 17

> Let $D = B(x, r)$ be a maximal disk in X or $\overline{X^c}$, intersecting the boundary of X in two distinct points a and b at least. Let ϕ be the angle $(\widehat{a, x, b})$. Let $B_i = B(x_i, r_i) \to D$ be a sequence of Delaunay disks corresponding to the triangulation of P_i. Then
>
> $$d(x_i, x) < k \times \frac{\cos \phi}{1 - \cos \phi} \times \theta_i^{-1}$$
>
> where k is a constant independent of X.

This proposition shows that, in fact, the convergence towards the skeleton is, outside a neighborhood of the endpoints, linear in θ_i^{-1}, because the ϕ angle is zero only at endpoints or at infinity.

3.3 Shape of the skeleton

In this paragraph we will restrict us to the planar case. If ∂X is at least C^3, Calabi [3] has shown that the skeleton of X is a locally finite planar graph. We have already seen the convergence of S_i, in the sense of closed sets. It does not imply that the different edges of the graph (called branches in the sequel) are respected. We are going to show that the branch points (the vertices of the graph) are effectively respected.

∂X is a C^3 loop or union of C^3 loops. The sampling points P_i can now be considered as a polygon or a union of polygons, because on each loop L, the sampling points $P_i \cap L$ are totally ordered and consequently organized as a polygon. For the sake of simplicity, we will speak of the polygon P_i even if ∂X is composed of many loops. How can we define a triple point on S_i?

Definition 18

> Let x_1^i and x_2^i be two points in S_i and τ_1 and τ_2 the associated triangles (the center of the circle circumscribed about τ_i is x_i). We say that x_1^i and x_2^i are <u>neighbors</u> if and only if τ_1 and τ_2 are adjacent according to an edge which is not an edge of the polygon P_i.

Definition 19

> x_1^i is a <u>triple point</u> if and only if it has three neighbors.

Remark: we only consider the case of triple points, which corresponds to the only stable singularities (with the endpoints) of the skeleton. We will then assume that for all i, four points of P_i are never cocircular. One way to avoid the difficulty, when more than four points are cocircular, is to triangulate these points in some arbitrary way. The centers of the Delaunay disks corresponding to these simplexes are identical, and the previous definitions can still be applied assuming S_i is a set such that some elements may appear several times.

Proposition 20
Let x be a triple point of the skeleton of X. For i large enough, there exists a sequence of Delaunay spheres $B_i = B(x_i, r_i)$ such that each x_i is a triple point converging towards x.

Proof: let $D = B(x, r)$ be the maximal disk whose center x is a triple point of the skeleton. This disk touches the boundary of X in at least three different points (a, b, c), because x is triple. Let d be a point interior to the triangle (a, b, c). For i great enough, d belongs to the convex hull of P_i, because the convex hull is continuous on $\mathcal{K}$. For all i, let t_i be a Delaunay triangle containing d (d can be on the boundary of t_i). Let t be a cluster point of t_i, and $t_{i_k} \to t$. Let (a_i, b_i, c_i) be the three vertices of t_i. If necessary, a subsequence of t_{i_k} can be extracted, so we will assume that a_{i_k}, b_{i_k} and c_{i_k} converge towards a', b' and c' respectively and that D_{i_k}, the Delaunay disk corresponding to t_{i_k} converges towards D'. Then, $d \in t = (a', b', c')$ and D' is a maximal circle which lies on a', b' and c'. Therefore t is the triangle (a, b, c), and $D = D'$. Because the triangle (a, b, c) is the only possible cluster point of t_i, $t_i \to t$. Since t is not a degenerated triangle, we can assume, by permuting if necessary a_i, b_i and c_i, $a_i \to a$, $b_i \to b$, $c_i \to c$ and $B_i \to D$.

For i great enough, the distances $d(a, b)$, $d(b, c)$ and $d(a, c)$ are all greater than $\frac{3}{\theta_i}$. Then $d(a_i, b_i)$, $d(b_i, c_i)$ and $d(c_i, a_i)$ are greater than θ_i^{-1}. Therefore, the three edges of the triangle t_i are not edges of the polygon P_i, and the point x_i has exactly three neighbors ∎

If we assume that the skeleton of X is a planar graph having a finite number of triple points and endpoints, the previous proposition proves that for a thin enough sampling of ∂X, the skeleton deduced from the Delaunay triangulation by linking the neighboring centers of the Delaunay disks has the same shape as the skeleton of X (the same topology in the sense of planar graph).

4 Random closed sets

The motivation of this section is the study of bucketing techniques as used in [2]. It is sometimes necessary to list all the intersections of the objects of one class with those of another. One simple way is to compute all the intersections, but this method is very time consuming. The principle of bucketing techniques is to divide the space into contiguous zones called buckets (often parallelepipedic buckets) and to associate to each the list of all the segments intersecting it. But, if two objects intersect, this intersection occurs in at least one zone. So we can work on each zone. What has been won is that a zone contains only a few objects.

In the worst case, nothing has been gained, since all the objects may intersect a given bucket. But this case does not occur very often so that an analysis of the average case is interesting. To achieve this goal, the objects are modeled with a random set X and we want to compute the probability that X intersects a given deterministic bucket B: $P(\{X \in \mathcal{F} : B \cap X \neq \emptyset\})$. Unfortunately such a set is generally non-measurable according to stochastic processes[1].

Bucketing techniques have already been studied (see for instance [1]), but the objects are always points. Our approach is well suited to study other objects consisting of more than a countable

---
[1] In particular, it is the case when B is not a countable set. However, X often can be parametrized by means of a finite number of variables, so that we have to compute probabilities on $\mathbb{R}^n$. We propose not to follow this way, in order to avoid to point out the parameters in each case.

set of points. We will first describe the σ-algebra provided by mathematical morphology and then show how to use it on real problems.

4.1 Definition of a σ-algebra and Choquet's theorem

When we have a topology on a set at hand, it is possible to associate the "natural" σ-algebra generated by the open sets (Borelian tribe). In the case of $(\mathcal{F}, \mathcal{T})$, this σ-algebra, which will be denoted by σ_f, is generated by the class $\mathcal{F}^K$ only or by the class $\mathcal{F}_G$ only for if $K_i \uparrow G$, then $\mathcal{F}_{K_i} \uparrow \mathcal{F}_G$. Analogously, if $(G_i)_{i \in \mathbb{N}}$ is a system of open neighborhoods of K and if $G_i \downarrow K$, $\mathcal{F}_{G_i} \downarrow \mathcal{F}_K$. So, for a given compact set K, $\{X : K \subset X^c\}$ is an element of σ_f, that is an <u>event</u> whose probability may be computed.

Definition 21
> We call <u>Random Closed Set</u> (or RACS) the triple $(\mathcal{F}, \sigma_f, P)$ where P is a probability on the measurable space $(\mathcal{F}, \sigma_f)$.

The existence of such probabilities is ensured by the compacity of $\mathcal{F}$.

Since σ_f is the Borelian tribe, any continuous and even semicontinuous mapping $\mathcal{A} \to \mathcal{F}$ (where $\mathcal{A}$ is a topological space provided with its Borelian tribe) is <u>measurable</u> and consequently define random variables. So we may speak of the union, intersection of two RACS, of the dilation or erosion of a RACS by a deterministic compact.

Just as a random variable is completely defined by its probability density function, a similar notion can be pointed out in the case of RACSs: it is the well-known Choquet's theorem [4] (see also [7]).

Theorem 22
> [Choquet's theorem] Let $T : \mathcal{K} \to \mathbb{R}$ be a functional. There exists a unique probability P on σ_f such that $P(\mathcal{F}_K) = P(\{X : X \cap K \neq \emptyset\}) = T(K)$ if and only if T is an alternating Choquet's capacity of infinite order such that $T(\emptyset) = 0$ and $0 \leq T \leq 1$.

We will not make clear the notion of Choquet's capacity because it is not necessary for our purpose. It is sufficient to know that there are conditions on T analogous to those of a probability density function: increasing and right continuous function.

In other words, it sufficient to know the functional $T(K) = P(\{X : X \cap K \neq \emptyset\})$ to entirely characterize a RACS.

4.2 The viewpoint of the space law

What the relationships between the σ-algebra σ_f and that which is used in stochastic process theory (i.e. in signal processing)? If X is a RACS, we can consider X as a random function f_X

from the angle of its support function: $f_X(x) = 1 \Leftrightarrow x \in X$. σ_f enables us to compute events of the form

$$\{X : K_1 \subset X^c \text{ and } K_2 \subset X\} = \{f_X : f_X(x) = 0 \text{ for } x \in K_1 \text{ and } f_X(x) = 1 \text{ for } x \in K_2\}$$

In stochastic process theory, only the sets where K_1 and K_2 are countable are measurable. It is then forbidden to study f_X on a ball which is a neighborhood of its center.

Let us examine this point in more details. To do this, let us use a common frame, namely the functions $\mathbb{R}^n \to \overline{\mathbb{R}}$. Let us show first how to transform a function into a set conversely.

Definition 23

> Let $f : \mathbb{R}^n \to \overline{\mathbb{R}}$ be an upper semicontinuous function. We call <u>undergraph</u> of f the closed set of $\mathbb{R}^n \times \overline{\mathbb{R}}$
>
> $$SG(f) = \{(x,t) \in \mathbb{R}^n \times \overline{\mathbb{R}} : f(x) \geq t\} \cup \{(x, -\infty), x \in \mathbb{R}^n\}.$$

It is worth noting that $SG(f)$ is a closed set because f is upper semicontinuous. Conversely, let us denote by $\mathcal{C}$ the set of all the closed sets F of $\mathbb{R}^n \times \overline{\mathbb{R}}$ verifying

$$((x,t) \in F \Longrightarrow (x,t') \in F \text{ for } t' < t) \text{ and } ((x,-\infty) \in F) \tag{1}$$

Then the function $f(x) = \sup\{t : (x,t) \in F\}$ is upper semicontinuous. There is therefore a bijective mapping between Φ_f, the space of all upper semicontinuous functions, and $\mathcal{C} \subset \mathcal{F}$, the space of all closed sets verifying condition 1.

Proposition 24

> The space $\mathcal{C}$ provided with the topology induced by $\mathcal{T}$ is compact.

All what was previously presented can be used again in the new frame of Φ_f. So we can define the notion of u.s.c. random function, the σ-algebra we use being generated by the events of the form

$$\{f \in \Phi_f : \sup\{f(x), x \in K\} \leq t\} \tag{2}$$

where K spans over the space $\mathcal{K}$.

In the case of stochastic processes, where the random functions have not to be u.s.c., the construction of a σ-algebra on an arbitrary product of measurable spaces generates the events for which K is countable.

Let be given a space law, that is a family of probability density functions

$$F_{x_1,\ldots,x_p}(u_1,\ldots,u_p) : \mathbb{R}^p \times \mathbb{R}^p \to [0,1]$$

which is equivalent to have the functional $T(K)$ (cf. Choquet's theorem) for all countable K.

Proposition 25

There exists <u>several</u> random functions s.c.s. (Φ_f, σ_f, P) such that

$$\forall p, (x_i)_{i=1}^p, (u_i)_{i=1}^p, \ P(f(x_1) < u_1, \ldots, f(x_p) < u_p) = F_{x_1,\ldots,x_p}(u_1, \ldots, u_p).$$

Only one of these functions, which is necessarily unique, is separable, i.e. there exists a countable set of $\mathbb{R}^n$ such that

$$f(x) = \limsup_{y \to x,\, y \in D} f(y).$$

By reasoning on sets, the separability property can be expressed by $SG(f) = \overline{SG(f) \cap (D \times \overline{\mathbb{R}})}$ where $SG(f)$ is the undergraph of f. (see [8] P 46 and [7] P 43).

To summarize, using the σ-algebra σ_f, the space law (i.e. $T(K)$ for all countable K) is not always sufficient to characterize a RACS. We have to know the functional $T(K)$ on $\mathcal{K}$ completely.

5 Application: study of bucketing algorithms in the average case

5.1 The Boolean model

Let us introduce now a special kind of RACS, namely the Boolean model, in order to model a "random" distribution of primary pattern, such as points, line segments, triangles, polygonal chains, etc. We shall carry out a somewhat heuristic presentation [13]. A more rigorous approach can be found in [7]. It is based on the notion of conditional RACS, which we want to avoid in this paper. The link between these two viewpoints is developed in [10].

Definition 26

Let $\theta \in \mathbb{R}^{+*}$ and X' be a RACS, which will be called <u>primary grain</u>. The <u>Boolean model</u> defined by θ and X' is the RACS X whose realizations are given by

$$X = \bigcup_{x \in \mathcal{P}} X'_x$$

where $\mathcal{P}$ is the realization of a stationary Poisson point process of density θ in $\mathbb{R}^n$ and X'_x the translation of the primary grain X' by vector $\vec{x}$, the $(X'_x)_{x \in \mathcal{P}}$ being chosen independently (see figure 7).

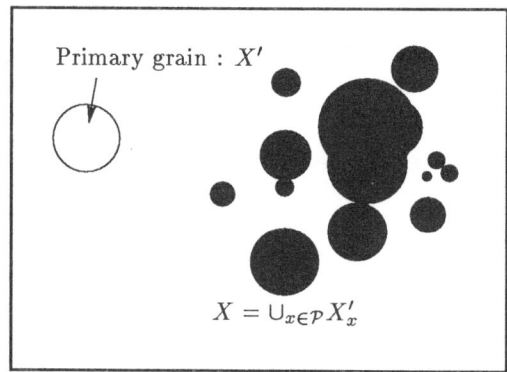

Figure 7: The Boolean model.

Remarks:

1. X' is a random closed set, so that it can model objects having different shapes according to a probability law.

2. This model does not take into account the side effects: the underlying Poisson point process is supposed to be of infinite extent into $\mathbb{R}^n$.

3. X' cannot be anything. X, the union of the X'_xs must be closed in order to speak of the RACS X. X is closed if and only if $E[V(X' \oplus K)] < \infty$ for K being the unit ball, $E[\cdot]$ the mathematical expectation and $V(\cdot)$ the Lebesgue measure in $\mathbb{R}^n$. Let us remark that this formula is meaningfull because $\cdot \oplus K : \mathcal{F} \to \mathcal{F}$ is a continuous mapping and $V(\cdot) : \mathcal{F} \to \mathbb{R}^+$ is measurable. For instance, we cannot take a straight line as primary grain.

Now let us state two theorems concerning the Boolean model which will be used in the estimation of the bucketing algorithms.

Theorem 27

The functional $T(K)$ corresponding to the Boolean model $X \equiv (\theta, X')$ is given by

$$1 - T(K) = P(K \cap X = \emptyset) = \exp\left(-\theta E[V(X' \oplus \check{K})]\right)$$

where $\check{K}$ is the symmetrical set of K with respect to the origin and $E[\cdot]$ the mathematical expectation.

Theorem 28

The random variable given the number of primary grains intersected by K follows a Poisson law of parameter $\lambda = \theta E[V(X' \oplus \check{K})]$:

$$P(N(K) = i) = \frac{\lambda^i}{i!} e^{-\lambda}$$

The Boolean model will be used in the following manner: X' will be a given pattern (for instance a segment of random orientation and mean length $\bar{l}$) and we shall deduce the probabilities concerning the intersections of these patterns with a deterministic bucket K.

5.2 Examples of computations

With the notations of the previous section, let us give some examples of special cases for X and K. A practical use of these results is given in [2]. We have seen that if X' represents the object whose intersection with the bucket B is under study, the number of objects hitting B follows a Poisson law of parameter $\theta E[V(X' \oplus \check{B})]$. In our case, the Lebesgue measure is the surface area in the plane and the volume in the 3D space and θ can be estimated by n/L^2 (resp. n/L^3) where n is the number of the primary grains falling in a square (resp. a cube) whose edge is of length L. So the average number of objects hitting B is precisely the parameter of the Poisson law.

5.2.1 Case of a line segment

Let us examine the case where X is a line segment whose midpoint is the origin, the distribution length l and mean length $\bar{l}$. We assume, for the sake of simplicity that its orientation is uniform. If B is a circle of radius ρ, the average number $\overline{ns}$ of segments intersecting B is given by

$$\overline{ns} = \theta \, \rho \left(2\bar{l} + \pi \rho \right)$$

In a similar way, in the 3D space, if B is a sphere of radius ρ we find

$$\overline{ns} = \theta \, \pi \, \rho^2 \left(\bar{l} + \frac{4}{3}\rho \right)$$

Suppose now that B is a square of edge length λ

$$\overline{ns} = \theta \lambda \left(\sqrt{2}\bar{l} + \lambda \right)$$

So all the computations amounts to estimate $V(X \oplus \check{B})$ for each given case and to take the mean value.

5.2.2 Case of a planar convex shape

If B is a planar square, it is possible to simplify the computations, because a square is the Minsowski sum of its two orthogonal edges S_x and S_y

$$B = [O, (\lambda, 0)] \oplus [O, (0, \lambda)] = S_x \oplus S_y.$$

The Minkowski sum being associative, the sum can be computed in two successive steps using only line segments

$$X \oplus B = (X \oplus S_x) \oplus S_y$$

If we denote by $\overline{A}$ the mean surface area of the random convex set X' and $\overline{x}$ (resp. $\overline{y}$) the mean length of its projection onto the x-axis (resp. y-axis), then the mean number of convex sets intersecting the square bucket B of edge length λ is (see figure 8)

$$\overline{nt} = \theta\left(\overline{A} + \lambda(\overline{x} + \overline{y}) + \lambda^2\right)$$

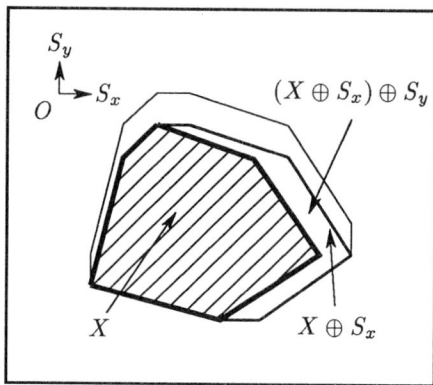

Figure 8: Minkowski sum of a planar convex shape and a square.

The numerical values of $\overline{A}$, $\overline{x}$ and $\overline{y}$ can now be computed using some special laws for the random convex set X', such as a triangle of fixed basis of a height following a given law and of uniform orientation...

We will not develop this point further. The interested reader is referred to the reference book about geometric probability [11]. Let us note that for a given size of the bucket, when the number of objects grows towards infinity in a bounded domain (i.e. $\theta \to \infty$), the complexity of the algorithms remains linear. But in practical cases, it is more reasonable to consider θ as a constant. Then we yield a precise estimation of the multiplicative constant and the size of the bucket will be chosen accordingly: not too small, because the pre-computation of all the lists of the objects intersecting the bucket will be too time consuming, and not too large, because nothing new will be reached by the bucketing techniques.

6 Conclusion

Mathematical morphology provides us with two often little known tools: a topology and a σ-algebra, whose application to the analysis of algorithms seems promising. In addition, many results can be found in the literature (statistical models, geometric probabilities) with may be used directly. These tools also provide an alternative to the usual Hausdorff distance and space law.

7 Notations

$\mathbb{R}^n$	Euclidean n-dimensional space
$\overline{X}$	topological closure of X
$\overset{\circ}{X}$	interior of X
∂X	boundary of X
K_x	translated set of X by vector $\vec{x}$
$\check{K} = \{k : -k \in K\}$	symetrical set of X with respect to O, the origin of $\mathbb{R}^n$
$B(x,\varepsilon)$	closed ball of radius ε centered at x

$X \oplus K = \{x + k : x \in X \text{ and } k \in K\}$
    Minkowski sum of X and K

$D_K(X) = X \oplus \check{K} = \{x : \check{K}_x \cap X \neq \emptyset\}$
    dilation of X by K

$E_K(X) = X \ominus \check{K} = \{x : K_x \subset X\}$
    erosion of X by K

$\varrho(K, K') = \max\left\{\sup_{x \in K} d(x, K'), \sup_{x \in K'} d(x, K)\right\} = \inf\{\varepsilon : \check{K} \subset K' \oplus B(O,\varepsilon), K' \subset K \oplus B(O,\varepsilon)\}$
    Hausdorff distance of K to K'

$\mathcal{F}(\mathbb{R}^n) = \mathcal{F}$	set of all the topological closed sets of $\mathbb{R}^n$
$\mathcal{K}(\mathbb{R}^n) = \mathcal{K}$	set of all the compact sets of $\mathbb{R}^n$
$\mathcal{K}' = \mathcal{K} \setminus \emptyset$	
$\mathcal{F}^K_{G_1,...,G_p} = \{F \in \mathcal{F} : F \cap K = \emptyset \text{ and } F \cap G_i \neq \emptyset\}$	
$\mathcal{T}$	topology on $\mathcal{F}$ generated by $\mathcal{F}^K_{G_1,...,G_p}$
σ_f	σ-algebra generated by $\mathcal{T}$

$Bmax(X) = Bmax_I(X) \cup Bmax_E(X)$
    set of the centers of the maximal balls inside X ($Bmax_I(X)$) and inside X^c ($Bmax_E(X)$)

$Sk(X) = Sk_I(X) \cup Sk_E(X)$
    skeleton of X, union of the interior ($Sk_I(X)$) and the exterior skeleton ($Sk_E(X)$) of X. $Sk_I(X) = \overline{Bmax_I(X)}$ and $Sk_E(X) = \overline{Bmax_E(X)}$

References

[1] T. Asano, M. Edahiro, H. Imai, and M. Iri. *Computational Geometry*, chapter Practical Use of Bucketing Techniques in Computational Geometry. Elsevier Science Publishers B.V. (North Holland), 1985.

[2] J.D. Boissonnat, O.D. Faugeras, and E. Le Bras-Mehlman. *Representing Stereo Data with the Delaunay Triangulation*. Technical Report 788, INRIA, February 1988.

[3] L. Calabi and J.A. Riley. *The Skeletons of Stable plane Sets.* Technical Report AF 19 (628-5711), Parke Math. Lab. Inc., December 1967.

[4] G. Choquet. Theory of capacities. *Ann. Inst. Fourier*, V:131–295, 1953-54.

[5] T. Lozano-Pérez and M.A. Wesley. An algorithm for planning collision free paths among polyhedral obstacles. *Comm. ACM*, 22:560–570, 1979.

[6] G. Matheron. *Image Analysis and Mathematical Morphology, Volume 2 : Theoretical Advances*, chapter Filters and Lattices. Academic Press, 1988.

[7] G. Matheron. *Random Sets and Integral Geometry.* Wiley, 1975.

[8] G. Matheron. *Théorie des ensembles aléatoires.* Les cahiers du Centre de Morphologie Mathématique, Ecole des Mines de Paris, 1969.

[9] F.P. Preparata and M.I. Shamos. *Computational Geometry : an Introduction.* Springer Verlag, 1985.

[10] F. Prêteux and M. Schmitt. *Analyse et synthèse de fonctions booléennes : théorèmes de caractérisation et démonstrations.* Technical Report 34/87/MM, CGMM, Ecole des Mines, 1987.

[11] L.A. Santalo. *Integral Geometry and Geometric Probability.* Addison Wesley, 1976.

[12] J. Serra. Eléments de théorie pour l'optique morphologique. Thèse d'Etat, Université de Paris VI, 1986.

[13] J. Serra. *Image Analysis and Mathematical Morphology.* Academic Press, 1982.

AN OPTIMAL ALGORITHM FOR THE BOUNDARY OF A CELL IN A UNION OF RAYS*

Panagiotis Alevizos [†], Jean-Daniel Boissonnat [‡]
and Franco P. Preparata [§]

Abstract

In this paper, we study a cell of the subdivision induced by a union of n half lines (or rays) in the plane. We present two results. The first one is a novel proof of the $O(n)$ bound on the number of edges of the boundary of such a cell, which is essentially of methodological interest. The second is an algorithm for constructing the boundary of any cell, which runs in optimal $\Theta(n \log n)$ time. A byproduct of our results are the notions of skeleton and of skeletal order, which may be of interest in their own right.

---

*This work was partly supported by the CEE ESPRIT Project P-940, by the Ecole Normale Supérieure, Paris, France, and by the NSF grant ECS-84-10902.

[†] Department of Mathematics, University of Patras, Greece and INRIA, Route des Lucioles, 06565 Valbonne, France.

[‡] INRIA, Route des Lucioles, 06565 Valbonne, France.

[§] University of Illinois at Urbana-Champaign, IL 61801, USA. This work was done in part while this author was visiting the Ecole Normale Supérieure, Paris, France.

1 Introduction

Given is a set $R = \{r_1 ..., r_n\}$ of n rays (i.e., half lines) in the plane. To avoid insignificant degeneracies, we assume that the rays are in general position, i.e., all termini are distinct and no three rays intersect in the same finite point. The union $F = \bigcup_{j=1}^n r_j$ partitions the plane into a collection of regions, some of them being unbounded. Let $D(R)$ be this subdivision induced by F.

In the special case where the termini of all rays lie on a single straight line, Chazelle, Guibas and Lee [4] and Edelsbrunner, O'Rourke and Seidel [7] have shown that the boundary E of the region of $D(R)$ containing all the ray termini consists of $O(n)$ edges (whereas the total boundary of F may have $O(n^2)$ edges) and have given an $O(n \log n)$-time algorithm to compute it.

In [4,7], the alignment of the ray termini serves the purpose of readily supplying the order of the termini along E. The result of [4,7] that E has $O(n)$ edges is of a topological nature, and thus it is easily extended to the more general case, provided that the order of the termini along E is known. A recent paper of Alevizos, Boissonnat and Yvinec [2] gives a method for computing this order (hereafter referred to as the ABY-order) when all ray termini are known to belong to a single region of $D(R)$. The ABY-order is based on the following binary relation $\prec$ on a set of rays, exhibiting the property that all intersect the line l of equation $y = 0$ and all ray termini have positive ordinates : given two rays r_i and r_j, let x_i and x_j be the abscissae of the intersections of r_i and r_j with l respectively. Ray r_i precedes r_j, denoted $r_i \prec r_j$, if either r_i and r_j do not intersect in the half-plane $y > 0$ and $x_i < x_j$ or r_i and r_j intersect in the half-plane $y > 0$ and $x_i > x_j$. It is shown in [2] that this relation is a total order if and only if all ray termini belong to a single region of $D(R)$ and contains cycles otherwise.

This problem of finding the boundary of any region of $D(R)$ is closely related to the problem of finding the upper envelope (i.e. pointwise maximum) of n functions. This problem was first studied by Atallah [1] and then by Sharir, Hart and Wiernik [5,10]. For the case of n line segments in the plane, none of which is vertical, Hart and Sharir have shown that the upper envelope consists of, at most, $O(n\alpha(n))$ edges where $\alpha(n)$ is an extremely slowly growing function related to the inverse Ackermann function. Their result extends immediately to the case of rays. Moreover, Wiernik and Sharir [10] proved that this bound is tight, in the worst-case, for line segments. Concurrently with the research leading to this paper, Edelsbrunner, Guibas and Sharir [6] have considered the problem of calculating one or several cells in an arrangement of line segments. In the case of one cell, their algorithm runs in time $O(n\alpha(n)\log^2 n)$. This result generalizes a previous result by Pollack, Sharir and Sifrony [9]. The results contained in this paper were obtained independently and thus several technical tools of the cited papers can be used to replace some of the arguments given here. It must be noticed however that all our algorithmic techniques are rather straightforward : we mainly use variants of the standard plane-sweep algorithm for computing intersections of line segments.

The paper is organized as follows. Section 2 contains structural preliminaries for important subsets of rays satisfying the so-called "half-plane constraint". Section 2 introduces the notion of skeletal order, central to our technique, and contains a novel proof of the $O(n)$ bound on the number of edges of the boundary of any region of $D(R)$. In Section 3, we present an optimal algorithm to compute any such portion of the boundary in $O(n\log n)$ time. The half-plane constraint is dropped in Section 4, where structural and algorithmic results are presented for general sets of rays. Final remarks are presented in Section 5.

2 Structural preliminaries

2.1 Bundles

Assuming without loss of generality, that R contains no horizontal ray, there is a real value M_0 such that for any $M > M_0$, R is partitioned into two subsets that are independent of the choice of M, one, R^+, consisting of all rays intersecting the line $y = -M$, the other, R^-, consisting of all rays intersecting the line $y = M$. Our approach consists in investigating in Sections 2 and 3 the problem for R^+

and R^- separately and in combining the results to solve the original problem in Section 4. Thus, with a minor abuse of notation, until further notice, we let R denote a set with the property of R^+ defined above (referred to concisely as the *half-plane constraint*) and we let l denote the line $y = -M$. Note that in this case there is a unique unbounded region lying above $y = -M$, called *external region*, whose boundary is referred to as *external* and denoted E. All other regions of $D(R)$ are called *internal*.

Set $R = \{r_1, ..., r_n\}$ is partitioned into classes $R_1, ..., R_h$ such that r_i and r_j are in the same class if and only if their respective termini lie in the same region of $D(R)$. The relation $\prec$ on a given class $R_h \subseteq R$ is a total order, by the result of [2]. We let t_i denote the terminus of r_i.

We consider the rays of a given class, say R_h, with $|R_h| = s$, and we re-index them so that $(r_1, ..., r_s)$ is their ABY-order. For convenience, we view this order as a string $\rho = r_1 r_2 ... r_s$. Given a substring $\alpha = r_i ... r_j$ of ρ, the external boundary of $\bigcup_{k=i}^{j} r_k$, called a *bundle*, consists of three portions : a polygonal chain C_I, called the *intermediate chain*, between the termini t_i of r_i and t_j of r_j, a polygonal chain C_L, called the *left chain*, between infinity and t_i, and a polygonal chain C_R, called the *right chain*, between infinity and t_j. It is immediate to realize that C_L and C_R are convex, and oppose their convex profiles. The bundle pertaining to α, denoted $b(\alpha)$, consists of chains C_L, C_I and C_R (see Figure 1); chains C_L and C_R are called the *external chains* of $b(\alpha)$. When necessary, an external chain will be assumed to be oriented from its terminus towards infinity.

We note that a single ray r with terminus t is itself a bundle, pertaining to a singleton string, with C_I degenerating to t and C_L and C_R both coinciding with the ray itself.

Two bundles are *disjoint* when they pertain to disjoint substrings. We now establish the following crucial property :

Lemma 1 *Any two external chains respectively belonging to two disjoint bundles intersect in at most one point.*

Proof : Let (C_L^i, C_I^i, C_R^i) be the chain of $b(\alpha_i)(i = 1, 2)$, with $\alpha_1 \cap \alpha_2 = \emptyset$. Referring to their ABY-order, let α_1 be the left bundle.

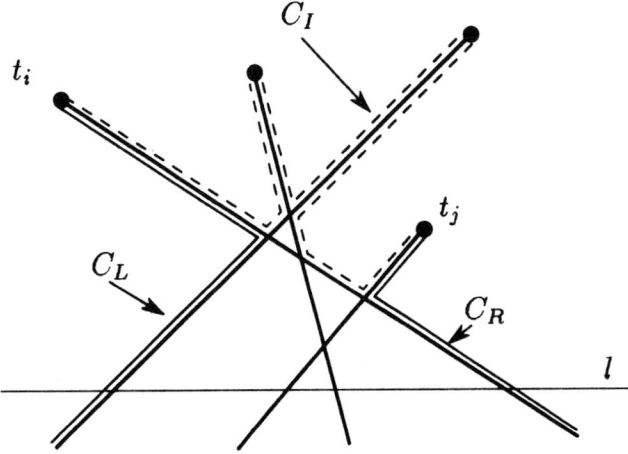

Figure 1: Illustration of the notion of $b(r_i, ..., r_j)$

First we observe that C_L^1 and C_R^2 intersect in at most one point. Indeed, suppose they intersect in more than one point. Then, due to their opposing convexities, they intersect in exactly two points, p_1 and p_2 with $y(p_1) > y(p_2)$ (see Figure 2a). Let r_1 and r_2 be the rays intersecting in p_2, with $r_1 \in \alpha_1$ and $r_2 \in \alpha_2$; according to the ABY-sorting rule given in the introduction, r_2 appears before r_1 in the ABY-order, contradicting the hypothesis.

Consider now the pair (C_R^1, C_L^2). Suppose they intersect in more than one point. Again, due to their opposing convexities, they intersect in exactly two points p_1 and p_2 with $y(p_1) > y(p_2)$ (see Figure 2b). Applying the same argument to the rays intersecting in p_1, we reach an analogous contradiction. Thus C_1^R and C_2^L intersect in at most one point.

Finally, consider the pair (C_L^1, C_L^2) (the pair (C_R^1, C_R^2) is treated analogously). Suppose that they intersect in more than one point. This means there are at least two intersections, and we consider two consecutive ones, q_1 and q_2. There are therefore four rays $r_{11}, r_{12}, r_{21},$ and r_{22}, such that r_{i1} and r_{i2} belong to $C_L^i (i = 1, 2)$, and r_{1j} intersects $r_{2j} (j = 1, 2)$. Moreover, we assume without loss of generality that r_{i1} and r_{i2} appear in this order when traversing $C_L^i (i = 1, 2)$ from its terminus to infinity (see Figure 2c).

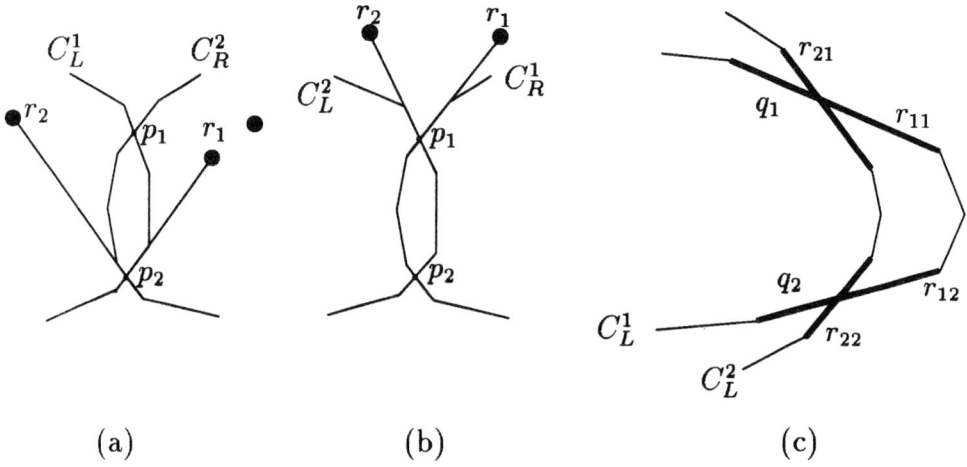

Figure 2: For the proof of Lemma 1 (Each pair of chains has at most one intersection point).

It is immediate to realize that these rays are interlaced either in the order $(r_{11}, r_{21}, r_{22}, r_{12})$, or in the order $(r_{21}, r_{11}, r_{12}, r_{22})$, contrary to the disjointness hypothesis. Thus C_L^1 and C_L^2 intersect in at most one point. □

2.2 Number of edges of the boundary of the regions of $D(R)$

2.2.1 Number of edges of the boundary of the external region

We make the conservative assumption that the boundary E of the external region passes through all ray termini. Indeed, let us suppose that a ray terminus is not on the external boundary; clearly, if we extend the ray so that it crosses the external boundary and its terminus appears on the external boundary, the number of edges of E increases. Thus the maximum number of edges of the external boundary is reached when all the ray termini are on the external boundary.

The edges of E are naturally ordered by a march along E from left to right. The attributes "before", "after" etc., as applied to edges of E, refer to this order. Due to the result of [2], the ray termini are encountered in their ABY-order, as E is traversed (this is also a variant of the consistency lemma of [8]). Let $\rho = r'_1...r'_n$

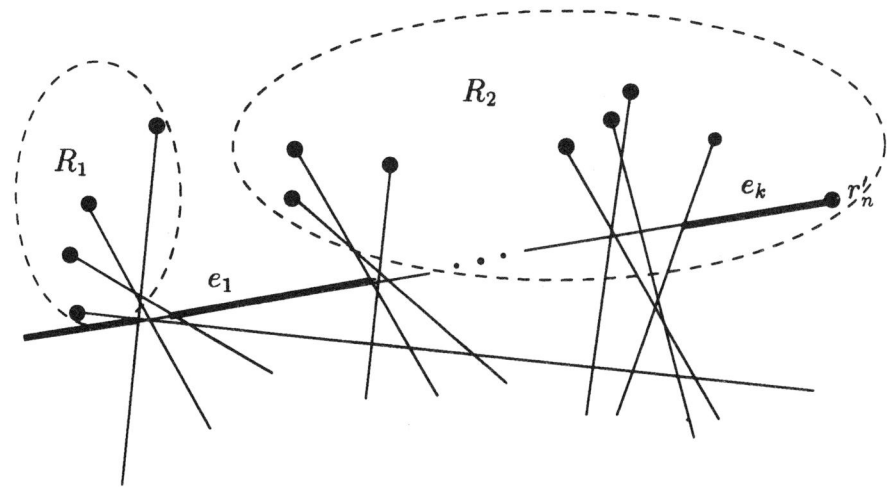

Figure 3: Subsets R_1 and R_2

be the string formed by the rays, which represents their ABY-order.

Ray r'_n is referred to as the *last* ray of set R. Ray r'_n contributes $k \geq 1$ line segments $e_1, ..., e_k$ of E (and possibly the initial half-line of E), in this order along E. We define R_1 as the subset of $R - \{r'_n\}$ consisting of all rays containing edges of E preceding e_1 (see Figure 3).

Due to the remark above, a substring $\alpha_1 = r'_1 ... r'_m$ of ρ is associated to R_1 and the complementary substring $\alpha_2 = r'_{m+1} ... r'_n$ is associated to the subset $R_2 \stackrel{\Delta}{=} R - R_1$. Thus the bundles $b(\alpha_1)$ and $b(\alpha_2)$ pertaining to α_1 and to α_2 are disjoint.

With this nomenclature, we are ready to prove the main theorem of this section :

Theorem 1 *The boundary of the external region of $D(R)$, for $R = \{r_1, ..., r_n\}$, has at most $4n - 2$ edges.*

Proof : The proof is by induction. The theorem obviously holds for $n = 1$; we assume that it holds for any positive integer $t \leq n - 1$.

Let E_i denote the external boundary of the union of the rays of set R_i, and $|E_i|$ the number of edges of $E_i (i = 1, 2)$. According to the inductive hypothesis,

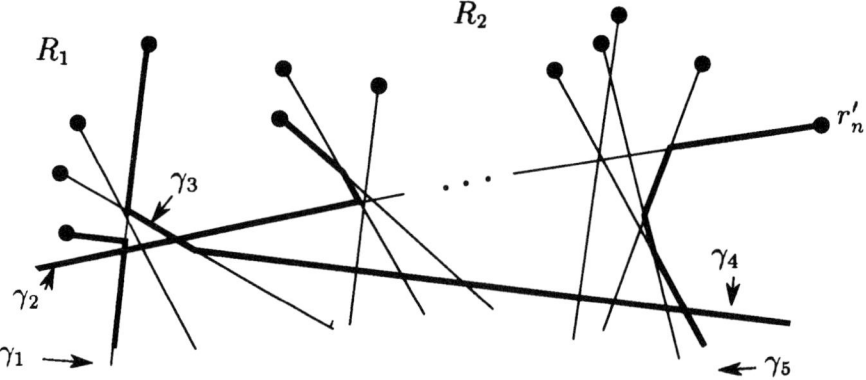

Figure 4: Intersections between E_1 and E_2

we have $|E_1| \leq 4m - 2$ and $|E_2| \leq 4(n - m) - 2$.

Let (C_L^i, C_I^i, C_R^i) denote the chains of $b(\alpha_i)(i = 1, 2)$. Note that C_I^1 is a portion of the final boundary E, since it precedes e_1 and e_1 is the first (bounded) segment of r'_n along E; thus, C_I^1 does not intersect R_2. Moreover, C_I^2 does not intersect any ray of R_1, since it occurs entirely between e_1 and e_k and lies above r'_n, whereas all rays of R_1 intersect r'_n to the left of e_1. Thus the only edges of E_1 and E_2 which intersect belong to right or left chains of the associated bundles.

Due to Lemma 1, these chains have at most three points of intersection. Specifically, we have at most (refer to Figure 4):

- an intersection between an edge of C_L^1, say γ_1, and the first edge of C_L^2, say γ_2, contained in r'_n.

- an intersection between an edge of C_R^1, say γ_3, and edge γ_2 of C_L^2.

- an intersection between an edge of C_R^1, say γ_4 (possibly coincident with γ_3), and an edge of C_R^2, say γ_5.

The external boundary E of the union of all the rays $(R_1 \bigcup R_2)$ is the external boundary of the union of the external boundaries E_1 and E_2. It is clear, from the above discussion, that the largest value of $|E|$ is attained with $\gamma_3 = \gamma_4$, in which case there is one edge of $E_1(\gamma_3)$ and one edge of $E_2(\gamma_2)$ such that each is cut into three pieces when merging the two bundles, two of them appearing as edges of E. Thus we have

$$|E| \leq |E_1| + |E_2| + 2 \leq 4n - 2$$

which completes the proof. □

2.2.2 Number of edges of an internal region of $D(R)$

Theorem 1 extends to the boundary of any internal region. This is obvious for those internal regions which do not contain a terminus. Let us then consider an internal region A whose boundary $E(A)$ contains ray termini. We now constructively define a polygon $C(A)$, which is the smallest convex polygon, bounded by rays and, if applicable, by the line at infinity, which encloses A (refer to Figure 5).

Let $R_1 \subset R$ be the set of rays with termini on the boundary of A, and denote by $D(R - R_1)$, the subdivision determined by $R - R_1$. Since the original subdivision is a refinement of this new subdivision, in the latter there is a unique region A_1, such that $A \subset A_1$. If in $D(R - R_1)$ region A_1 does not contain any ray terminus on its boundary, then it is a convex polygon and we let $C(A) = A_1$. Otherwise let $R_2 \neq \emptyset$ be the set of rays whose termini are on the boundary of A_1. We remove them from $R - R_1$ to obtain a unique region A_2 of $D(R - R_1 - R_2)$ containing A_1. This process is iterated, and we obtain a sequence of planar regions $A \subset A_1 \subset A_2 \ldots \subset A_s$ where $A_s = C(A)$ is a (possibly unbounded) convex polygon. The boundary of $C(A)$ partitions the set R of rays into $\{R^{(1)}, R^{(2)}\}$, where $R^{(1)}$ consists of the rays whose termini are internal to A_s (that is, $R^{(1)}$ is the union of the rays that have been successively removed in the constructions of the $A_1, A_2, \ldots A_s$). This also shows that the ray termini on $E(A)$ in $D(R)$ appear on the boundary of the external region in $D(R^{(1)})$, and that $E(A)$ is the union of connected components of the following alternating two types:

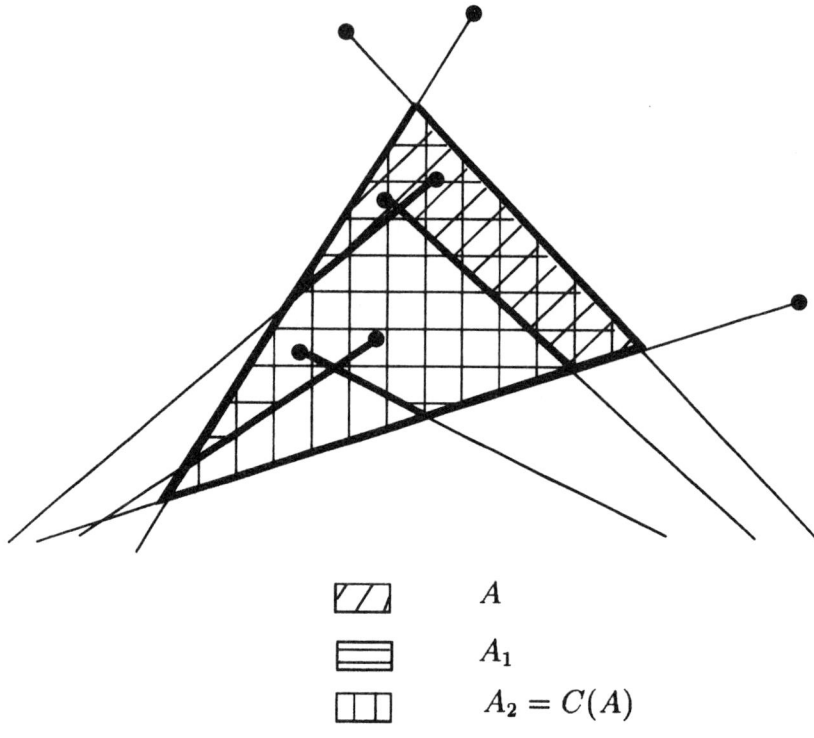

Figure 5: Illustration of the construction of $C(A)$ for an internal region A

Type (i) a connected portion of the intersection of the boundary of $C(A)$ with the external region of $D(R^{(1)})$.

Type (ii) a connected portion of the intersection of the boundary of the external region of $D(R^{(1)})$ with the interior of $C(A)$.

Since each edge of the boundary of $C(A)$ belongs to a distinct ray of $R^{(2)}$, the boundary of $C(A)$ has at most $|R^{(2)}|$ edges. We partition the set of rays $R^{(1)}$ into disjoint subsets, each consisting of the rays of $R^{(1)}$ intersecting the boundary of $C(A)$ between two successive components of Type (i). Let $m \geq 1$ be the number of such subsets and n_i the number of rays pertaining to subset i ($i = 1, ..., m$). We have $\sum_{i=1}^{m} n_i \leq |R^{(1)}|$. Disjoint bundles are associated to these subsets and we have

$$|E(A)| \leq \sum_{i=1}^{m}((4n_i - 2) + |C(A)| + m),$$

since Theorem 1 holds for each bundle, and merging $C(A)$ and the bundles adds at most m edges to $E(A)$. With $\sum_{i=1}^{m} n_i \leq n - |C(A)|$, we conclude $|E(A)| \leq 4n - m$. This yields the following result:

Theorem 2 *The boundary of any internal region of $D(R)$, for $R = \{r_1, ..., r_n\}$, has at most $4n - 1$ edges.*

2.3 Legal descriptions and skeletons

Let $R = \{r_1, ..., r_n\}$. In alternative cases, we shall augment R with an additional ray, either r_L – the line l with its terminus at $-\infty$ – or r_R – the same line l with its terminus at $+\infty$. Note that each ray r_i, $i = 1, ..., n$, intersects both r_L and r_R. We say that a string $\rho = r'_1 r'_2 \ldots r'_n$ (where $r'_1 r'_2 \ldots r'_n$ is a permutation of $r_1 \ldots r_n$) is a *legal description* of R or, briefly, *legal* if the subsequence of the rays belonging to a given class R_h of the partition of R (as defined in Section 2.1) yields their ABY-order. For example, the set R displayed in Figure 6 is partitioned into three classes: $\{r_6, r_8\}, \{r_5, r_4, r_{10}, r_9, r_1, r_2\}$ and $\{r_3, r_7\}$; $\rho = r_6 r_8 r_5 r_4 r_{10} r_9 r_7 r_3 r_1 r_2$ is a legal string for R.

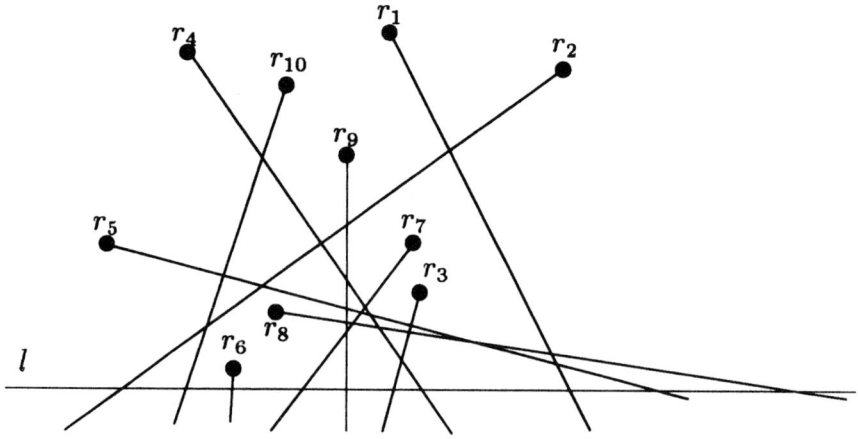

Figure 6: A set of rays

We now introduce the notion of "skeleton" of a set R of rays. Let $r'_1 r'_2 \ldots r'_n$ be an arbitrary indexing of the members of R. We then have:

Definition 1 *The skeleton G of R pertaining to indexing $r'_1 r'_2 \ldots r'_n$ is constructively defined as follows: $G_0 = \{r_0\}$ (in the sequel, r_0 will always be either r_L or r_R); G_i is obtained from G_{i-1} by adding to G_{i-1} the segment between the terminus of r'_i and its first intersection with G_{i-1}. Then, $G = G_n$.*

The skeleton associated with the set of rays in Figure 6, indexed as shown, is illustrated in Figure 7. Given an indexing $r'_1 r'_2 \ldots r'_n$, Definition 1 constructively divides a ray r'_j into two portions, the *active* one (which is added to the skeleton), and its complement (the *inactive* portion, which may be empty). With this notion, we have the following alternative and equivalent definition of skeleton:

Definition 1′ . *The skeleton G of R pertaining to indexing $r'_1 r'_2 \ldots r'_n$ is constructively defined as follows: $G_0^* = F$ (the union of the rays, including r_0); G_i^* is obtained from G_{i-1}^* by removing from G_{i-1}^* the inactive portion of r'_i. Then, $G = G_n^*$.*

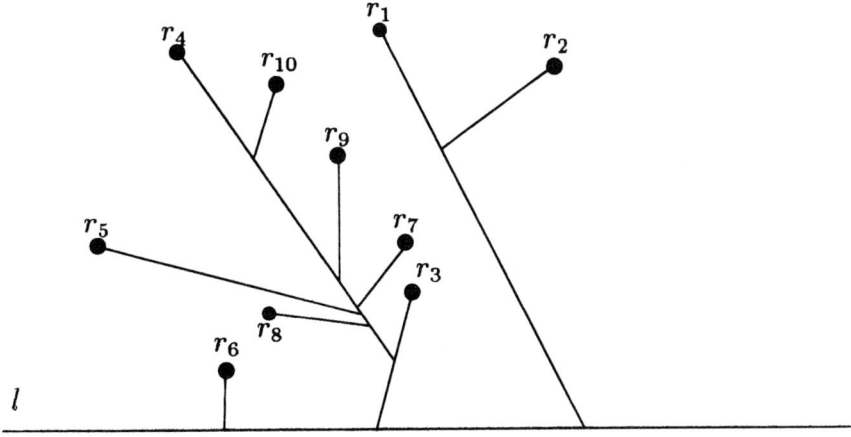

Figure 7: The skeleton associated with the set of rays in Figure 6

It is also convenient to view a ray r as an arbitrarily thin rectangle $\epsilon(r)$ (called the ϵ-*thickening* of r) obtained by an arbitrarily small translation of r orthogonal to r itself (see Figure 8). The boundary of $\epsilon(r)$ is conventionally directed as a *clockwise* circuit. This interpretation is naturally extended to the structures $G_0, G_1, \ldots, G_n$ and $G_0^*, G_1^*, \ldots, G_n^*$ introduced in Definitions 1 and $1'$; we let $\epsilon(G_i)$ denote the ϵ-thickening of G_i. (We recognize that the boundary of $\epsilon(G_n) = \epsilon(G_n^*) = \epsilon(G)$ is also the directed boundary of $\bar{G}$, the polygon obtained by removing from the plane the points of the skeleton G.) Examining in detail the transformations taking from G_{i-1}^* to G_i^*, we note that $\epsilon(G_i^*)$ is obtained by removing from $\epsilon(G_{i-1}^*)$ a collection of trapezoids (see Figure 9), specifically, the points of $\epsilon(r_i')$ corresponding to the inactive portion of r_i' and not shared by any other $\epsilon(r_j')$, $j > i$. Finally, we note that the boundary of $\epsilon(G_0^*)$ is the collection of the directed boundaries of all the regions of $D(R)$ while $\epsilon(G)$ consists of a single circuit. It is a simple exercise to verify that G is a tree and that $\epsilon(G)$ has $4n + 3$ arcs (note that there are $n + 1$ rays).

Given a skeleton G, the *skeletal order* of G is the sequence of the ray termini in the left-to-right traversal of the boundary of $\bar{G}$. In the example of Figure 6, the skeletal order is $r_6 r_8 r_5 r_4 r_{10} r_9 r_7 r_3 r_1 r_2$.

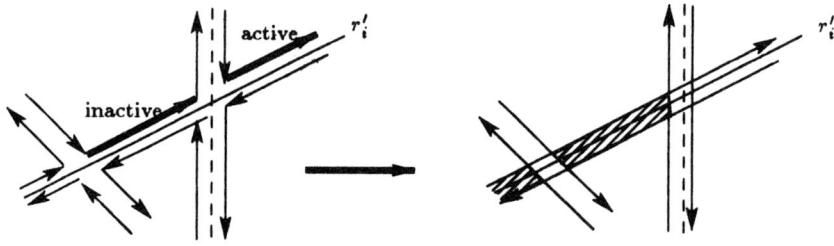

Figure 8: Illustration of the transformation from G_{i-1}^* to G_i^* in terms of their ϵ-thickenings.

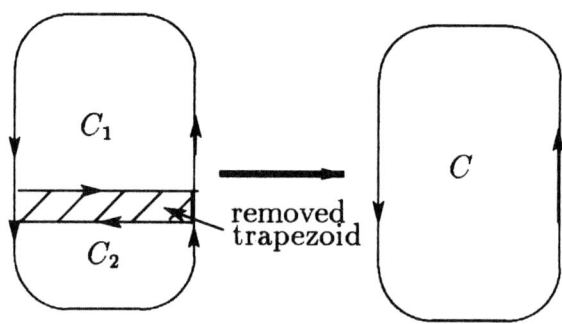

Figure 9: For the proof of Lemma 2

We now link the notions of legal description and of skeletal order.

Lemma 2 *Let r' and r'' be two rays belonging to the same class R_h. r' precedes r'' in the ABY-order if and only if r' precedes r'' in a skeletal order.*

Proof: Referring to the concepts introduced above, removal of an inactive portion trapezoid of $\epsilon(r_i')$ (while forming G_i^* from G_{i-1}^*) simply merges two boundary circuits C_1 and C_2 into a single circuit C as illustrated in Figure 9.

Thus the removal of the trapezoid in question preserves the order of the termini, i.e., if r' preceded r'' in either C_1 or C_2, the same will hold in the resulting circuit C. Since the process of globally transforming G_0^* to G_n^* consists of the successive merging of pairs of boundary circuits, while preserving the order of the termini, the lemma follows. □

We then have:

Theorem 3 *The skeletal order corresponding to an arbitrary indexing of the rays of R is a legal description of R.*

Of particular interest are the two following skeletons. Let $(r'_1, r'_2, ..., r'_n)$ denote a labelling of the rays corresponding to increasing angle with respect to the x-axis, oriented in the usual manner (note that this is also the left-to-right order of the intersections of the rays with the horizontal line l previously defined). We define the *leftist* skeleton G_L as the one corresponding to the order $(r_L, r'_1, ..., r'_n)$, and the *rightist* as the one corresponding to the order $(r_R, r'_n, ..., r'_1)$. The leftist (resp., rightist) skeleton has the property that the unique directed path from a ray terminus t to infinity is a polygonal line, monotone with respect to the vertical and containing only right (resp., left) turns. For the set of Figure 6, the leftist and rightist skeletons are shown in Figure 10.

3 Algorithmic issues

3.1 Lower bounds

We observe that $\Omega(n \log n)$ is a lower bound for the time complexity of any algorithm for either the leftist or rightist order of a set $R = \{r_1, ..., r_n\}$. This result follows from transforming sorting to the problem in question. Indeed, let $x_1, ..., x_n$ be n real numbers. For each x_i ($i = 1, ..., n$), we construct the vertical ray with terminus $(x_i, 1)$ and extending to $y = -\infty$. The resulting set of rays is of the same type as R^- (introduced in Section 2.1) and its leftist and rightist orders are identical and give the natural order of $x_1, ..., x_n$.

An analogous argument transforms sorting of n real numbers to the construction of the boundary of any region of $D(R)$, thereby establishing $\Omega(n \log n)$ as a lower bound also for the latter problem.

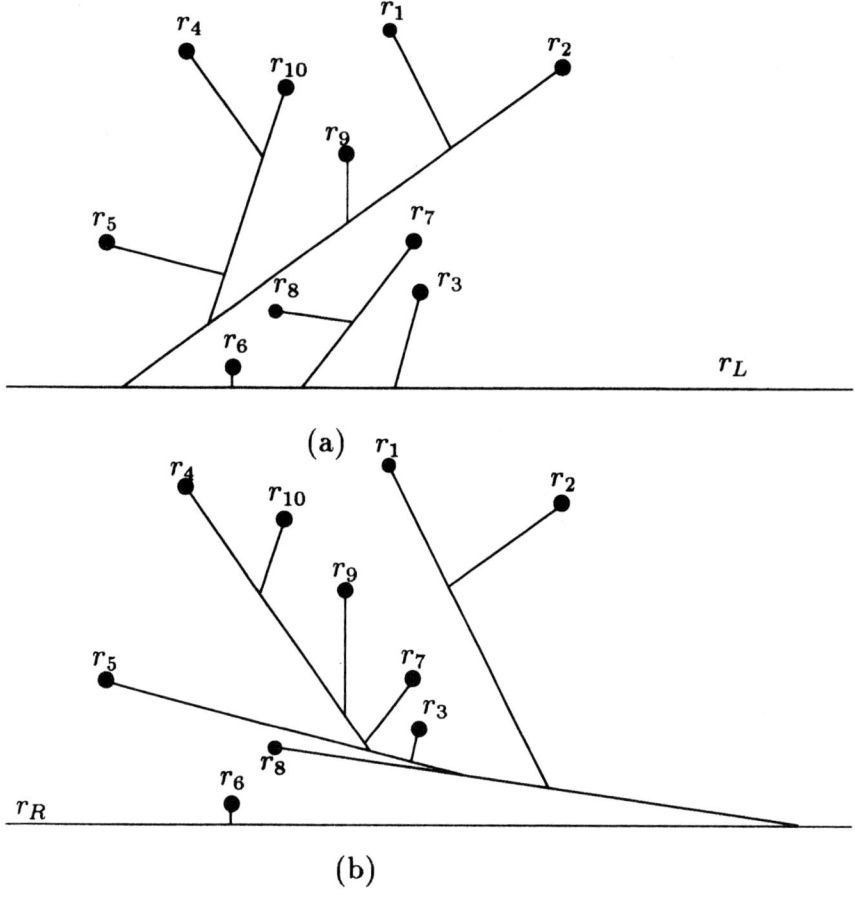

Figure 10: Leftist (a) and rightist (b) skeletons.

3.2 Computation of the leftist and rightist orders

We first note that the leftist (resp., rightist) order is readily obtainable from the corresponding skeleton. Indeed, we have shown in Section 2.3, that for any skeleton G, the corresponding skeletal order is obtained by a left-to-right traversal of the boundary of its ϵ-thickening $\epsilon(G)$. Since, for a set of n rays, the latter has $4n + 3$ arcs, the order is obtained in linear time from a skeleton.

The leftist and rightist skeletons are easily computable by a sweep algorithm by decreasing ordinate over the set of termini. The algorithm for constructing the leftist (resp., rightist) skeleton is a simple modification of the classical algorithm for intersecting segments. Indeed, the only difference with respect to the original algorithm is that any time the sweep-line reaches an intersection q of two segments r' and r'', rather than reversing the order of r' and r'' on the sweep-line, we just delete from the segment set the ray branch below q which lies to the right (resp., left). Clearly this task visits at most $(n-1)$ intersection points, and spends $O(\log n)$ work at each visited intersection. The time performance of the technique is therefore $O(n \log n)$ and the correctness is trivially established.

Theorem 4 *The leftist and rightist skeletons of $R = \{r_1, ..., r_n\}$ can be constructed in time $O(n \log n)$, and this is optimal.*

3.3 Construction of the external boundary

The construction of E is obtained from the leftist and rightist skeletons, G_L and G_R respectively. Let us consider the union $G_L \bigcup G_R$.

Theorem 5 *The external boundary of $G_L \bigcup G_R$ is identical to E.*

Proof: To prove the theorem, it is sufficient to prove that each edge of E belongs to an edge of G_L or to an edge of G_R (or both). Assume for a contradiction that e is an edge of E that does not belong to G_L nor to G_R. Let t be the terminus of the ray r containing e. If e does not belong to G_L, there necessarily exists a ray r' intersecting r between t and e whose terminus t' lies on the right side of the line containing r' (see Figure 11). Analogously, if e does not belong to G_R, there

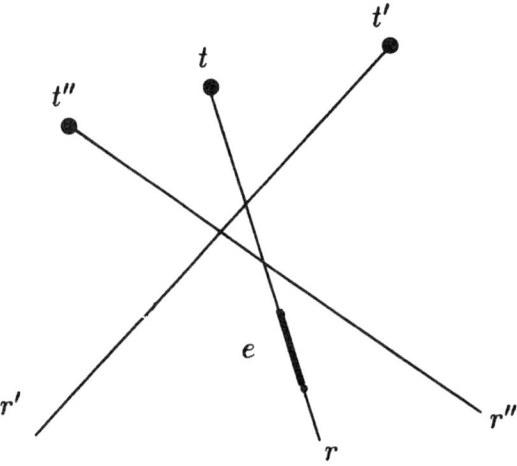

Figure 11: For the proof of Theorem 6

exists a ray r'' intersecting r between t and e whose terminus t'' lies on the left side of the line containing r''. Thus e belongs to the wedge formed by r' and r'' and so cannot be an edge of the external boundary E, a contradiction. □

To construct the boundary of $G_L \bigcup G_R$ we use a plane sweep. This algorithm – closely related to the one used to compute either G_L or G_R – is again a natural modification of the classical algorithm for computing the intersections of segments.

The set S of segments to be swept is defined as follows. Consider G_L and G_R as plane graphs and let V be the union of the vertex set of G_L and of the vertex set of G_R. Each edge of either G_L or G_R gives rise to one or two segments depending upon whether it contains a vertex of V in its interior. Thus there are *two* segments incident on each ray terminus. Moreover, it is convenient to assign a direction to each segment in S: upward if the segment is due to G_L, downward otherwise; this direction is called the *label* of a segment and belongs to the binary set $\{\uparrow, \downarrow\}$, with obvious meanings.

The horizontal sweep-line status δ maintains the left-to-right sequence of the intersected segments of the boundary E (a subset of S). If we introduce two dummy sentinel vertical segments at $-\infty$ and $+\infty$, with respective labels $\downarrow$ and $\uparrow$, then the sequence of labels in δ is of the form "$\downarrow (\uparrow\downarrow)^s \uparrow$", for some $s \geq 0$. A pair "$\uparrow\downarrow$" defines a horizontal interval (called *internal* interval) contained within the (unbounded) polygon P bounded above by E; correspondingly, a pair "$\downarrow\uparrow$" defines an *external* interval. Note that an internal interval may have width zero, and that δ is initialized with just the two sentinels to a single external interval.

The event-point schedule is initialized as V, but it is dynamically augmented during the sweep with a subset of the intersections of the segments of G_L and G_R, specifically, exactly those intersections that lie on E (whose number is obviously $O(n)$, as a consequence of Theorem 1). We guarantee that no other intersection is encountered by the sweep by systematically avoiding the insertion into δ of segments known to lie in the interior of the polygon P defined above (one such segment is called *internal*). A segment found to be internal is also removed from set S.

For each event point v (either a vertex or an intersection) we define $\text{up}(v)$ as the *current* left-to-right sequence of segments incident upon v from above; the sequence $\text{down}(v)$ is analogously defined. During the sweep, the current event point v is at first classified as either internal or external. In the first case, the only action needed is the removal of the members of $\text{down}(v)$ from S (since they are internal segments) and δ is left unaltered. Indeed, by not inserting any internal segments into δ we maintain the invariant that $|\text{up}(v)| = 0$ for an internal point.

If v is external, then the members of $\text{up}(v)$ are deleted from δ. In addition, we distinguish the following subcases (refer to Figure 12):

1. v is a vertex of both G_L and G_R. In this case $|\text{down}(v)| = 2$ and either $|\text{up}(v)| = 0$ or $|\text{up}(v)| = 4$ depending upon whether v is a terminus or not, respectively. The label sequence of $\text{down}(v)$ is "$\uparrow\downarrow$", and both terms of $\text{down}(v)$ are inserted into δ.

2. v is vertex of G_L but not of G_R.

    2.1. $|\text{up}(v)| = 3$ and $|\text{down}(v)| = 2$. Let $\text{up}(v) = (e_1, e_2, e_3)$ and $\text{down}(v) = (e'_1, e'_2)$, with respective label sequences "$\uparrow\downarrow\uparrow$" and "$(\uparrow\downarrow)$". After deletion

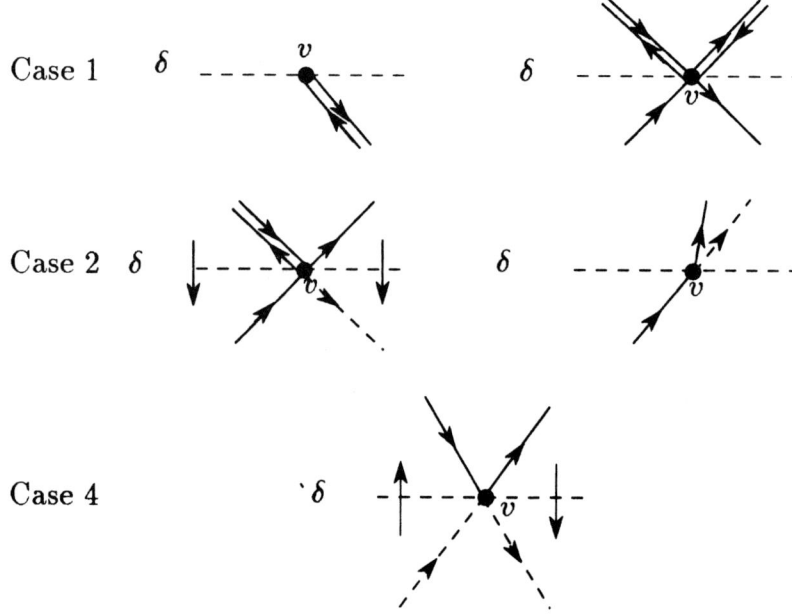

Figure 12: Illustration of the distinct cases of external vertices encountered by the plane-sweep

of (e_1, e_2, e_3) from δ, the label sequence of δ remains legal only by inserting the edge with label "↑", i.e., e_1'. Thus, e_2' is found to be internal and is removed from S.

2.2. $|\text{up}(v)| = 1$ and $|\text{down}(v)| = 1$. Let $\text{up}(v) = (e_1)$ and $\text{down}(v) = (e_1')$, with identical labels "↑". Note that $|\text{up}(v)| = 1$ because of the two segments originally incident upon v from above in G_L, the rightmost one has been previously removed from S, since it is internal. Therefore we correctly insert e_1' into δ (after deleting e_1).

3. v is a vertex of G_R but not of G_L. Analogous to Case 2 above.

4. v is an intersection. In this case $\text{up}(v) = (e_1, e_2)$ with label sequence "↓↑" and $\text{down}(v) = (e_1', e_2')$ with label sequence "↑↓". Clearly neither e_1' and e_2' can be inserted into δ without violating the legality of the label sequence; indeed, both e_1' and e_2' are internal and are removed from S.

These actions are applied to the event points lying above the horizontal line l; the latter is handled by specific straightforward processing.

Clearly, the above actions indicate that the plane sweep maintains as an invariant the legality of the label sequence, i.e., the only edges of $G_L \bigcup G_R$ which appear in the sweep-line status are those that appear on E and the only vertices of $G_L \bigcup G_R$ which are inserted into the event-point schedule are those that appear on E. The set of event points is the union of the vertices of G_L and G_R (two sets of size $O(n)$) and of the intersections between G_L and G_R that lie on E; by Theorem 1, the latter are also $O(n)$ in number.

Let us consider the performance of the above algorithm. We first need to sort by decreasing ordinates the $O(n)$ vertices of G_L and of G_R and to insert them into the event-point schedule, which can be done in $O(n \log n)$ time. Since processing an event point requires $O(\log n)$ time, we conclude :

Theorem 6 *The external boundary E can be computed in time $O(n \log n)$, and this is optimal.*

Remark. The construction of E from G_L and G_R could also be achieved by adapting an algorithm developed by Pollack, Sharir, and Sifrony[9]. However, we feel that the above technique is noteworthy for its simplicity of implementation.

3.4 Construction of the boundary of an internal region

We recall from Section 2.2.2, that for any internal region A we can define a unique convex polygon $C(A)$, which encloses A and is bounded by rays and, possibly, by the line at infinity. The boundary of $C(A)$ consists of two monotone chains C_L and C_R sharing a vertex v with largest ordinate in $C(A)$. The construction of C_L and C_R is our next objective.

Recall from Section 2.2.2 that $R^{(2)}$ is the set of rays whose termini lie outside $C(A)$. Vertex v is the intersection of two rays r' and r'', both in $R^{(2)}$. In particular, v is the minimum-ordinate intersection for all pairs of rays in $R^{(2)}$ whose lower wedge contains A, as is immediately shown by contradiction.

Let p be a point known to be in the interior of A (typically, p is a ray terminus). Given p and a ray $r \in R$, we say that r is a *left ray* (with respect to p) if p lies in the half-plane to the left of the line containing r, directed from the terminus t to infinity on r, and is a *right ray* otherwise.

Clearly, all edges of C_L belong to left rays, and all edges of C_R belong to right rays. Therefore, let $\{R_L, R_R\}$ be the partition of R into left and right rays (this partition is obtainable in $O(n)$ time), and let $r' \in R_L$ and $r'' \in R_R$ be the two rays intersecting at vertex v. It is clear that C_L belongs to the path originating at the terminus t of r' in $G_R(R_L)$ and C_R belongs to the path originating at the terminus of t'' of r'' in $G_L(R_R)$.

Therefore to construct the boundary of the region containing a given point p we proceed as follows:

1. Partition R into $\{R_L, R_R\}$ with respect to p; (* $T = O(n)$ *)

2. Construct $G_R(R_L)$ and $G_L(R_R)$. (* $T = O(n \log n)$ *)

3. Determine the lowest-ordinate intersection v with $y(v) > y(p)$ between

$G_R(R_L)$ and $G_L(R_R)$. From v, construct C_L and C_R; (* $T = O(n \log n)$, as shown below*)

4. Partition R into $\{R^{(1)}, R^{(2)}\}$; (* $T = O(n \log n)$ *)

   If $R^{(1)} = \emptyset$, A is a trivial region, identical to the polygon (C_L, C_R). Else we perform two more steps :

5. Construct the boundary of the external region for set $R^{(1)}$, consisting of rays with terminus internal to the polygon (C_L, C_R). (* $T = O(n \log n)$ from Theorem 7 *)

6. Construct the intersection between the convex polygon (C_L, C_R) and the external region of $D(R^{(1)})$. (* $T = O(n \log n)$ *)

We now show that Step 3 can be carried out in time $O(n \log n)$. We recall the definition of the envelope of a set of segments. The left shadow of a segment s is the portion of the horizontal slab determined by the extremes of s and lying to its left; the left shadow of a set of segments is the union of the left shadow of its members; the right envelope of a set of segments is the right boundary of its left shadow. The left envelope is analogously defined (refer to Figure 13).

Since the line $y = -M$ intersects all rays of R and leaves all termini on the same side, to determine v we just need to consider the right envelope of $G_R(R_L)$ and the left envelope of $G_L(R_R)$. A vertical plane-sweep by increasing ordinate constructs these envelopes in time $O(n \log n)$. Indeed, consider $G_R(R_L)$, and assume inductively that the sweep-line status contains the order of the intersections of $G_R(R_L)$ with the horizontal sweep-line. An event is either a terminus (when we suppress a term from the sweep-line status) or a branching (when we insert a term). The sequence of the rightmost terms appearing on the sweep-line status provides the right envelope. Clearly, this computation runs in time $O(n \log n)$.

Finally, the intersection of the two envelopes is analogous to merging two sorted sequences and is completed in additional $O(n)$ time (for $y \geq y(t)$). We conclude that the construction of the boundary of the region containing a given point p takes $O(n \log n)$ time. In particular, when p is a ray terminus, we have :

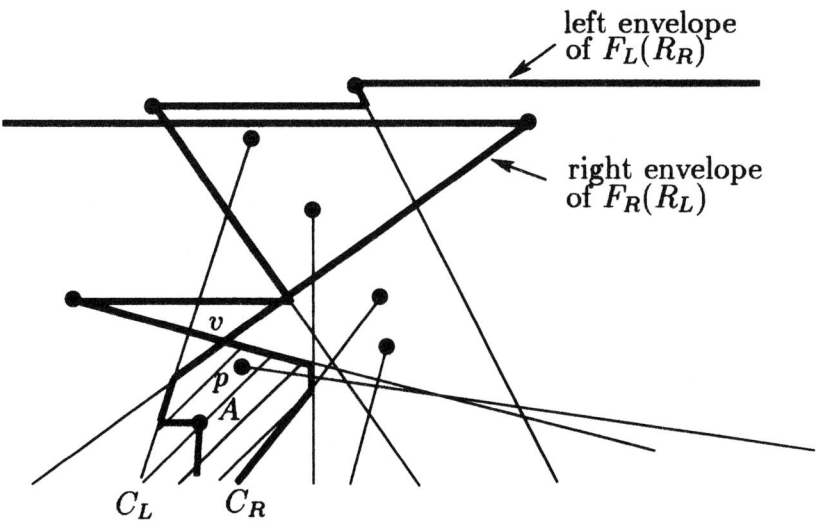

Figure 13: Left and right envelopes

Theorem 7 *The boundary of any internal region of $D(R)$ can be computed in time $O(n \log n)$, and this is optimal.*

Note that the described method agrees with the original procedure for computing the external boundary. Indeed, in the latter case, v does not exist, and the role of the pair (C_L, C_R) is taken by the line at infinity.

4 The general case: structural and algorithmic results

In this section, we remove the half-plane constraint adopted at the beginning of Section 2 and held throughout Section 3, and consider an arbitrary set of rays.

Let us consider a region of the subdivision $D(R)$, identified by a point t in its interior, typically a ray terminus. Then, letting $n = |R|$, in time $O(n \log n)$ we can construct the boundaries E^- and E^+ of the regions $D(R^-)$ and $D(R^+)$, respectively, which contain the chosen t. These are the boundaries of two simple polygons.[1] The desired boundary E^* is the boundary of the connected component of the intersection of these two polygons which contains t.

In order to show that E^* has $O(|R|)$ edges, we prove the following lemma :

Lemma 3 *Let P and Q be two simple polygons with a total number of n edges and let $D(P \bigcup Q)$ be the subdivision induced by the union of the edges of P and Q. The boundary of any region of $D(P \bigcup Q)$ has at most $2n$ edges.*

Proof: : Let E^* be the boundary of a region of $D(P \bigcup Q)$. An edge of E^* is either contained in an edge of P or in an edge of Q. If an edge of E^* is contained in an edge e of P or Q, we label it with e. Thus we associate to E^* a circular sequence L of labels.

An edge of either P or Q will be called of type C_i if it contains i edges of E^*. Let γ_i be the number of such edges and m the maximum of i.

The following observation is due to Pollack, Sharir and Sifrony [9] : for any two edges e and f, L cannot contain a subsequence of the form $efefe$.

From this observation, we deduce that among the edges of P (resp.Q) containing at least one edge of E^* and intersecting a given edge e of Q (resp., P), the only ones which may be of type $C_i, i > 1$, are the first one and the last one (when marching along E^*). Moreover these "extreme" edges cannot intersect e twice.

As a consequence, for each edge $e \in Q$ (resp., P) of type $C_j, j \geq 3$, P (resp., Q) contributes at most two edges of type $C_i, i > 1$ (the first and the last ones), and at least $2j - 4$ edges of type C_1 (the intermediate ones). Thus

$$\gamma_1 \geq \sum_{j=3}^{m}(2j-4)\gamma_j.$$

Besides we have
$$\sum_{i=1}^{m} \gamma_i = n$$

---

[1] No extension of the usual definition of simple polygon is in order if one adopts the fiction, set forth in Section 2.3, of replacing each ray with its ϵ-thickening.

and
$$\sum_{i=1}^{m} i\gamma_i = |E^*|.$$

Hence we conclude that

$$|E^*| \leq 2n - \frac{\gamma_1}{2} \leq 2n.$$

$\square$

It should be noticed that Lemma 5 is a variant of the combination lemma of [6].

Boundary E^* can be computed in time $O(n \log n)$. Indeed, Pollack, Sharir and Sifrony have shown that the contour of the union of two polygons can be obtained by applying the ray-shooting technique of Chazelle and Guibas [3] in time $O(\log n)$ per contour edge, i.e. in total $O(n \log n)$ time [9]. Their algorithm can be easily extended to compute a given connected component of the intersection of two simple polygons in the same time bound. Another solution, which does not make use of the rather intricate ray-shooting technique, is the "blue-red merge" algorithm of [6] which is a combination of two simple plane-sweep algorithms. Thus, we can state the main result of this paper :

Theorem 8 *The boundary of any region of the planar subdivision induced by n rays can be constructed in optimal time $\Theta(n \log n)$.*

5 Concluding remarks

In conclusion, the main result of this paper is an optimal time algorithm for the construction of the boundary of any region of the subdivision determined by a set of n rays in the plane.

The fact – established in its full generality also in this paper – that the boundary of any such region has $O(n)$ edges draws a sharp conceptual distinction between sets of rays and sets of segments. Indeed, Wiernik and Sharir [10] have proved that the upper-envelope of the union of n line segments may have $\Omega(n\alpha(n))$

edges. This bound is not tight for some special classes of line segments, as shown by the following proposition, which is a direct consequence of Theorems 2 and 9.

Proposition 1 *The boundary of any region of the planar subdivision induced by a set of n line segments, each intersecting a given line l, has $O(n)$ edges and can be constructed in $O(n \log n)$ time.*

Proof : Suppose without loss of generality that l is the x axis. Each line segment e_i has an end-point p_i^+ with positive ordinate and the other p_i^- with negative ordinate. Segment e_i can be considered as the intersection of two rays r_i^+ with terminus p_i^+ and r_i^- with terminus p_i^-. Let $R^+ = \{r_i^+ : i = 1, ..., n\}$ and $R^- = \{r_i^- : i = 1, ..., n\}$. Each region of the planar subdivision determined by the n segments is the disjoint union of two (not simultaneously empty) regions, respectively of $D(R^+) \bigcap (y \geq 0)$ and $D(R^-) \bigcap (y < 0)$. The boundary of any such region has $O(n)$ edges (by Theorem 2) and can be constructed in time $O(n \log n)$ by Theorem 9. This establishes the claim. □

Further research is needed to see if it is possible to extend our technique, most notably the notion of skeletal order, to the case of line segments in general position and to design an algorithm for computing the boundary of any region of the resulting arrangement.

Acknowledgments

Jean-Pierre Merlet is acknowledged for supplying to us his interactive drawing preparation system JPdraw . Thanks are due to the referees for their constructive criticism.

References

[1] M. Atallah: Dynamic Computational Geometry, *Proc. 24th IEEE Symp. on Foundations of Computer Science*, 92-99; Oct. 1983.

[2] P. Alevizos, J.D. Boissonnat, and M. Yvinec: An optimal $O(n \log n)$ Algorithm for Contour Reconstruction from Rays, *Proc. 3rd ACM Symposium on Computational Geometry*, Waterloo, 162-170; June 1987.

[3] B. Chazelle and L. Guibas: Visibility and Intersection Problems in Plane Geometry, *Proc. 1st ACM Symposium on Computational Geometry*, Baltimore, 135-147; June 1985.

[4] B. Chazelle, L. Guibas, and D.T. Lee: The Power of Geometric Duality, *BIT* 25, 76-90; (1985).

[5] S. Hart and M. Sharir: Non Linearity of Davenport-Schinzel Sequences and of Generalized Path Compression Schemes, *Combinatorica* 6(2), 151-177 (1986).

[6] H. Edelsbrunner, L.J. Guibas, and M. Sharir: The Complexity of Many Faces in Arrangements of Lines and Segments, *Prof. 4th ACM Symposium on Computational Geometry*, Urbana, 44-56; June 1988.

[7] H. Edelsbrunner, J. O'Rourke, and R. Seidel: Constructing Arrangements of Lines and Hyperplanes with Applications, *SIAM J. Comp.* 15, 341-363 (1986).

[8] L. J. Guibas, M. Sharir, and S. Sifrony: On the General Motion Planning Problem with Two Degrees of Freedom, *Prof. 4th ACM Symposium on Computational Geometry*, Urbana, 319-329; June 1988.

[9] R. Pollack, M. Sharir, and S. Sifrony: Separating Two Simple Polygons by a Sequence of Translations, *Discrete Comp. Geom.* 3:123-136 (1988).

[10] A. Wiernik and M. Sharir: Planar Realizations of Nonlinear Davenport - Schinzel Sequences by Segments, *Discrete Comp. Geom.* 3:15-47 (1988).

Triangulation in 2D and 3D space.

M. Yvinec
LIENS _ URA CNRS 1327
Ecole Normale Supérieure
45 rue d'Ulm
75230 Paris FRANCE

Abstract

This paper is a review of some problems related to the triangulation of polygons or point sets in 2D and 3D space. It includes in particular a proof of the equiangularity properties of the Delaunay triangulation in 2D space and a short review on the different algorithms for the triangulation of polygons. The (still open) question of the intrinsic complexity of the triangulation problem for a simple polygon is also raised. As far as 3D space is concerned, the combinatory relations which provide bounds on the number of tetrahedra which appear in the triangulation of a set of points are given. A divide and conquer algorithm for triangulating arbitrary set of points is also presented. This algorithm is based on a splitting theorem which has been proved independently by Avis and ElGindy on one side and by Edelsbrunner, Preparata and West on the other side.

1 Introduction

The triangulation of a set of points is one of the basic tools in the domain of numerical analysis to deal with, for instance, interpolation problems for multivariate functions or the solution of partial differential equations by the finite element method. The triangulation methods are also largely used in domains such as Robotics or Computer Vision, for instance to reconstruct solid objects from a set of measured points ([Boi84], [Boi86]) or else in the domain of Computer Graphics or Image Synthesis to solve the clipping problems or to the interpolation of rendering functions ([Rog85]). In the field of Computational Geometry, triangulating a set of points, a planar graph, a polygon, a polyhedra or any kind of geometric object is often a prerequisite treatment to the execution of another algorithm. For instance, the localisation algorithm of Kirkpatrick ([Kir83]) assumes that the planar regions are triangulated and the algorithms for visibility or shortest path problems within a given polygon begin with a triangulation of the polygon ([GHL*87]).

In fact, some questions are still unanswered, and, at the present time, the triangulation problems are among the main concern of many researchers which is obvious from the high number of papers recently appeared on that subject. This paper presents a survey on the present state of the art in the field of triangulation. It was impossible here to be exhaustive on that topic and we have rather choosen to restrict to a few aspects of the problem. The second part of this paper concerns planar triangulations while triangulations in 3D space are dealt with in the third part.

One of the main problems which appears in the domain of numerical analysis is to obtain *good* triangulations, that is as regular as possible. As far as this is concerned, Part 2 begins with the equiangularity properties of the Delaunay triangulations. The next subsection of Part 2 deals with the (still unsolved) question of knowing what is the intrinsic complexity of the triangulation problem for a simple polygon. At last, the last subsection of Part 2 is a survey of the different methods used in triangulation algorithms.

In 3D space the triangulation problem is made much more complicated by the fact that the number of tetrahedra which is to appear is unknown a priori. The combinatorial facts, which allow to bound the number of tetrahedra, are given in the first subsection of Part 3. The next subsection of Part 3 presents a triangulation algorithm for an arbitrary set of points in 3D space. This algorithm uses the *divide and conquer* method and is based on a splitting theorem which was proved independently by Avis and ElGindy on one hand and Edelsbrunner, Preparata and West on the other hand. At last, the concluding Part includes essentially a list of unsolved problems.

2 Planar Triangulation

2.1 Definition and Complexity

In 2D space, the triangulation of a set P of n points can be defined as a planar maximal graph on the set of points P. Such a graph necessarily includes the $n'(n' \leq n)$ edges of the convex hull $CH(P)$ of P and partitions the interior of $CH(P)$ in triangles whose interiors are disjoint.

For every triangulation of a set P of n points, the number of triangles is known as a function of the number n' of extremal points of P. Indeed, like any planar graph, a triangulation satisfies the Euler relation :

$$n - e + t = 1 \tag{1}$$

where e is the number of edges and t the number of triangles. Furthermore, each triangle is adjacent to three edges and each edges, except the convex hull boundary edges, is adjacent to two triangles :

$$2e - n' = 3t \qquad (2)$$

thus :

$$e = 3n - n' - 3 \qquad (3)$$
$$t = 2n - n' - 2 \qquad (4)$$

The complexity of the triangulation problem in 3D space is $O(n \log n)$. Indeed, on one hand there are triangulation algorithms in time $O(n \log n)$ and on the other hand the problem of sorting n real numbers is linearly reducible to the problem of triangulating a set of n points (cf [PS85, page 188]).

2.2 The Delaunay Triangulations

Generally, especially in the domain of numerical analysis, the aim is to obtain a triangulation as regular as possible, avoiding the appearance of too thin (nearly degenerate) triangles. Several criteria can be used to define what is actually a *good* triangulation. For instance, one sometimes tries to minimize the total length of the edges or to maximize the smallest of the angles. With respect to this last criterium, the Delaunay triangulations are particularly interesting because any Delaunay triangulation actually achieves the fact that its smallest angle is maximum.

Definition

A Delaunay triangulation is a triangulation which satisfies the following property:
the circumscribed circle to any triangle in the triangulation includes no points of the set P in its interior.

If the n points of the set P are in general position, that is if there are no four cocircular points, the Delaunay triangulation is unique and can be obtained as the dual of the Voronoi diagram of the set of points. (This dual is a graph defined on the set P by adding one edge for each pair of points whose Voronoi cells share an edge) (cf [PS85]).

If the points of the set P are not in general position, the dual of the Voronoi diagram partitions the interior of the convex hull in convex faces. Every triangulation, obtained from this graph by adding some edges to triangulate the faces adjacent to more than three edges, is a Delaunay triangulation.

Equiangularity of Delaunay triangulations

A triangulation is said to be globally equiangular if the smallest angle is maximum. R. Sibson [Sib78] proved that every Delaunay triangulation is globally equiangular.

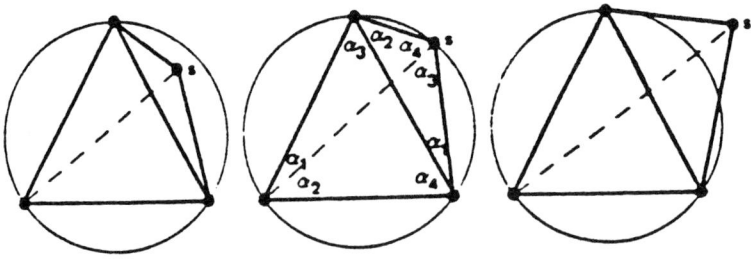

Figure 1: Local equiangularity ([Ede87])

The proof uses the notion of local equiangularity. An edge e, adjacent to the two triangles t_1 and t_2 whose union is a convex quadrilater, is said to be legal if the smallest angle in t_1 and t_2 is greater than the smallest angle of the triangles t'_1 and t'_2 obtained if the edge e is replaced by the other diagonal e' of the quadrilater $t_1 \cup t_2$. A triangulation is locally equiangular if all its edges are legal. Straightforward elementary geometry (cf Figure 1 from [Ede87]) shows that the edge e is legal if and only if the circle $b(t_1)$ circumscribed to the triangle t_1 does not include the vertex of t_2 which is not adjacent to e. Thus the property of circumscribed circles implies that any Delaunay triangulation is locally equiangular.

Conversely, any locally equiangular triangulation is a Delaunay triangulation. Indeed, assume that this is wrong and let T be a locally equiangular triangulation which fails to satisfy the Delaunay property. There is a triangle t_1 and a vertex s such that s is included in the circle $b(t_1)$ circumscribed to t_1. Since T is a triangulation s is not included in the interior of t_1 and let e be the edge of t_1 such that the triangle t^*, convex hull of $t \cup e$, is disjoint from the interior of t_1. Among all the possible triplets (triangle,vertex,edge) which can thus be defined, let (t_1, s, e) be the triplet which maximise the angle of the triangle t^* at the vertex s. Let t_2 be the other triangle of T adjacent to the edge e and let s_2 be the vertex of t_2 which is not adjacent to e. (cf Figure 2). From equiangularity s_2 is not included in the interior of $b(t_1)$ and thus t_2 is different from t^*. The circle $b(t_2)$ circumscribed to t_2 intersects $b(t_1)$ at both ends of the edge e, and as s_2 is not included in the interior of $b(t_1)$, the interior of $b(t_2)$ includes the interior part of $b(t_1)$ which is disjoint from the interior of t_1 and limited by e as cord segment. Thus, the vertex s is included in the interior of $b(t_2)$. Let e_2 be the edge of t_2 such that the convex hull of $e_2 \cup s$ is disjoint from the interior of t_2. The triplet (t_2, s, e_2) forms an angle greater than the angle formed by (t_1, s, e), which is contradictory with the hypothesis.

To complete Sibson's proof, we just have to show that any locally equiangular triangulation is at the same time globally equiangular. (The converse

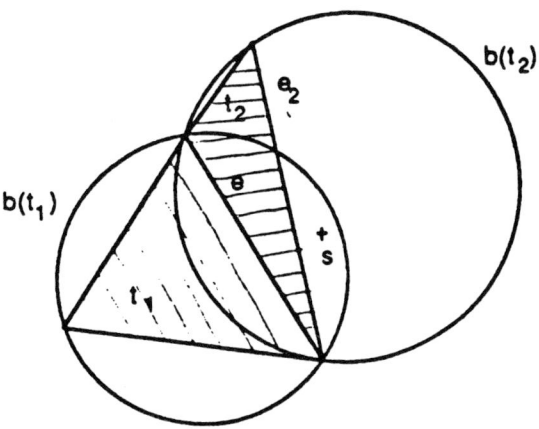

Figure 2: proof for the locally equiangularity of a Delaunay triangulation

proposition is not true). First, we notice that all the locally equiangular triangulations or Delaunay triangulation of a given point set P have the same smallest angle. Furthermore, any triangulation T' can be turned into a locally equiangular triangulation by an algorithm which, as long as there are some illegal edges, picks up an illegal edge and replaces it by the other diagonal of the same convex quadrilater. Each edge exchange replaces two triangles by two different ones which have both strictly smaller circumradii than the both they replace. Thus, cycling cannot occur and since there is only a finite number of triangulations, this algorithm is guaranteed to terminate. Applied to a locally equiangular triangulation T', this algorithm yields a globally equiangular triangulation T whose smallest angle cannot be less than the smallest angle of T'. Both triangulations T and T' have thus the same smallest angle which achieves to prove that any locally equiangular triangulation is also globally equiangular.

2.3 Triangulation of polygons

In many cases, the triangulation problem is constrained, that is a set of edges are prescribed to belong to the triangulation. Typically, this is the case when the problem is to triangulate the interior of a single polygon. Then, the problem is set as follows :
Let P be a simple polygon given by the list of its vertex $(p_0, p_1, \ldots, p_n = p_0)$. The objective is to add to the planar graph formed by this polygon a set of $n - 3$ new edges (diagonals), lying entirely inside P, and without intersections between each other. These new edges partition the interior of P in $n-2$ triangles whose interiors are disjoint.

The triangulation of polygons is one of the main problems of Computational

Geometry. Indeed, on one hand triangulation is a basic prerequisite operation for many other algorithms like the algorithms to find shortest paths or visibility graphs inside a polygon and on the other hand the intrinsic complexity of this problem is not yet exactly known.

2.3.1 Complexity of the triangulation of a simple polygon

The well known triangulation algorithms as the algorithms proposed by Garey et al. [GJPT78], Hertel and Mehlhorn [HM83], Fournier and Montuno [FM84] and also the algorithm of B. Chazelle [Cha82] perform the triangulation of a simple polygon with n vertex in time $O(n \log n)$ and a linear memory space. Many other algorithms have been developped to triangulate, in linear time, restricted classes of polygons such that convex, star-shaped [SVL80], monotone polygons [GJPT78] or edge-visible polygons[TA82]. Other algorithms can adapt to the actual shape of the polygon and thus run faster in special cases. For instance, Mehlhorn's algorithm [HM83], in its sophisticated version, performs the triangulation of a simple polygon in time $O(n+r \log r)$ where r is the number of concave vertex of the polygon. The algorithm of Chazelle and Incerpi [CI84] has a running time of $O(n \log s)$, where the sinuosity s is a parameter depending on the shape of the polygon. This parameter is seldom over a few units even for very complicated polygons and in practice this algorithm is quasi-linear ; furthermore, it is shown to run in time $O(n \log \log n)$ for a very large class of polygons. At last, recently Tarjan and Van Wyk [TVW86] have published an algorithm running in time $O(n \log \log n)$ for any simple polygon. Despite from being simple to implement, this algorithm proves that the knowledge of a simple polygonal path through the points lowers the complexity of the triangulation problem. This algorithm has not yet been proved to be optimal.

2.3.2 A survey of the main triangulation algorithms for simple polygons

This subsection gives some ideas about the various methods used in triangulating polygons. Of course, it was not possible in here to be exhaustive or to go into the details of each algorithm.

Triangulation of a convex polygon. The triangulation of a convex polygon is a trivial operation : for instance, one of the vertex can be simply joined to all the other.

Linear triangulation of a monotone polygone [GJPT78]. A polygon is said to be monotone, with respect to the y-axis for instance if its boundary is the concatenation of two monotone chains , a chain $(p_0, p_1, \ldots, p_k)$ with increasing y-coordinates and a chain $(p_k, p_{k+1}, \ldots, p_n = p_0)$ with decreasing y-coordinates.

The Garey and al. algorithm first sorts the vertex of the polygon according to decreasing y-coordinates, simply merging the two monotone chains whose concatenation is the polygon boundary. Let $(q_0 = p_0, q_1, \ldots, q_n)$ be this ordered list, the vertices are then considered one at a time in that order. In the course of the algorithm, some edges are added to the graph of the polygon. Each added diagonal separates a triangle from the remaining part of the polygon which is still a monotone polygon with one less vertex. The algorithm holds in a stack all the vertices which have already been considered and belong to the boundary of the remaining polygon. At each step, the stack content $(x_1, x_2, \ldots, x_i)$ fulfills the following invariant :

- $x_1, x_2, \ldots, x_i$ are ordered according to decreasing y-coordinate,

- $x_1, x_2, \ldots, x_i$ form a chain on the boundary of the remaining polygon,

- for all j, $2 \leq j \leq i - 1$, the vertex x_j is a concave vertex of the remaining polygon,

- the next vertex to be considered x is adjacent to either x_1 or to x_i or, else, to both of them.

As a first step, the two first vertex q_0 et q_1 are pushed on the stack . At general step, let $(x_1, x_2, \ldots, x_i)$, be the content of the stack and let x be the next vertex to be considered :

1. If x is adjacent to x_1 but not to x_i (cf Figure 3a), the diagonals $xx_1, xx_2, \ldots, xx_i$ are added. The stack content is replaced by (x_1, x).

2. If x is adjacent to x_i but not to x_1 (cf Figure 3b), then, while $i > 1$ and while the angle of the remaining polygon at vertex x_i is less than π, repeat the following instructions :
{ add the diagonal xx_{i-1}
pop x_i from the stack
decrement i. }

3. If x is adjacent to x_1 and x_i (cf Figure 3c), the diagonals $xx_1, xx_2, \ldots, xx_{i-1}$ are added and the processing is complete.

The following algorithms triangulate any simple polygon.

The Garey-Johnson-Preparata-Tarjan algorithm [GJPT78]. First, this algorithm turns the graph of the polygon into a *regular* graph by adding some edges. A graph is said to be *regular* if the following conditions are fulfilled :

- each vertex, except the ones with largest y-coordinate, is joined directly to at least one vertex with larger y-coordinate,

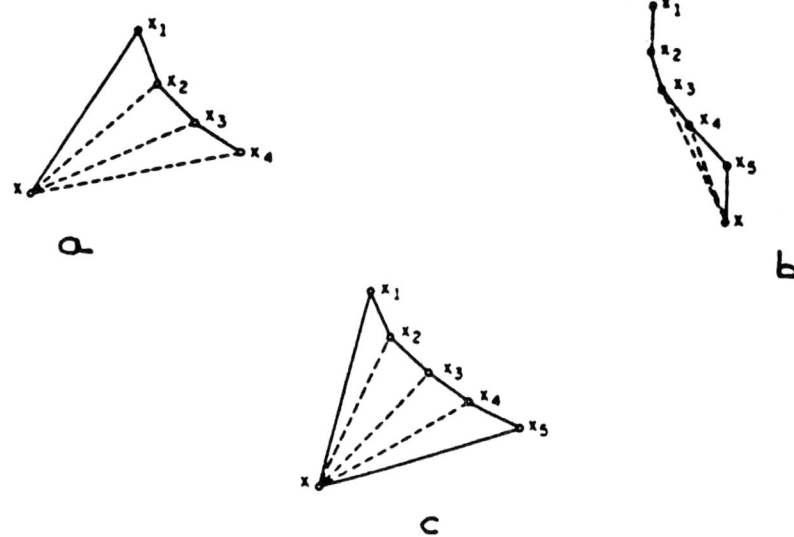

Figure 3: Triangulation of a monotone polygone ([GJPT78])

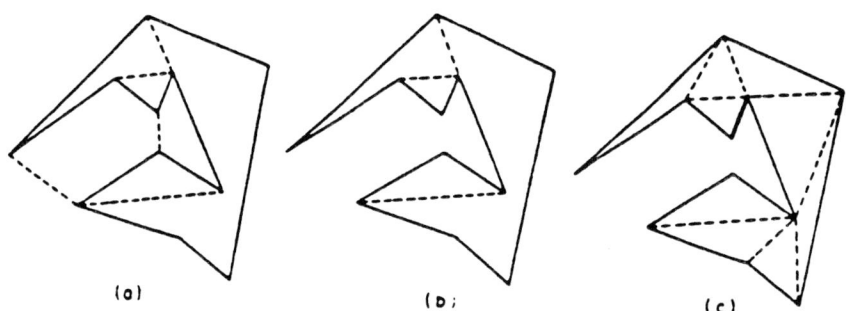

Figure 4: a) regularization process; b)splitting into monotone polygons c)triangulation ([GJPT78])

- each vertex, except the vertex with smallest y-coordinate, is joined directly to at least one vertex with smaller y-coordinate.

The regular graph which is obtained partitions the interior of polygon P into monotone polygons which are then triangulate using the above algorithm. (cf Figure 4 from [GJPT78]). The regularization process is performed in time $O(n \log n)$ through a double plane sweep algorithm, and the monotone polygons are then triangulated in time $O(n)$, thus the complexity of the whole algorithm is $O(n \log n)$.

The algorithm of K. Mehlhorn [HM83] and [Meh85, pages 160-172]. In its simple version, this algorithm looks like the preceeding one, although the three operations of regularization, splitting and triangulation are performed together in the course of a single sweep line process. In the elaborate version,

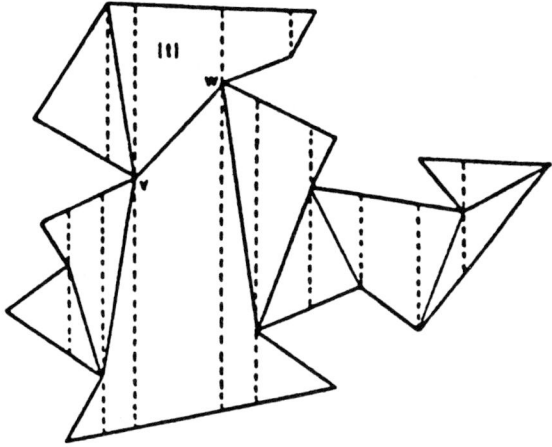

Figure 5: Vertical decomposition of a simple polygon ([CI84])

the sweeping stops only at some particular vertices, whose number is $2r$ where r is the number of concave vertices of the input polygon, which lowers the running time of the algorithm to $O(n + r \log r)$.

The algorithm of A. Fournier et Y. Montuno [FM84]. This algorithm performs a *vertical decomposition* of the input polygon, which means that the polygon is partitioned into trapezoids or slabs.

Each trapezoid has two edges parallell to the y-coordinate axis and is determined by two vertices of the input polygon. This decomposition is performed in time $O(n \log n)$ by a sweep line algorithm.

The resulting trapezoids fall in two classes : trapezoids of class A are determined by two vertices of the polygon which are not adjacent either on the trapezoid or on the input polygon while trapezoids of class B are determined by two adjacent vertices. For each trapeziod of class A, it is possible to create one of the edge of the triangulation by adding the diagonal joining the two input vertices. When this is done for all trapezoids of class A, we are left with monotone polygons which are then triangulated in linear time. (cf Figure 5 from [CI84])

The algorithm of B. Chazelle and J. Incerpi [CI84]. The algorithm of B. Chazelle and J. Incerpi uses the same idea of vertical decomposition of a polygon. B. Chazelle and J. Incerpi have given a very efficient algorithm to perform this decomposition in time $O(n \log s)$ where s is a parameter characteristic of the polygon shape which they called the *sinuosity* : any simple polygon can be regarded as the concatenation of alternatively *spiralling* and *antispiralling* polygonal lines, and s is the number of polygon lines included.

The $O(n \log \log n)$ algorithm of R. Tarjan and C. Van Wyk [TVW86].
A. Fournier et Y. Montuno [FM84] have shown that the triangulation of a simple polygon is a linearly reducible to the vertical decomposition of this polygon, a problem which is itself reducible to the problem of finding all the visible edge-vertex pair for a given sight direction. Assume wlog that the sight direction is the vertical direction (y-direction), the visible edge-vertex pairs are those edge vertex pairs which can be connected by a vertical segment lying entirely within the polygon.

The algorithm proposed by R. Tarjan and C. Van Wyk yields all the visible edge-vertex pairs of a simple polygon in time $O(n \log n)$. This algorithm uses basically the *divide and conquer* method associated to the Jordan sorting algorithm and a few sophisticated data structures such as *finger trees*.

Triangulation using the *divide and conquer* paradigm [Cha82]. At last, we have to mention a straightforward algorithm which triangulates a simple polygons through recursive splittings based on the *polygon cutting theorem* proved by B. Chazelle [Cha82]. Here is a simplified statement of this theorem :

Let P be a simple polygon with n vertices $(p_0, p_1, \ldots, p_n = p_0)$, it is possible to find in $O(n)$ time a pair of vertices $p_i p_j$ such that the segment $p_i p_j$ lies entirely inside polygon P and partitions it into two simple polygons P_1 and P_2 whose respective sizes n_1 and n_2 satisfy :

$$n_1 \leq n_2 \leq 2n/3 + 2 \tag{5}$$

The recursive splitting of a simple polygon using this theorem leads to an $O(n \log n)$ triangulation algorithm.

3 Triangulation in 3D space

3.1 Definition and combinatorial facts

In a 3D space, the triangulation problem can be set as follows :
Let P be a set of n points. A triangulation of P is a partition of the interior of the convex hull $CH(P)$ of P into tetrahedra such that

- the set P includes the four vertices of each tetrahedra and no other point.

- the intersection of two tetrahedra is either empty or a face of each. (The term *face* is here to be understood in its general meaning : a face of a tetrahedron is the convex hull of a subset of its vertices, thus it may be a vertex, an edge or a facet.)

The main problem arising in 3D space comes from the fact that the number of tetrahedra involved in the triangulation of a given point set P is not known a priori.

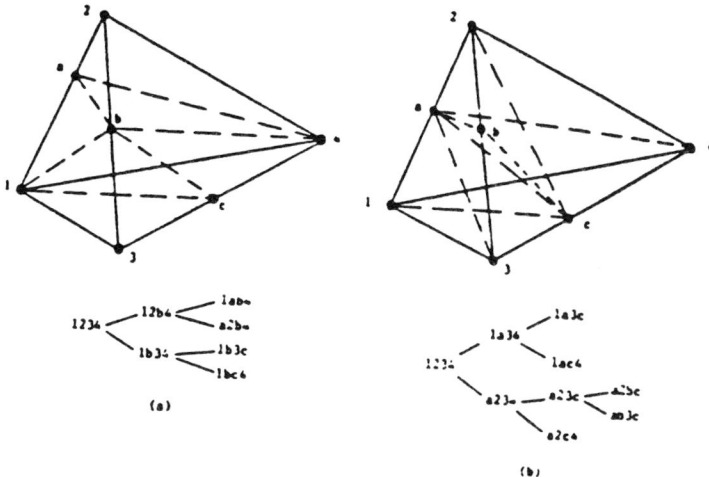

Figure 6: Some examples of 3D triangulations ([AE87])

1. The same point set can be triangulated using different numbers of tetrahedra as appears on Figure 6 from [AE87].

2. There are points sets which can be triangulated with a linear number of tetrahedra.

   For instance, it is easily seen that any simplicial set P. of n points in general position has a triangulation of size $3(n-4)+1$.(The size of a triangulation denote the number of tetrahedra used. A set of points in a space of dimension d is said to be simplicial if it has the dimension d and $d+1$ extreme points. A set of points P is said to be in general position if any subset of $d+1$ points has the dimension d.)Indeed, let us think of a triangulation which is built incrementally, adding one at a time the point of P which are inside the convex hull. Each added internal point falls in the interior of a tetrahedra which has to be splitted in four tetrahedra, which increases by 3 the total number of tetrahedra. Moreover, it is shown later that any set of points P without 3 colinear points has a triangulation of linear size.

3. There exist point sets which have only triangulations of quadratic size. Let us for instance consider the set of $n+2$ points formed by the four points $1, 2, 3, 4$ in general position and $n-2$ points on the edge 43 as shown on Figure 7 from [AE87]. This point set is triangulated using $n-1$ tetrahedra sharing all the edge 12. Any additional point on this edge 12 increases by $n-1$ the number of tetrahedra and, if $n-2$ points are added on the edge 12, we get a point set P with $2n$ points and a unique possible triangulation of size $(n-1)^2$.

4. There exist point sets which have both linear and quadratic triangulations

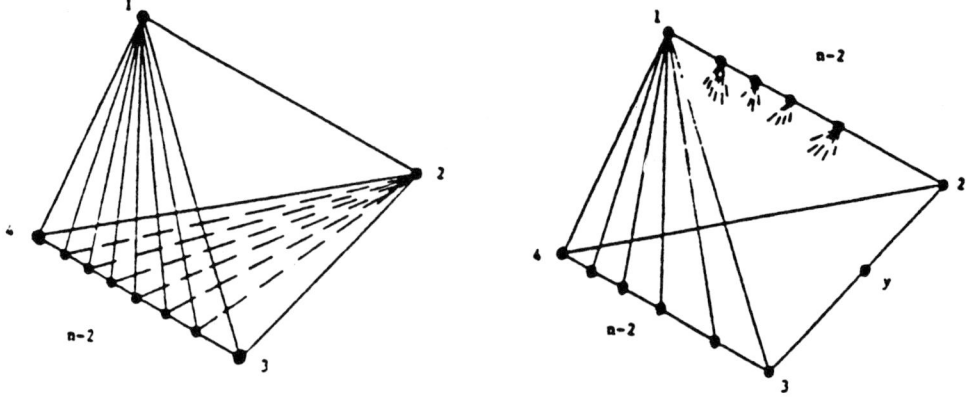

Figure 7: Quadratic Triangulations ([AE87])

Let us for instance consider the set P' formed by the $2n$ points of the above set P with one additional point y on the edge 23 (cf Figure 7). On one hand, if we ignore that point y and triangulate the remaining points, we obtain as before $(n-1)^2$ tetrahedra ; inserting point y creates one additionnal tetrahedra which yields a triangulation of size $(n-1)^2+1$. On the other hand the tetrahedron $\{1,2,3,4\}$ can be first partitionned into the tetrahedra $\{1,y,3,4\}$ and $\{1,y,2,4\}$. Each of these tetrahedra is similar to the above example and can be triangulated using $n-1$ tetrahedra, which yields, on the whole, $2(n-1)$ tetrahedra.

Bounds on the number of tetrahedra

Let P be a set of n points and let T be a triangulation of P. We start with the Euler formula which for any 3D triangulation reads :

$$n - e + f - t = 1 \qquad (6)$$

where n is the number of points
$\quad$ e is the number of edges
$\quad$ f is the number of facets
$\quad$ and t is the number of tetrahedra.
Let n' be the number of points on the convex hull $CH(P)$ of P and n'' be the number of internal points (points which are inside the convex hull)

$$n' + n'' = n \qquad (7)$$

The vertices, edges and facets of T which lie on the boundary of the convex hull form a maximal planar graph. Thus there are $e' = 3n'-6$ edges and $f' = 2n'-4$ facets on $CH(P)$. Each tetrahedra has four facets and all facets, except the hull facets belong to 2 tetrahedra. Thus :

$$4t = 2f - (2n' - 4) \qquad (8)$$

which yields for the size of T :

$$t = e - n - n' - 3 = e - 2n' - n'' - 3 \tag{9}$$

There are at most $\binom{n}{2}$ edges and at least $3n' - 6$ hull edges. In addition every interior vertex belongs to at least 4 internal edges, so there are at least $2n''$ internal edges :

$$(3n' - 6) + 2n'' \leq e \leq \binom{n}{2} \tag{10}$$

Hence, we get the following trivial bounds :

$$n - 3 \leq t \leq \binom{n-1}{2} - n' + 2 \tag{11}$$

The next question concerns the accuracy of these trivial bounds. Partial answers to this question are found in [EPW86]. In particular, it is shown there that, for $n \geq n' \geq 4$, there are point sets P with n points and n' extremal points which have a triangulation of size t larger than $\binom{n-1}{2} - (n'-4)n'' - n' + 2$ and other sets P which have a triangulation of size less than $n - 3 + 2n''$.

It is shown in the next paragraph that any point set P in general position has a triangulation of size $t = 2n' - 4 - \delta + 3n''$ where δ is the maximum degree for a vertex on the convex hull, and it is conjectured that this triangulation has the minimal size.

3.2 3D Triangulation using *divide and conquer*

Let P be a set of n points in general position (there are no four coplanar points) in a 3D space . There is an algorithm to triangulate this set using $t = 2n' - 4 - \delta + 3n$ tetrahedra and a linear running time. This algorithm uses the method *divide and conquer* and is based on a splitting theorem proved independently by D. Avis and H. ElGindy [AE87] on one side and H. Edelsbrunner, F. Preparata and B. West [EPW86] on another side. D. Avis and H. ElGindy also provide an algorithm to deal with arbitrary sets of points (instead of sets in general position) in time $O(n \log n + t)$ where t is the size of the triangulation.

The splitting theorem

Theorem 1 *Every simplicial set P of n points in d dimensions with n" interior points can be partitioned into $d+1$ simplices, none of which contains more than $dn"/(d+1)$ interior points. The partitionning point is contained in P and is called a $d/(d+1)$-splitter. It can be found in time $O(d^4 + nd^2)$.*

The proof of this fact, given in [AE87] and [EPW86], is a constructive proof which yields an algorithm to find a $d/(d+1)$-*splitter* in linear time. Besides, the ratio $d/(d+1)$ is optimal, indeed some simplicial point sets do not have better splitter.

Triangulation of a point set in general position

In the initial step, the convex hull $CH(P)$ of P is constructed and the vertex v with highest degree δ on the convex hull is found. The triangulation is initialized by the set of tetrahedra spanned by v and a hull facet that does not contain v. There is $2n' - 4 - \delta$ such tetrahedra. Each of the $n"$ interior point is localized in one of those initial tetrahedra and then, each initial tetrahedra is triangulated through recursive splittings using at each step a $d/(d+1)$-*splitter*. This algorithm yields a triangulation of P with $2n' - 4 - \delta + 3n"$ tetrahedra.

From the splitting theorem, a $d/(d+1)$-*splitter* can be found in linear time and the recursive phase of this algorithm runs in time $O(n \log n)$. The convex hull of the set P is constructed in time $O(n \log n)$ using for instance the algorithm of Preparata and Hong [PH77]. The localization of the internal points of P can also be performed in time $O(n \log n)$ using the localization method of Kirkpatrick [Kir83] on a central projection of the convex hull $CH(P)$ of P from the point v. Thus, as a whole, this triangulation algorithm runs in time $O(n \log n)$.

Triangulation of an arbitrary set of points

The arbitrary sets of points with possible cases of coplanar or colinear points could be handled using the perturbation method. This method has the disadvantage to form degenerated tetrahedra whose vertices are coplanar or colinear. D. Avis and H. ElGindy [AE87] derived a special algorithm to handle the cases of coplanar or colinear points. At each step of the partitionning process, a list of points which are found interior to the new facets or the new edges of the triangulation is kept and the process of recursive splitting is continued, forgetting about the points on the lists. After the last step of recursive splitting we are thus left with tetrahedra without internal points but possibly with some points on their facets or edges. D. Avis and H. ElGindy provide an $O(t)$ algorithm to triangulate these tetrahedra (t is the number of used tetrahedra) : first, each non empty facet is triangulated without taking into account the points which possibly are on its edges. From these 2D triangulations, a 3D triangulation of the tetrahedra is formed, all the facets are now empty but there are possibly some points on the edges. Simplices such as $\{r, s, t, u\}$ with many points on their edges are then triangulated as follows : as long as there is a face, such as face $\{r, s, t\}$ with at least two non empty edges, that face is triangulate and the vertex u is joined to each vertex of this triangulation. Then, if the points internal to the edges appear on two opposit edges such as rs and tu, a quadratic number of tetrahedra are created as shown in the example of Figure 7. Thus, it is to be noticed that only cases of colinear points compell to use a quadratic number of tetrahedra. The complexity of the last phase is linear with respect to the number of created tetrahedra and the overall complexity, including the initial step and the recursive splitting process is $O(n \log n + t)$.

3.3 The Delaunay triangulation in 3D space

The above triangulation method is likely to yield very thin tetrahedra which is usually to avoid especially in numerical analysis applications. Therefore, Delaunay triangulations are usually performed. Unfortunately, in 3D space, the Delaunay triangulation of a point set is possibly of quadratic size. Moreover, no generalization of the equiangularity properties of the 3D Delaunay triangulation is presently available for 3D space.

The Delaunay triangulation of a point set can be obtained from the projection of the lower convex hull of a set of points in 4d space. This assumes the classical relation between a 3D point $m(x, y, z)$ and the 4d point $M(x, y, z, w = x^2 + y^2 + z^2)$ lying on the unity paraboloid. This method yields an $O(n^2)$ algorithm.

There is also an iterative method to construct the Delaunay triangulation in 3D space. The points of the set are added one at a time in the triangulation. For each newly added point, all tetrahedra whose circumscribed sphere contain that point are destroyed and the region of space spanned by all the distroyed tetrahedra is triangulated again taking the new point into account. This algorithm runs in time $O(n^3)$ in the worst case because all the existing tetrahedra can be destroyed for each new point. In practice, this algorithm is much more efficient, especially if the triangulation is described using the hierarchical structure introduced by J. D. Boissonnat [BT86].

4 Conclusion

The triangulation problems are now well known in 3D space but Computational Geometry is still at its beginning as far as 3D triangulation is concerned. For instance, to our knowledge there does not exist an algorithm to triangulate a non convex polyhedra. As a conclusion, here is a list of unsolved questions given in [EPW86] :

1. The sets of n points with n' extremal points which achieve the triangulations of maximal and minimal size are not the same. What is the maximum ratio between the maximal and minimal sizes of the triangulations of the same point set ?

2. What are the minimum and maximum sizes of Delaunay triangulations for sets of n points with n' extremal points ?

3. Can the region between two non overlapping convex polytopes always be triangulated with a linear number of tetrahedra ?

4. Is there an algorithm for the Delaunay triangulation for sets of n points with n' extremal points that runs in time $O(n(\log n)^\alpha + t)$ where t is the size of the triangulation ?

References

[AE87]   D. Avis and H. ElGindy. Triangulating point sets in space. *Discrete Comput. Geom.*, 2, 1987.

[Boi84]   J. D. Boissonnat. Geometric structures for three-dimensional shape representation. *ACM Trans. on Graphics*, 3, 1984.

[Boi86]   J. D. Boissonnat. *Shape Reconstruction from Planair Cross-sections.* Technical Report 546, INRIA, Rocquencourt, France, 1986.

[BT86]   J. D. Boissonnat and M. Teillaud. An hierarchical representation of objects : The delaunay tree. In *Proc. of the 2d annual Symposium of Computational Geometry*, pages 260–268, Yorktown Heights, New York, June 86, 1986.

[Cha82]   B. Chazelle. A theorem on polygon cutting with applications. In *Proc. of the 23rd IEEE Annual Symp. on Found. of Comp. Sci.*, IEEE, New York, Chicago, Nov. 3-5 1982, 1982.

[CI84]   B. Chazelle and J. Incerpi. Triangulation and shape complexity. *ACM Transactions on Graphics*, 3(2), 1984.

[Ede87]   H. Edelsbrunner. *Algorithms in Combinatorial Geometry.* EATCS Monographs on Theoretical Computer Science, Springer-Verlag, 1987.

[EPW86]   H. Edelsbrunner, F. P. Preparata, and D. B. West. *Tetrahedrizing point sets in three dimension.* Technical Report UIUC DCS-R-86-1310, Department of Computer Sciences, Univ. of Illinois, Urbana, Champaign, 1986.

[FM84]   A. Fournier and D. Y. Montuno. Triangulating simple polygons and equivalent problems. *ACM Transactions on Graphics*, 3(2), 1984.

[GHL*87] L. Guibas, J. Hershberger, D. Leven, M. Sharir, and R. E. Tarjan. Linear-time algorithm for visibility and shortest path problems inside triangulated polygons. *Algorithmica*, 2, 1987.

[GJPT78] A. R. Garey, D. S. Johnson, F. P. Preparata, and R. E. Tarjan. Triangulating a simple polygon. *Inform. Process. Lett.*, 7, 1978.

[HM83]　S. Hertel and K. Mehlhorn. Fast triangulation of simple polygons. In *Proc. of the Conference on Foundations of Computing Theory*, Springer-Verlag, Borgholm, Sweden, 1983.

[Kir83]　D. G. Kirkpatrick. Optimal search in planar subdivision. *SIAM J. Comput.*, 12, 1983.

[Meh85]　K. Mehlhorn. *Data Structures and Algorithms 3 : Multidimensional Searching and Computational Geometry*. McGraw-Hill, 1985.

[PH77]　F. P. Preparata and S. J. Hong. Convex hulls of finite sets in two and three dimensions. *Comm. ACM*, 20, 1977.

[PS85]　F. P. Preparata and M. I. Shamos. *Computational Geometry*. Springer-Verlag, New York, 1985.

[Rog85]　D. F. Rogers. *Procedural Elements for Computer Graphics*. EATCS Monographs on Theoretical Computer Science, Springer-Verlag, 1985.

[Sib78]　R. Sibson. Locally equiangular triangulations. *Comput. J.*, 21, 1978.

[SVL80]　A. A. Schoone and J. Van Leeuwen. *Triangulating a star shaped polygon*. Technical Report RUV-CS-80-3, Univ. of Utrecht, 1980.

[TA82]　G. T. Toussaint and D. Avis. On a convex hull algorithm for polygons and its application to triangulation problems. *Pattern Recogn.*, 15, 1982.

[TVW86]　R. E. Tarjan and C. J. Vam Wyk. *An O(nloglogn) Time Algorithm for Triangulating Simple Polygons*. Technical Report CS-TR-52-6, Dept. Comput. Sci., Princeton Univ., NJ, 1986.

Hamiltonian cycles in Delaunay complexes

by Henry Crapo (*) and Jean-Paul Laumond (**)
(*) Bât 10, INRIA, B.P. 105, 78153 Le Chesnay Cedex, France
(**) LAAS / CNRS, 7, av du Colonel Roche, 31077 Toulouse Cedex, France

Abstract

We restate a conjecture concerning the existence of Hamiltonian cycles in graphs resulting from the Delaunay triangulation of point sets in the plane R^2. We introduce the notion of Delaunay complex, the natural completion of a Delaunay triangulation. We show that Delaunay complexes are necessarily 3-connected. It remains an unsolved problem to prove that Delaunay complexes have Hamiltonian cycles, or to provide a counterexample. On the other hand, using the methods of I. Rivin and W. Thurston to specify a Delaunay triangulation by its list of hyperbolic dihedral angles, we settle a related conjecture. We provide an example to show that the Hamiltonian cycles in a Delaunay complex may not generate all non-degenerate geometric realizations of Delaunay complexes. That is, there are geometric realizations of Delaunay complexes that are not convex sums of Hamiltonian cycles.

1. Delaunay complexes and triangulations

Definitions. Let T be a finite set of points in real d-dimensional space, not confined to any (d-1)-dimensional subspace. We define the *Delaunay complex* of T by specifying its faces, or cells of dimensions 0, 1, ... d. A subset $A \subseteq T$ of the set T of points is a *cell* in the Delaunay complex D(T) if and only if there is a (d-1)-sphere C such that $C \cap T = A$, and such that the sphere C does not split the set T (such that the entire set T is either in the closure of the "inside" of C, or in the closure of the "outside" of C, both such regions being closed topological balls). The *graph* G(T) of the Delaunay complex is its 1-skeleton, the incidence structure of its 0- and 1-dimensional cells. The *Delaunay sphere* S(T) of T, or *geometric realization* of the Delaunay complex, is the geometric complex whose vertices are the points in T, whose cells are points, line segments, polygonal faces, polyhedral cells, ... whose combinatorial structure is given by the Delaunay complex, and whose geometric locations are determined by the locations of the points in the set T.

Figure 1 gives an elementary example of a Delaunay sphere. This drawing was generated by a programme written in Postscript, using only the set of points as input. The numbers displayed on the edges of the complex are the hyperbolic dihedral angles (for which see below) here recorded as *percent* of 2π. The maximum permissible value on an edge is thus 50.

Proposition. The Delaunay complex of a point set T in R^d can be constructed as follows:
(1) Select a d-sphere S in R^{d+1}, tangent at its "south" pole to the given space R^d, and project the point set T onto the sphere S by inverse stereographic projection [9] with respect to its "north" pole as centre of projection.
(2) Form the convex hull H(T') of the resulting set T' of points on S.
(3) Project the boundary complex of H back down into the space R^d, by stereographic projection.

Proof: The reason this gives the Delaunay complex is as follows. Each cell of the convex polytope H(T') is exactly the intersection Q∩T' of a hyperplane Q in R^{d+1} that does not split T' (such that the entire set T' is contained in one of the two closed half-spaces associated with Q). The intersection Q∩S, for such a hyperplane, is a (d-1)-sphere that projects to a (d-1)-sphere in R^d that does not split the set T. ♦

There are two types of spheres in R^d that define faces of the Delaunay complex: those that contain all points of T in their "interior" closed ball, and those that contain no points of T in their "interior" open ball. The former, which we call spheres of *type -1*, arise from supporting hyperplanes in R^3 that separate the set T from the "north" pole; the latter, which we call spheres of *type +1*, arise from supporting hyperplanes of which one closed half space contains both the set T and the "north" pole. See **Figure 2**, a vertical section of the stereographic projection from R^3 to R^2.

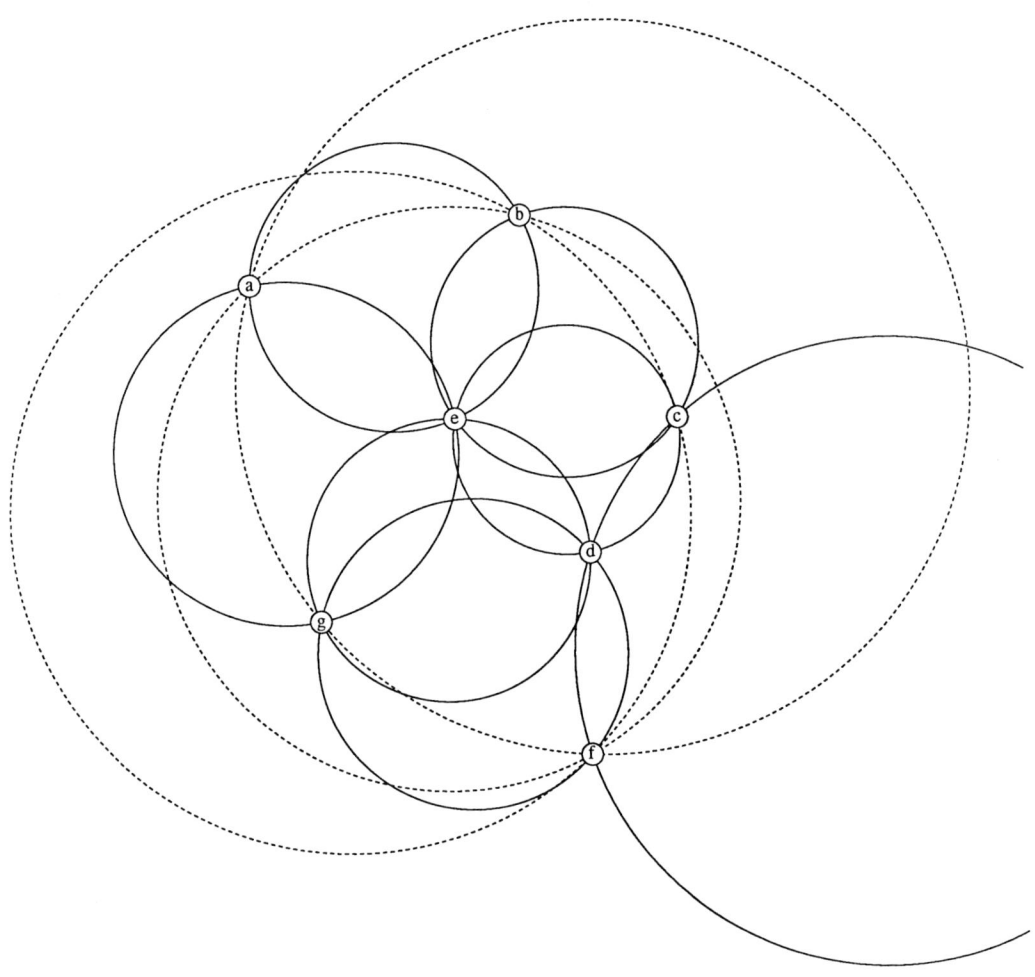

Figure 1a

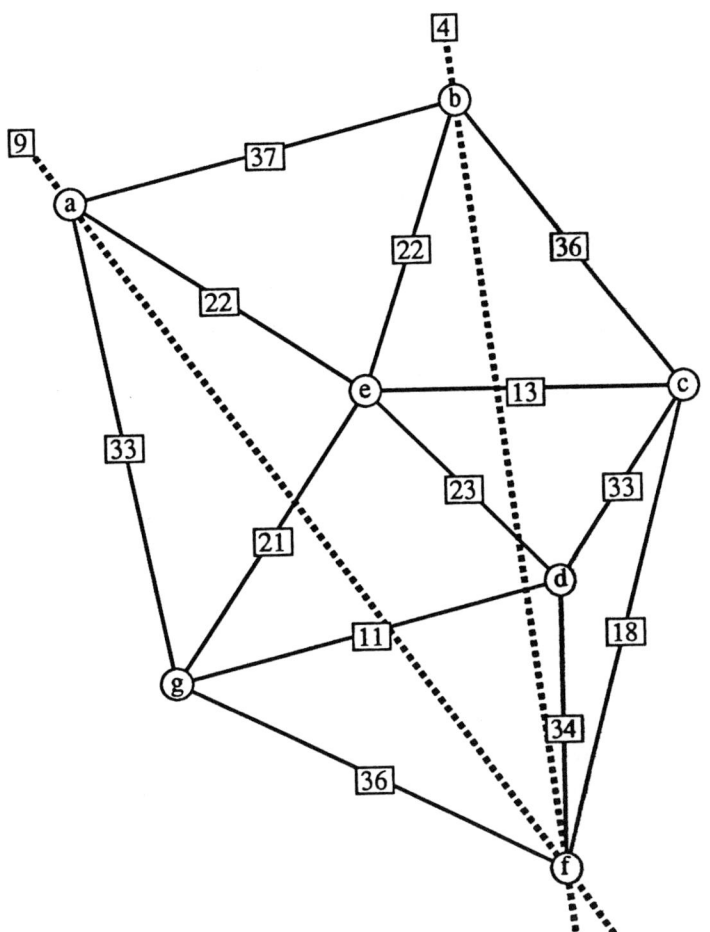

Figure 1b

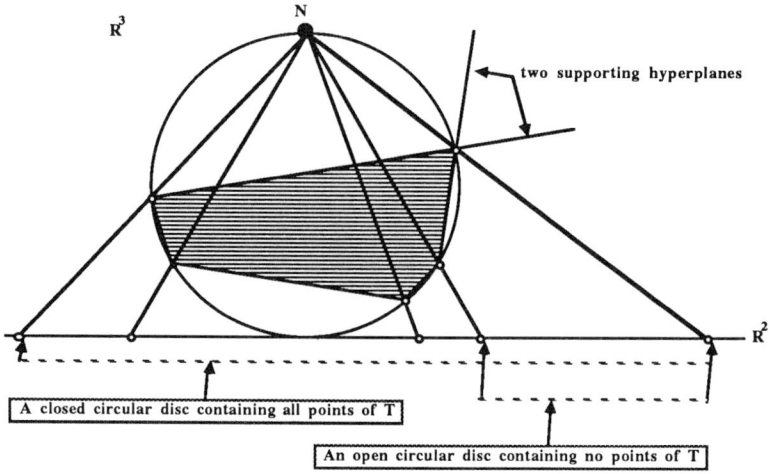

Figure 2

Definition. The "lower half" of a Delaunay complex, generated in the plane by only those circles of type +1, is called a *Delaunay triangulation*.

The Delaunay triangulation of a set T of points in the plane is topologically a disc, and for points in general position (no four being cocircular) it provides a triangulation of the convex hull of the point set T. To emphasize the distinction between Delaunay complexes and triangulations, recall that a Delaunay complex is a topological *sphere*, while a Delaunay triangulation is a topological *ball* (thus a disc, for d=2).

Take d=2. For an arbitrary set T of points in the real plane, the inverse stereographic projection T' of T is an arbitrary finite set of points on a real 2-sphere, none of these points being at the "north" pole. What may seem strange is that not all 3-connected spherical polyhedra can be obtained in this way, despite the fact that all such polyhedra have convex realizations in R^3. The fact that the point set T' is *cospherical* on the sphere S has a strong influence on the structure of the convex polytope H(T').

2. Historical comments

Several conjectures concerning Delaunay complexes and triangulations arose from work in robotics and 3-D solid modeling with J-D Boissonnat [1] in 1985. Boissonnat's aim was to construct a triangulation (and thereby to reconstruct a model of the boundary) of a 2- or 3-dimensional concave object, starting from a discrete set of points on its boundary, points obtained by laser scanning and optical detection. His method was to construct the Delaunay triangulation of the given set of points, then to "sculpture" the triangulation until all the original points lay on the surface of the resulting solid. For point sets in the real plane, Boissonnat found that it seemed to matter not at all that the points were taken on the boundary of a single connected object; even for a randomly selected set of points in the plane, his efficient, rather straight-forward algorithm always seemed to produce a triangulated concave region with no interior vertices. In 2-D, the boundary of such a region is of course a Hamiltonian cycle in the graph of vertices and edges of the Delaunay triangulation. In 3-D, it is a topological sphere that passes through all the points. What was missing in the description of this algorithm was some assurance that the process would always result in a properly sculpted solid model. It was thus in the interest of providing solid theoretical foundations for some possible modification of Boissonnat's algorithm that we posed the question: given an arbitrary finite set of points, is there always at least *some* bounding surface (in 3-D) or Hamiltonian cycle (in 2-D) that is a subcomplex of their Delaunay triangulation, and that passes through all the points?

In 1986, Crapo observed [3] that the usual Delaunay triangulation of the convex hull of a set of points should be completed to a Delaunay complex, topologically a d–sphere when the points are in d-dimensional space. In this way, he was able to show that Boissonnat's "sculpture" is precisely the better-known process of *shelling* [8] of the boundary complex of a convex polytope. It was in terms of this complex that Crapo stated (at the ACM meeting on Computational Geometry, Yorktown Heights, 1986) a number of conjectures concerning the existence of Hamiltonian cycles. William Thurston [15] had worked with Ivan Rivin, to "coordinatize" Delaunay spheres (the geometric realizations of Delaunay complexes) by listing a hyperbolic dihedral angle for each edge. Rivin's thesis [12] includes a proof that the angle-lists for Delaunay spheres are precisely the solutions of certain *linear* inequalities.

Thurston conjectured that the Hamilton cycles, which are certain of the extreme points of this set of solutions, might be sufficient to generate all Delaunay spheres, by the formation of convex sums. He soon convinced himself to the contrary, by observing that the only algorithm likely to "push" a sphere toward one of those extreme points would certainly run into topological difficulties.

In 1987, Dillencourt [4] provided an example of a Delaunay triangulation that does not possess a Hamiltonian cycle. This shows among other things that the Boissonnat algorithm, as currently stated, will not, under some circumstances, produce a satisfactory result. Dillencourt's example is, however, not three-connected, and its extension to the complete Delaunay complex has a Hamiltonian cycle.

3. Conjectures [1]

We state four conjectures concerning the existence of Hamiltonian cycles in Delaunay complexes.

Conjecture 1. (Crapo) The Delaunay complex of any finite point set in R^d has a Hamiltonian cycle.

Conjecture 2. (Crapo) The Delaunay complex D(T) of any point set T in R^d, not confined to any hyperplane in R^d, contains as a subcomplex a "Hamiltonian sphere", a topological (d–1)-sphere passing through all points in T. Furthermore, there is a shelling of the complex that produces, at some intermediate stage, a topological ball whose boundary is topologically equivalent to that (d-1)-sphere.

Conjecture 3. (Laumond) If the Delaunay triangulation of a finite point set in R^2 is 3-connected, it has a Hamiltonian cycle.

Conjecture 4. (Thurston) Every Delaunay sphere (geometric realization of a Delaunay complex) is a convex sum of trivial realizations, those corresponding to Hamiltonian cycles.

The first three conjectures remain unsettled. We give a counterexample to conjecture 4, below.

4. Connectivity and the Hamiltonian problem

Theorem. The graph G(T) of the Delaunay complex of a finite point set in R^d, not contained in any hyperplane of R^d, is (d+1)-connected.

Proof: By the above proposition, the topological structure of the graph G(T) of the Delaunay complex is inherited from that of the convex (d+1)-polytope H(T). By a theorem of Balinski, quoted in Grünbaum [8] (page 213, theorem 2), such a 1-skeleton is (d+1)-connected. Δ

---

[1] The third following conjecture has recently been disproved since the meeting of Toulouse : indeed, in a recent report (*An upper bound on the shortness exponent of incribale polytopes* Technical Report CS-TR-1868, Comp. SC. Dpt, Un. of Maryland, 1988), Dillencourt has build a Delaunay triangulation in the plane which is triconnected and non Hamiltonian. Moreover the Delaunay complexe associated with this example is non Hamiltonian; that disproves the conjecture 1 for d=2.

Even if Delaunay complexes of point sets in R^d are (d+1)-connected, the same does *not* hold for the corresponding Delaunay triangulations [5, 11] (of their convex hulls). The example of 9 points in the plane, given by Dillencourt [4], has a Delaunay triangulation that is not 3-connected. The non-Hamiltonian character of this graph is related to its lack of 3-connectivity. The Delaunay triangulation of any point set in the plane is, however, a planar graph. The problem of finding Hamiltonian cycles in a planar graph is known to be NP-complete [6], but every 4-connected planar graph is Hamiltonian [16]. Gouyou-Beauchamps [7] and Chiba [2] give algorithms that are of order $O(n^3)$, $O(n)$, respectively, for finding Hamiltonian cycles in such graphs. Laumond [10][1] analyzed the connectivity of Delaunay triangulations, and proposes a linear-time test for 3- and for 4-connectivity. He thus extracts a sufficient (and linear) condition for deciding that a Delaunay triangulation is Hamiltonian. His algorithm relies on a characterization of triplets of articulation. In the process, he shows there exist non-degenerate Delaunay triangulations that are 3-connected but not 4-connected.

Figure 3 shows the Delaunay complex and one of its hamiltonian cycles, for the example constructed by Dillencourt [4]. Because of the symmetry of this example, there is one quadrilateral face; a small displacement of one of these points would, however, produce a Delaunay sphere that is triangulated.

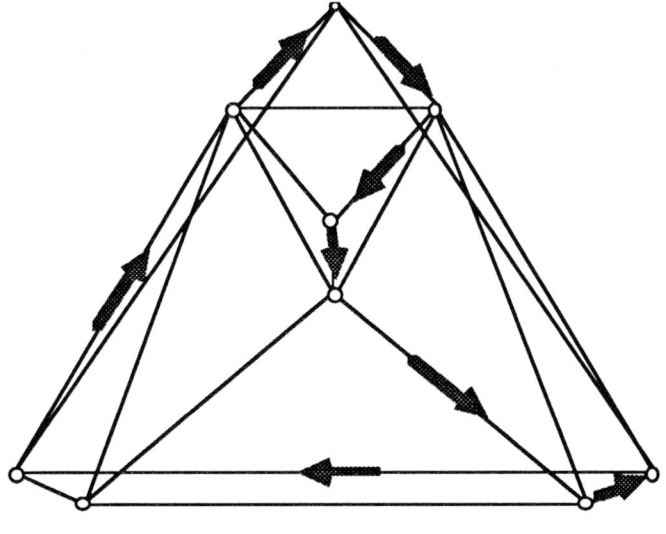

Figure 3

---

[1] A more recent version of this paper appears as *Connectivity of plane triangulations* Technical Report LAAS/CNRS November 1988. In particular, it establishes that : the articulation sets containing k vertices (k=3, 4 or 5), of a plane triangulation are circuits in the triangulation which are not faces.

For point sets in R^d, $d \geq 3$, all Delaunay complexes have graphs G(T) that are at least 4-connected. Whenever such a graph happens to be planar, it is necessarily Hamiltonian. We propose a stronger conjecture for such complexes: that there exists a (d-1)-sphere passing through all points of the complex. See conjecture 2, above.

5. The Poincaré model of a Delaunay complex

5.1 Definition.

Take d=2. Let $A' \subseteq T'$ be a facet of the convex hull H(T') of the set T' obtained by inverse stereographic projection (onto a fixed sphere S) of a point set T in the real plane. The set A' is coplanar on a plane P_A in R^3. The plane P_A meets the sphere S in a circle D_A, and there is a unique sphere S_A such that $S \cap S_A = D_A$ and such that S_A is perpendicular to the sphere S along this circle D_A. The portion of the sphere S_A interior to the sphere S is a called a *plane* in the Poincaré model of hyperbolic 3-space [9] (pages 256-259).

For each pair A, B of adjacent facets of the convex hull H(T), we can measure the hyperbolic dihedral angle (inside the sphere S_A, outside the sphere S_B, along the circle $C_{AB} = S_A \cap S_B$). This angle is easily measured as the angle between two tangents to circles in the plane tangent to the sphere S at either of the two points $D_A \cap D_B$ (which are vertices of the convex hull H(T), points in the set T'). Since stereographic projection preserves angles, this angle, which we call α_{AB}, is equally well measured as the angle between (the tangents to) the two Delaunay circles C_A, C_B in the real plane (see **Figure 4**). It is useful to carry this simplification one step further. For each Delaunay circle of *type 1* (which contains no points of the complex in its open interior) list the vertices around that face in clockwise order; for each Delaunay circle of *type -1* (which contains all points of the complex in its closed interior) list the vertices in counter-clockwise order. Then every ordered pair **ab** of points on an edge of the complex will occur as an adjacent ordered pair in exactly *one* of these lists; we call that face *the face to the right* of the oriented edge **ab**. It will also occur exactly once in the opposite order, for the face that we call *the face to the left* of the edge **ab**. For an edge **ab**, both of whose incident Delaunay circles are of type 1, the hyperbolic dihedral angle is the angle at **a**, measures counterclockwise from the centre of the circle to the right of **ab** to the centre of the circle to the left of **ab**. The same rule applies equally well to edges incident with circles of type -1, if the centre **p** of each such circle is first reflected in the point **a**.

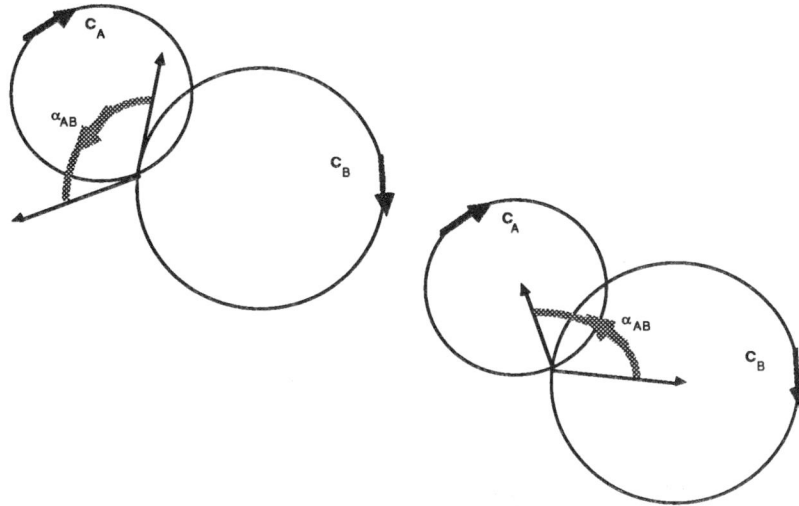

Figure 4

5.2 Rivin's characterization of Delaunay spheres

William Thurston [15] and Ivan Rivin [12] showed that the list of hyperbolic dihedral angles for convex hulls H(T') of point sets T' on a 2-sphere (that is, for hyperbolic polyhedra whose vertices lie on the "absolute" sphere of hyperbolic space) satisfy the following necessary and sufficient conditions, which we call the *Rivin inequalities*:

(1) For every pair of adjacent facets A, B,
$$0 \leq \alpha_{AB} \leq 2\pi.$$

(2) For every cycle A, B, ... E, A of facets adjacent about a vertex, the sum
$$\alpha_{AB} + \alpha_{BC} + \ldots + \alpha_{DE} + \alpha_{EA}$$
is *equal to* 2π.

(3) For every cycle A, B, ... E, A of adjacent facets, *not* a cycle about a vertex, the sum
$$\alpha_{AB} + \alpha_{BC} + \ldots + \alpha_{DE} + \alpha_{EA}$$
is *greater than* 2π.

For any point set T in the plane, let L(T) be the corresponding list
$$\{\alpha_{AB} \, ; \, A, B \text{ adjacent facets in the complex } D(T)\}$$
of hyperbolic dihedral angles.

The set of Delaunay spheres realizing a given graph G is determined (uniquely, up to an angle preserving map, or Möbius transformation, of the sphere) by the solutions L = {α_{AB},

...} of the Rivin inequalities. These solutions form a convex set **L**(G) (possibly empty) in the space $[0, \pi]^m$, where m is the number of edges in the graph G. Every list L can be expressed as a convex linear combination of extreme points of the set **L**. If all the angles in a list L lie *strictly* between 0 and π, we say the list L is a *non-degenerate* realization of the graph G.

5.3 Delaunay spheres by the simplex method

It is easy, using the Rivin inequalities, together with the simplex method of linear programming, to determine whether a given 3-connected planar graph is the 1-skeleton of some non-degenerate geometric realization of a Delaunay complex. For graphs of reasonable size, the process can even be carried out by hand.

Example. If a 3-connected planar graph G* has exactly as many faces as vertices, then the graph G obtained by placing a new vertex in the middle of every face of G*, joined to all vertices around that face, has the property that any solution to the Rivin inequalities would have to be equal to zero on all edges of the subgraph G*. A similar argument shows that any Hamiltonian cycle in G uses no edge of G*. If, on the other hand, the graph G* has more faces than vertices, the analogous argument shows that the graph G is neither Delaunay, nor Hamiltonian. This construction goes back to Steinitz [14], in his attack on Steiner's problem, "Does every combinatorial type of 3-polytopes have representatives all vertices of which belong to a sphere?" [13, 8].

Example. Consider the cube graph, with one vertex truncated, as in **Figure 5**.

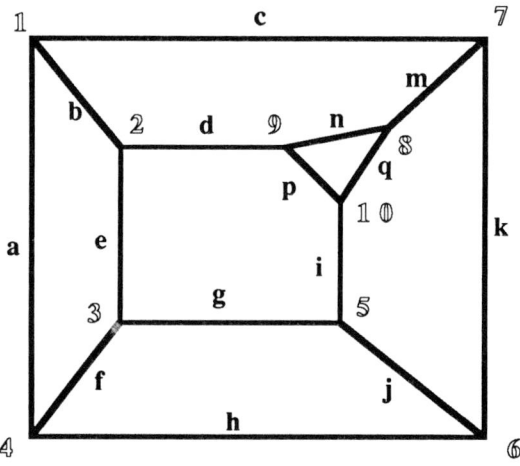

Figure 5

We compute a parametric representation of the family of solutions of the Rivin inequalities, as follows:

$$\alpha_a = s, \alpha_b = t, \alpha_e = u, \alpha_f = v.$$
$$\alpha_c = 2\pi - s - t, \text{ to satisfy condition (2) at vertex 1.}$$
$$\alpha_d = 2\pi - t - u, \text{ to satisfy condition (2) at vertex 2.}$$
$$\alpha_g = 2\pi - u - v, \text{ to satisfy condition (2) at vertex 3.}$$
$$\alpha_h = 2\pi - s - v, \text{ to satisfy condition (2) at vertex 4.}$$
$$\alpha_a i = w.$$
$$\alpha_j = u + v - w, \text{ to satisfy condition (2) at vertex 5.}$$
$$\alpha_k = s - u + w, \text{ to satisfy condition (2) at vertex 6.}$$
$$\alpha_m = u - w + t, \text{ to satisfy condition (2) at vertex 7.}$$

But now $\alpha_d + \alpha_i + \alpha_m = 2\pi - t - u + w + u + u - w + t = 2\pi$, in violation of condition (3). This (algebraic) fact that the sum is forced to 2π reflects the (geometric) fact that there is no way of placing vertices 8, 9, 10 on the sphere. In fact, if 7 vertices of a projective cube are cospherical, the 8th vertex lies also on that sphere. The lines (2, 9), (5, 10), (7, 8) are defined in space as the intersections of known face planes, and those lines meet in a single point that lies on the sphere. No other points along those three lines lie on the sphere, so there is no place on the sphere to install the three distinct vertices 8, 9, 10.

5.4 Convex sums of Hamiltonian cycles

Each Hamiltonian cycle in a Delaunay complex yields a particularly degenerate model of a Delaunay sphere. For each pair A, B of faces adjacent across an edge in the Hamiltonian cycle, assign the hyperbolic dihedral angle $\alpha_{AB} = \pi$, and for all other pairs A, B of adjacent faces, assign $\alpha_{AB} = 0$. We verify easily that these values satisfy the Rivin inequalities, with "≥" rather than "≥" in inequality 3. The corresponding "polyhedron" is flattened to a disc, with all its vertices lying along a single circle on the sphere S. The corresponding list L is an extreme point of the convex set L(G). As explained in the introduction, we once thought that all non-degenerate Delaunay spheres with a given graph might be convex sums of degenerate realizations associated with Hamiltonian cycles. The following example shows that this is not so.

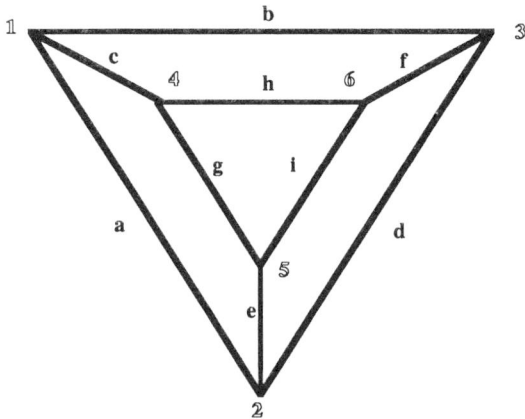

Figure 6

Example. Let G be the graph of the triangular prism, as shown in **Figure 6**, and let L be the list with assignments $.8\pi$ on the edges of the two triangular faces, $.6\pi$ on the remaining 3 edges. This list L satisfies the Rivin inequalities. There are three distinct Hamiltonian cycles H_i in the graph G; their lists are given below:

edge	a	b	c	d	e	f	g	h	i
L	$.6\pi$	$.6\pi$	$.8\pi$	$.6\pi$	$.8\pi$	$.8\pi$	$.6\pi$	$.6\pi$	$.6\pi$
H_1	0	π	π	π	π	0	0	π	π
H_2	π	0	π	π	0	π	π	0	π
H_3	π	π	0	0	π	π	π	π	0

Any convex sum of the lists H_i will yield the *same* values for "opposite" edges a, f, whereas these sums are supposed to be .6, .8, respectively.

With the aid of a micro-computer program furnished by Komei Fukuda, we have computed the polytope of all angle-lists of Delaunay spheres with this graph. It is a tetrahedron. Three of its extreme points are the lists H_i for the three Hamiltonian cycles. The remaining extreme point is

edge	a	b	c	d	e	f	g	h	i
F	$.5\pi$	$.5\pi$	π	$.5\pi$	π	π	$.5\pi$	$.5\pi$	$.5\pi$

So the three Hamiltonian cycles generate a plane of solutions, all of which possess the symmetry $\alpha_a = \alpha_f = \alpha_g$, $\alpha_b = \alpha_e = \alpha_h$, $\alpha_c = \alpha_d = \alpha_i$. Any solution of this plane requires a non-zero coefficient for F in its (unique) expression as a convex sum of extreme points.

Bibliography

[1] Jean-Daniel Boissonnat, *Geometric Structures for 3-dimensional Shape Representation*, **ACM Transactions on Graphics 3**, no. 4, 1984.

[2] Norishige Chiba and Takao Nishizeki, *The Hamiltonian Cycle Problem is Linear-time solvable for 4-connected Planar Graphs*, **Proceedings of ISCAS 85**, 961-964.

[3] H. Crapo, *Delaunay triangulations,* with Jean-Daniel Boissonnat, (extended abstract) Séminaire du Centre de Mathématique, Ecole Polytechnique, Palaiseau, January 1987.

[4] Michael B. Dillencourt, *A non-Hamiltonian, non-degenerate Delaunay Triangulation,* **Information Processing Letters 25** (1987), 149-151.

[5] H. Edelsbrunner, **Algorithms in Combinatorial Geometry**, Springer-Verlag, 1987.

[6] M.R. Garey, D.S. Johnson and R.E. Tarjan, *The Planar Hamiltonian Circuit Problem is NP-complete*, **SIAM J. of Computing 5**, (1976), 704-714.

[7] D. Gouyou-Beauchamps, *The Hamiltonian Circuit Problem is Polynomial for 4-connected Planar Graphs,* **SIAM J. of Computing 11** (1982), 529-539.

[8] Branko Grünbaum, **Convex Polytopes**, John Wiley & Sons, Inc., 1967.

[9] David Hilbert and S. Cohn-Vossen, **Geometry and the Imagination,** Chelsea Publ. Co., New York, 1952.

[10] Jean-Paul Laumond, *Connectivity and the Hamiltonian Circuit Problem for Delaunay Triangulations*, LAAS (CNRS) internal report #87414, Toulouse, December 1987.

[11] F.P. Preparata and M.I. Shamos, **Computational Geometry, an Introduction**, Springer-Verlag, 1985.

[12] Ivor Rivin, *On the Geometry of Convex Polyhedra in Hyperbolic 3-space*, Ph.D. thesis, Princeton University, June 1986, supervised by W. Thurston.

[13] J. Steiner, **Gesammelte Werke** (2 volumes), Berlin 1881, 1882.

[14] E. Steinitz, *Über isoperimetrische Probleme bei konvexen Polyedern*, **J. reine angew. Math.** (Crelle), **158** (1927), 129-153, **159** (1928), 133-143.

[15] William Thurston, private communication, June 1985.

[16] W. T. Tutte, *A Theorem on Planar Graphs,* **Trans. Amer. Math. Soc. 82** (1956), 99-116.

Models of Robot Manipulators

B.GORLA, M.RENAUD
Groupe Robotique et Intelligence Artificielle
LAAS-CNRS
7 Avenue du Colonel Roche, 31077 Toulouse Cedex, France

1 A model for a robot manipulator

A robot manipulator is an active system featuring links and joints organized around either a simple chain structure (see Fig. 1a), a tree chain structure (see Fig. 1b) or a complex chain structure (see Fig. 1c) that supports the end effector required to accomplish a task.

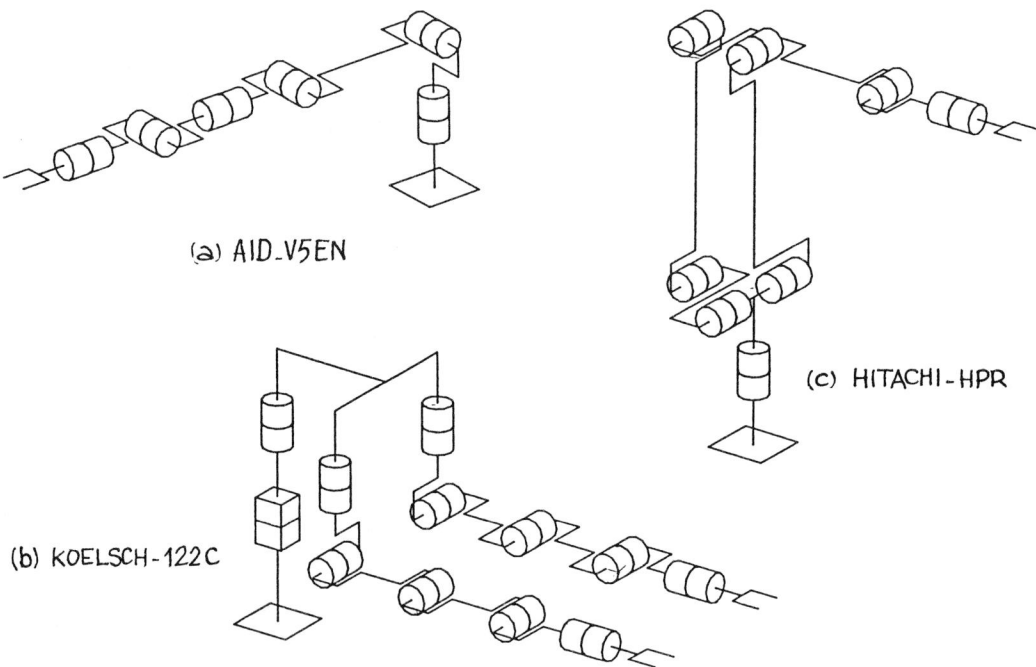

Fig.1 The three types of structure of a robot manipulator

If the links that form the robot manipulator are viewed as rigid solids and if the joints used are n and with one degree of freedom, that is, if they permit elementary translational or rotational motion, then the configuration, i.e., geometry of the mechanical structure, is defined by n parameters referred to as generalized cooordinates, components of a column matrix q of dimension nx1 belonging to the generalized workspace D_q, of dimension n, contained in the generalized space R^n:

$$q = (q_1 q_2 ... q_n)^t$$

Changes in the robot manipulator configuration bring about a new location involving a different position and orientation of the robot manipulator's end effector. As shown in the following, this location is painstainkingly defined with the aid of m parameters referred to as operational coordinates, components of a column matrix of dimension mx1 belonging to the operational workspace D_x, of dimension m, contained in the operational space R^m:

$$x = (x_1 x_2 ... x_m)^t$$

The objective of modeling is to establish the relations that allow mapping from the generalized space R^n to the operational space R^m (direct models) or in the opposite direction (inverse models) in order to achieve robot manipulator control with the help of a numerical computer.

In fact, there are five models, viz:

- the direct geometric model allowing computation of the end effector location x, as a function of the robot manipulator configuration q:

$$x=f(q)$$

- the inverse geometric models allowing computation of the configurations q as a function of the location x, because there are several inverse geometric models since f is non linear:

$$q=g(x)$$

- the direct differential model allowing computation of the differential of location dx as a function of the differential of configuration dq. It involves, explicitly or implicitly, the Jacobian matrix of the robot manipulator J(q,x) of dimension mxn:

$$dx = J(q,x)dq$$

- the inverse differential model allowing computation of the differential of configuration dq, from a given configuration, as a function of the differential of location dx. It involves, explicitly or implicitly, a generalized inverse G(q,x) of dimension nxm of the matrix J(q,x):

$$dq = G(q,x)dx$$

- the dynamic model allowing computation of the generalized forces exerted by the actuators on the joints as a function of the generalized coordinates, velocities and accelerations they produce:

$$\Gamma = \Gamma(q, \dot{q}, \ddot{q})$$

with:

$$\Gamma = (\Gamma^1 \Gamma^2 ... \Gamma^n)^t$$

Today, methods for obtaining these models are efficient; most of them deal with simple or tree chain structure robot manipulators. In this case, they are quasi-systematic except for those involving inverse geometric models and, in order to achieve real time control, they rely on computations involving a quasi-minimal number of arithmetic operations.

While these results may be adequate for industrial robot manipulators, some difficult mathematical problems remain unsolved.

2 Advantage of the notion of minimal computation

The real time constraint imposed by the control system of a robot manipulator is such that numeric computation of the models must be minimal (in the order of a few milliseconds). This supposes that a minimal number of elementary arithmetical operations (+ , - , *, /) be involved implicitly or explicitly in the computation of certain functions (e.g., cosine and sine trigonometrical functions or square root function). Unfortunately, this issue is difficult to address. This is evidenced by the computation of the n-th power of a given real number for which no optimal algorithm exists [REN84]. For example, the binary algorithm which, for n=15, requires a series of six successive multiplications is not optimal and can be refined, providing computation is effected with the aid of the following five successive multiplications:

$$x^2 = x * x, x^3 = x^2 * x, x^6 = x^3 * x^3, x^{12} = x^6 * x^6, x^{15} = x^{12} * x^3$$

Similarly, the matrix product, frequently employed in modeling, is always effected in a classical manner, whereas STRASSEN's method permits minimization of the number of multiplications involved in this product. However this last method is achieved at the cost of additions whose total number is increased. Thus, the product C of two matrices A and B of order 2 can be effected with the help of 7 multiplications and 18 additions in lieu of the usual 8 multiplications and 4 additions:

$$M_1 = (A_{12} - A_{22}) * (B_{21} + B_{22})$$
$$M_2 = (A_{11} + A_{22}) * (B_{11} + B_{22})$$
$$M_3 = (A_{11} - A_{21}) * (B_{11} + B_{12})$$
$$M_4 = (A_{11} + A_{12}) * B_{22}$$
$$M_5 = A_{11} * (B_{12} - B_{22})$$
$$M_6 = A_{22} * (B_{21} - B_{11})$$
$$M_7 = (A_{21} + A_{22}) * B_{11}$$
$$C_{11} = M_1 + M_2 - M_4 + M_6$$
$$C_{12} = M_4 + M_5$$
$$C_{21} = M_6 + M_7$$
$$C_{22} = M_2 - M_3 + M_5 - M_7$$

Notice, however, that STRASSEN's method may not be so interesting from a computational time point-of-view. Indeed, it all depends upon the relative time required to compute each elementary arithmetical operation. The use of slice processors seems to bring execution time for multiplications closer to that of additions, disqualifying thus STRASSEN's method.

3   Advantage of the notion of intrinsic computation

The final form of the models obtained utilizes only the scalars which are components of vectors, tensors, quaternions, etc... in specific reference frames. These scalars are extrinsic mathematical entities lacking the generality of the intrinsic mathematical entities they represent. It is therefore clumsy to compute models directly with these extrinsic entities and it is more judicious to compute them first with the initial intrinsic entities. Then, the particular reference frame in which the components take on their simplest form must be selected.

This approach may be illustrated by the extraction of a system of equations which is carried out while computing either the inverse geometric models or the Jacobian matrix involved in the direct differential model of a robot manipulator with a simple chain structure. In both cases, the extrinsic computation requires that, following the intrinsic computation, a "preferential" frame related to the link situated "about the middle " of the structure be selected as particular reference frame.

4   Criterion for selecting a formalism

Computing a model involves a particular formalism that must permit systematic computation for a large class of robot manipulators.

Thus, the first four models that deal with the robot manipulator's geometry can be obtained through use of:

- homogeneous transformation matrix formalism [DEN55], [PAU81], [GOR84], [CRA 86], [KHA 86],

- hypercomplex dual quaternion formalism [CAS86],

- etc...

and the last model regarding dynamics may be obtained with the aid of :

- NEWTON-EULER's formalism [BRO 73],

- LAGRANGE's formalism [LUR 68],

- LIE's algebra and group formalism [ARN78],

- etc...

The objective comparison of the efficiency of the different formalisms requires to take into consideration the number of operations involved in the numeric computation of the models, from the same set of data.

5 Advantage of the notion of analytic computation

The model must be expressed in an analytic or literal form before the numeric calculation is effected. In this way, unnecessary operations, e.g., additions or subtractions with zero, subtraction or division of two equal quantities, multiplication by zero or one, division by one, etc... can be avoided. This is particularly interesting when calculating matrix products.

6 Advantage of the notion of iterative computation

While effecting the computation, any new quantity involving at least one multiplication, division, addition or subtraction must be given a name so that it can then be used under this name. By accomplishing this, auxiliary equations are generated, leading to an iterative computation of the model that avoids the expanded model whose complexity exponentially increases as a function of the number n of joints, as in the case of a robot manipulator with a simple chain structure.

For example, it is more judicious to compute:

$$y_1 = x_1 * x_2$$

$$y_2 = y_1 * x_3$$

than to compute:

$$y_1 = x_1 * x_2$$

$$y_2 = x_1 * x_2 * x_3$$

7 Difficulties arising from the computation of the geometric models

7.1 Definition of orientation

Whatever the type of robot manipulator envisaged, it is always possible to consider only those prismatic and revolute joints that make up a complete group for the set of mechanical joints.

Then, according to a procedure developed by DENAVIT and HARTENBERG [DEN55] and modified by KHALIL and KLEINFINGER [KHA86], a direct orthogonal frame can be attached to each link that forms the structure (some of these links may be virtual on account of the preceding remark).

This systematic procedure conveniently allows introduction, for each link, of two shape parameters and two joint parameters, one of which, that corresponds to either a translation for a prismatic joint or a rotation for a revolute joint, constitutes one of the n generalized coordinates. Thus, the definition of the robot manipulator's configuration is unambiguous. Such is not the case, however, of the definition of the end effector location; this location is theoretically determined with the aid of six independent parameters; three of which correspond, for example, to the position of a reference point of the end effector and three to the orientation of a frame attached to this effector, relative to a fixed frame attached to the base. Interestingly, it may be noticed, before anything else, that it is relatively easy for the designer to evaluate rather accurately the position of the reference point but sensing orientation is much more difficult. This is why the choice of orientation parameters still remains such a difficult task. This is aggravated by the fact that singularities and redundancy cannot be avoided at the same time.

Generally, all systems of parameters for the definition of orientation can be assimilated to one of the three following classes:

- the three classical EULER angles, BRYANT angles, etc... for which a singularity inevitably exists whenever two rotations become linearly dependent,

- the nine direction cosines which define orientation in a one-to-one way but for which the six orthogonality relations introduce a major redundancy (in practice, these relationships give rise to the difficult question of maintaining the orthogonality of the direction cosine matrix due to rounded off errors that appear in the numeric computations),

- the four EULER or RODRIGUES or RODRIGUES-HAMILTON parameters which constitute a reasonable tradeoff. Indeed, a redundancy relationship allows avoidance of any singularity (note, however, that these parameters are devoid of any physical significance that could be directly interpreted by the designer).

These are not the only difficulties that particularize orientation relative to position. A major difference lies in the "periodic" nature of orientation. A difference in the angle of rotation of $2k\pi$ about any axis, k being any relative integer, does not bring about any change in the end effector location before and after rotation is completed. Hence, the classical notion of transformation matrix cannot mathematically account for this difference in rotation. It may therefore be necessary to take into account the number of end effector twists, as in the case of screwing. Mathematical concepts are then lacking to tackle this problem.

Finally, orientation which is not easily sensed by the designer is also difficult to measure in practice because of the absence of location sensors, at least for large position and orientation variations. However, for reasons of accuracy, these sensors would be needed to integrate the sensory system to the control system (to take account of clearances, distortions etc... of the mechanical structure) and to simplify the geometric reasoning in the operational space R^m.

7.2 The inverse geometric models of a robot manipulator with a simple chain structure

The computation of the direct geometric model of a robot manipulator with a simple chain structure is a systematic operation as far as the four phases that lead to the result are concerned.

These phases are:

- introduction of the frames R_i attached to the robot manipulator links,

- extraction of the modified DENAVIT-HARTENBERG parameters $a_{i-1}, \alpha_{i-1}, r_i, \theta_i$,

- computation of the homogeneous transformation matrix T_{on} of the robot manipulator,

- determination of the operational coordinates x as a function of the components t_{ij} of the preceding matrix.

Conversely, the analytic computation of the inverse geometric models has not been addressed yet in all its generality because in principle it consists of inverting the direct model which is a non-linear function: the theorems of existence and unicity are then lacking.

However, the next two computational phases are systematic:

- evaluation of the components of the homogeneous transformation matrix T_{on} of the robot manipulator as a function of the operational coordinates x.

- extraction of nine dependent scalar equations from the identification of the "preferential" homogeneous transformation matrices T_{pn} (see Section 3) computed either as a function of $q_1, q_2, ..., q_p$ for one of them, or as a function of $q_n, q_{n-1}, ..., q_{p+1}$ and of the operational coordinates x, for the other.

The third phase that allows effective computation of the generalized coordinates q must be done manually. But if the structure is fairly complex, that is, if an excessive number of modified DENAVIT-HARTENBERG parameters are not simple, computation may fail, as occurred in the case of the R.M.S. robot manipulator carried on board the American space Shuttle.

In this case, iterative numeric methods, that cannot be used in a real time context because of the long computational delays involved, are utilized. These allow to address a general issue, resulting in the determination of one configuration only, that is, one inverse geometric model, although several configurations exist in principle (of course, the configuration obtained depends on the initial evaluation used in the iterative numeric method).

A few theoretical and partial results have been demonstrated by ROTH [ROTH 76] who introduced the important notion of solvability. This concerns the possibility for a system of nine equations with n unknowns to be simplified to a single equation of the form $F(cos\theta, sin\theta) = 0$ where F is a polynomial with two variables, usually of high degree. This operation remains highly complex and may even fail to produce a solution. If it is successful, an equation of the form $G(cos\theta) = 0$ or $H(sin\theta) = 0$ may finally be used. Determining only the number of solutions would be a major achievement. It seems, a priori, that ROUTH's criterion or any equivalent, can be used for this purpose, but in use, computation cannot be effected because of the complexity involved. If, conversely, an iterative numeric method is utilized, it is necessary to arrive at a computation algorithm which yields the number of solutions through variation of the initial condition. However, this seems impractical without dramatically increasing the computational burden.

7.3 Direct geometric model of certain robot manipulators with a complex chain structure

The STEWART's platform [FIC86], depicted in Fig. 2, constitutes a system with six degrees of freedom which is frequently used in robotics.

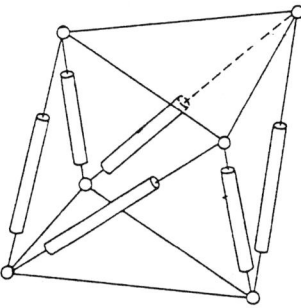

Fig . 2 STEWART's platform

However, the analytic computation of its direct geometric model remains unknown to this day, whereas the analytic computation of its inverse model does not pose unsurmountable obstacles. It is therefore reasonable to think that a mechanical duality may exist between a simple chain structure and a complex one.

As previously noted, computation of the direct model reduces to that of the roots of a complex polynomial equation of the form $F(cos\theta, sin\theta) = 0$, which, unfortunately, is obtained in a non-systematic manner.

8 Generalized workspace D_q and operational workspace D_x

8.1 Definitions

Generally, the variation interval of a generalized coordinate q_i, is bounded either for mechanical reasons, or as a result of the presence of position sensors that can deliver only a finite information. Then q_i belongs to the closed segment $[q_{im}, q_{iM}]$.

Thus, the generalized workspace D_q of a robot manipulator is the hyperparallelopiped of the generalized space R^n, cartesian product of the preceding intervals:

$$D_q = \prod_{i=1}^n [q_{im}, q_{iM}]$$

and the operational workspace D_x is the image by the function f -direct geometric model of the robot manipulator- of the generalized workspace D_q:

$$D_x = f(D_q)$$

The operational workspace D_x, usually of dimension six, cannot be visualized. This is why the representation is limited to that of the operational workspace in position D_{x_p} at each point of which one may envisage to associate an index representing the robot manipulator's capability to orient the end effector locally. Unfortunately, this index is not easy to define, this is why some authors define the notions of primary workspace -locus of the points of D_{x_p} without orientation limitation- and secondary workspace - complementary to the preceding one relative to D_{x_p}- [KUM 81] [ROT 76] and compute, if required, the ratio of the volume of the primary workspace over the volume of the workspace D_{x_p} [GUP86]. In fact, the value of this index is not very significant because it is often small. Other authors compute, for a given position of the wrist "centre ", the ratio of the surface swept out by a reference point of the end effector (with respect to the motion allowed by wrist joints) to the surface of the sphere of same radius. Unfortunately, this index does not enable us to take into account the rotation provided by a last revolute joint whose axis is superposed with a radius of the reference sphere [HAN 83]. Other authors have utilized the notion of solid angle which entails the same limitations but which may allow for a generalization.

Nevertheless, the problems posed by the generation and the computation of the volume of the operational workspace in position D_{x_p} must first be addressed in the general case of robot manipulators with n revolute and/or prismatic joints and limited motions. Given the complex form of the workspace D_{x_p}, its analytic generation seems extremely difficult to achieve. So various numeric computation algorithms have already been proposed. They are fairly costly in terms of computational time or memory use and usually concern robot manipulators using n revolute joints permitting complete 360^0 rotation or robot manipulators reduced to a basic simple chain structure consisting of the first three joints. Studies leading to the topologic and geometric characterization of the operational workspace in position D_{x_p}, and to the definition of analytic computational methods for the volume of this domain would be extremely useful in pratice for the geometric design and the optimal implementation of the robot manipulator's mechanical structures.

8.2 Notions of aspect

In this respect, the notion of aspect proposed by BORREL [BOR 86] is of paramount importance.

An aspect A_i is primarily a connected open set of points of the generalized workspace D_q. It is such that each minor of order m of the Jacobian matrix J of the robot manipulator is nonzero in any point of the aspect, except if it is zero everywhere (the m operational coordinates being supposed independent).

If a configuration q^0 belonging to an aspect A_i is then considered and if it corresponds to a location x^0, the theorem of implicit functions allows to state that there exists locally a single solution q, close to q^0, that corresponds to a location x, close to x^0, by assigning arbitrary values to (n-m) generalized coordinates close to those analogous to q^0.

This result holds whatever the point q^0 in the aspect A_i. Hence, the notion of aspect allows extension of the preceding neighbourhood of q^0 to the whole aspect A_i. It may therefore be stated that the number of configurations which correspond to an imposed location, by assigning arbitrary values to (n-m) generalized coordinates, is at the most equal to the number of aspects.

This method for determining the number of inverse geometric models can be applied to specific robot manipulators only. Determining the number of aspects for any robot manipulator would lead to the establishment of an upper bound for the number of its configurations corresponding to a given location of its end effector.

These results are illustrated in Appendix for a robot manipulator with three revolute joints.

In a first case, the mechanical structure is non redundant and is used to position the reference point O_4 and to orient the end effector simultaneously. The Jacobian matrix is of order 3 and one finds that:

$$det J = a_1 a_2 S_2$$

The singularity surface $S_2 = 0$ defines two aspects in the generalized workspace D_q.

In the second case, the structure is redundant since only the position of the reference point O_4 is considered. The three minors of the Jacobian matrix of dimension 2x3 are then as follows:

$$m_1 = a_1(a_2 S_2 + a_3 S_{2+3}), \; m_2 = a_3(a_2 S_3 + a_1 S_{2+3}), \; m_3 = a_2 a_3 S_3$$

The singularity surfaces $m_1 = 0$, $m_2 = 0$ and $m_3 = 0$ define six aspects in the generalized workspace D_q.

Another application of the notion of aspect concerns the generation of the operational workspace D_x. Indeed, it can be shown that the image by f of the boundary ∂A_i of an aspect A_i constitutes the boundary of the image by f of aspect A_i. This allows to envisage the analytic characterization of the closure $\overline{f(A_i)}$ of image $f(A_i)$ of one aspect. The study of the intersections of the closures of the images of the different aspects could then lead to the safe determination of the number of configurations which correspond to a particular location of the end effector.

Finally note that if the generalized workspace D_q consists of p aspects $A_1, A_2, ..., A_p$ then:

$$D_q = \cup_{i=1}^{p} \overline{A_i}$$

By denoting $f_p(\overline{A_i})$ the projection of $f(\overline{A_i})$ into the operational workspace in position D_{x_p}, it is shown that:

$$D_x = \cup_{i=1}^{p} f(\overline{A_i})$$

$$D_{x_p} = \cup_{i=1}^{p} f_p(\overline{A_i})$$

Figs. A.4 in Appendix show the results obtained in the two cases studied.

8.3 Computation of the volume of the operational workspace in position D_{x_p}

The volume of the operational workspace in position D_{x_p} is a global characterization that allows for an interesting classification of robot manipulators. The numeric methods presented in [BOR 86] permit computation in most cases but no analytical method is operational at this stage. This is because the dimension of the generalized workspace D_q, equal to n, does not coincide with that of the operational workspace in position D_{x_p}. It would therefore be interesting to compute first the volume of the image of an aspect $f_p(A_i)$ from the knowledge of topology and geometry of the aspect A_i. This could then serve as a criterion for choosing the inverse geometric model that allows placement in the image of the largest volume and avoidance of too frequent reconfigurations. In addition, the sum of the volumes of the images of the aspects which is an upper bound of the volume of operational workspace in position D_{x_p}, would give an insight into the robot manipulator's positioning capability .

8.4 Bibliography

[ARN 78] V.I.ARNOLD. Mathematical methods of Classical Mechanics. SPRINGER-VERLAG. New-York. Heidelberg. Berlin. 1978.

[BOR 86] P.BORREL. Contribution à la modélisation géométrique des robots manipulateurs. Application à la conception assistée par ordinateur. Thèse de Doctorat d'Etat.Université de Montpellier. July 1986.

[BRO 73] P.BROUSSE. Cours de mécanique. A. COLIN Publishers. Paris. 1973.

[CAS 86] J.M.CASTELAIN. Application de la méthode hypercomplexe aux modélisations géométrique et différentielle des robots constitués d'une chaîne cinématique simple. Thèse de Doctorat d'Etat. Université de Valenciennes et du Hainaut-Cambrésis. December 1986.

[CRA 86] J.CRAIG. Introduction to Robotics. ADDISON-WESLEY Publishing Co. 1986.

[DEN 55] J.DENAVIT and R.S.HARTENBERG. A kinematic notation for lower-pair mechanisms based on matrices. ASME. Journal of Applied Mechanics. Vol 17. June 1955.

[FIC 86] E.F.FICHTER. A STEWART platform-based manipulator: general theory and practical construction. Int. Jour. of Robotics Research. Vol. 5. No. 2. Summer 1986.

[GOR 84] B.GORLA and M.RENAUD. Modèles des robots manipulateurs. Application à leur commande. CEPADUES Publishers. Toulouse. May 1984.

[GUP 86] K.C. GUPTA. On the nature of the robot workspace. Int. Jour. of Robotics Reasearch. Vol. 5. No. 2. Summer 1986.

[HAN 83] J.A.HANSEN and K.C.GUPTA. Generation and evaluation of the workspace of a manipulator. 6th. IFToMM Congress on Machines and Mechanisms. New-Delhi. December 1983.

[IRI 86] J.IRIGOYEN EIZMENDI. Commande en position et force d'un robot manipulateur d'assemblage. Thèse de l'Université P.SABATIER. Toulouse. October 1986.

[KHA 86] W.KHALIL and J.F.KLEINFINGER. A new geometric notation for open and closed loop robots. IEEE Int. Conf. on Robotics and Automation. San Francisco. 1986.

[KUM 81] A.KUMAR and K.J. WALDRON. The dextrous workspace. ASME paper No. 80-DET-108. 1981.

[LUR 68] L.LUR'E. Mécanique Analytique. Tomes 1 and 2. MASSON Publishers. Paris. 1968.

[PAU 81] R.PAUL. Robot manipulators: mathematics, programming and control. MIT Press. Cambridge, Massachussets and London, England. 1981.

[REN 84] M.RENAUD. Calcul du modèle géométrique des robots manipulateurs en utilisant un nombre quasi-minimal d'opérations arithmétiques. Journée d'étude AFCET: La réduction des modèles. Chatou(France). May 1984.

[ROT 79] B.ROTH. Performance evaluation of manipulator from a kinematic viewpoint. Cours de Robotique IFTIM. September 1976.

Appendix

The robot manipulator depicted in Fig. A.1(a) possesses three revolute joints with parallel axes and its end effector moves parallel to the plane $(O_0, \underline{x}_0, \underline{y}_0)$. The latter's position defined for example by the cartesian coordinates X and Y of point O_4 and its orientation defined, for example, by angle θ, measured algebraically between $\underline{x}_0$ and $\underline{x}_3$, can vary independently of each other.

If we consider simultaneously the position and orientation of the end effector, that is if the location is defined by X, Y and θ, the robot manipulator is said to be non redundant: the variations of the three generalized coordinates generate those of an equal number of operational coordinates (see fig. A.1 (b)). A finite number of configurations (usually two) permits placement of the end effector in a given location.

If we consider only the position of the end effector, that is if the location is defined by X and Y, the robot manipulator is said to be redundant: the variations of the three generalized coordinates generate those of the two operational coordinates (see Fig. A. 1(c)). An infinity of configurations allow placement of the end effector in a given location.

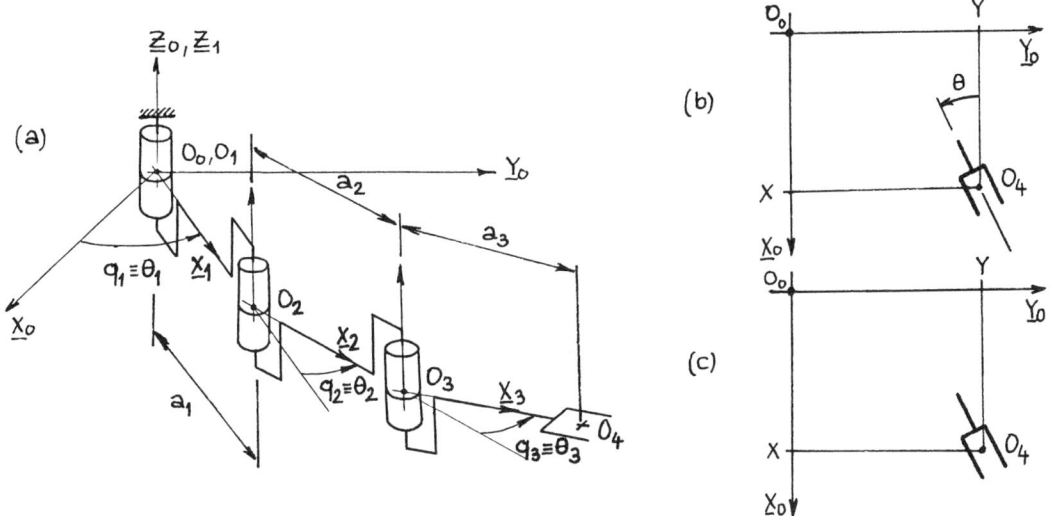

Fig. A.1 R.R.R. Robot manipulator

1 Direct geometric model

1.1 Step 1

Introduction of the reference frames R_o thru R_3 attached to the links of the robot manipulator.

See Fig. A.1

1.2 Step 2

Extraction of the modified DENAVIT-HARTENBERG parameters

	1	2	3
α_{i-1}	0	0	0
a_{i-1}	0	a_1	a_2
θ_i	q_1	q_2	q_3
r_i	0	0	0

Table A.2 Modified DENAVIT-HARTENBERG parameters

1.3 Step 3

Computation of the elementary homogeneous matrices $T_{i-1,i}$

$$T_{01} = \begin{pmatrix} C_1 & -S_1 & 0 & 0 \\ S_1 & C_1 & 0 & 0 \\ 0 & 0 & 1 & 0 \\ 0 & 0 & 0 & 1 \end{pmatrix}$$

$$T_{12} = \begin{pmatrix} C_2 & -S_2 & 0 & a_1 \\ S_2 & C_2 & 0 & 0 \\ 0 & 0 & 1 & 0 \\ 0 & 0 & 0 & 1 \end{pmatrix}$$

$$T_{23} = \begin{pmatrix} C_3 & -S_3 & 0 & a_2 \\ S_3 & C_3 & 0 & 0 \\ 0 & 0 & 1 & 0 \\ 0 & 0 & 0 & 1 \end{pmatrix}$$

1.4 Step 4

$$C_{23} = C_2 C_3 - S_2 S_3$$

$$S_{23} = S_2 C_3 + C_2 S_3$$

$$D_1 = C_2 a_2 + a_1$$

$$D_2 = S_2 a_2$$

$$t_{11} = C_1 C_{23} - S1 S_{23}$$

$$t_{14} = C_1 D_1 - S1 D_2$$

$$t_{21} = S1 C_{23} + C_1 S_{23}$$

$$t_{24} = S1 D_1 + C_1 D_2$$

1.5 Step 5

Determination of the operational coordinates x as a function of the elements t_{ij} of the homogeneous transformation matrix T_{03}

FIRST CASE: Non-redundant robot manipulator

$$x = (XY\theta)^t$$

X,Y are the cartesian coordinates of point O_4 with respect to frame R_o

θ is the algebraic angle $(\underline{x}_o, \underline{x}_3)$

$$X = t_{14} + a_3 t_{11}$$

$$Y = t_{24} + a_3 t_{21}$$

$$\theta = \text{Atan2}(t_{21}, t_{11})$$

SECOND CASE: Redundant manipulator

$$x = (XY)^t$$

$$X = t_{14}$$

$$Y = t_{24}$$

In this case, it is not necessary to evaluate t_{11} and t_{21} to compute the direct geometric model only.

2 Inverse geometric models

2.1 Step 1

Evaluation of the elements of the homogeneous transformation matrix T_{on} as a function of the imposed operational coordinates x.

FIRST CASE:

$$t_{11} = C\theta$$

$$t_{21} = S\theta$$

$$t_{14} = X - a_3 t_{11}$$

$$t_{24} = Y - a_3 t_{21}$$

SECOND CASE:

$$t_{14} = X - a_3 t_{11}$$

$$t_{24} = Y - a_3 t_{21}$$

with any t_{11} and t_{21} such that $(t_{11})^2 + (t_{21})^2 = 1$

2.2 Step 2

Extraction of the scalar equations derived from the identification of the "preferential" homogeneous transformation matrices :

FIRST CASE

$$C_1 t_{11} + S_1 t_{21} = C(2+3)$$

$$-S_1 t_{11} + C_1 t_{21} = S(2+3)$$

$$C_1 t_{14} + S_1 t_{24} = a_2 C_2 + a_1$$

$$-S_1 t_{14} + C_1 t_{24} = a_2 S_2$$

which yields:

$$F1 = (t_{14})^2 + (t_{24})^2$$

$$C_2 = (F1 - (a_1^2 + a_2^2))/2a_1 a_2$$

If $|C_2| > 1$ then location not reachable, End.

Else

$$S_2 = \epsilon_2 Sqrt(1 - C_2^2) \text{ with } \epsilon_2 = +/-1$$

$$q_2 = Arctg2(S_2, C_2)$$

Endif

$$D_1 = C_2 a_2 + a_1$$

$$D_2 = S_2 a_2$$

If $F1 = 0$ then q_1 arbitrary

Else

$$C_1 = (D_1 t_{14} + D_2 t_{24})/F1$$

$$S_1 = (D_1 t_{24} - D_2 t_{14})/F1$$

$$q_1 = Arctg2(S_1, C_1)$$

Endif

$$C_{23} = C_1 t_{11} + S_1 t_{21}$$

$$S_{23} = -S_1 t_{11} + C_1 t_{21}$$

$$q_{23} = Arctg2(S_{23}, C_{23})$$

$$q_3 = q_{23} - q_2$$

End.

SECOND CASE

$$F1 = (t_{14})^2 + (t_{24})^2$$

$$C_2 = (F1 - (a_1^2 + a_2^2))/2a_1a_2$$

If $|C_2| > 1$ then location not reachable, End.

Else

$$S_2 = \epsilon_2 Sqrt(1 - C_2^2) \text{ with } \epsilon_2 = +/-1$$

$$q_2 = Arctg2(S_2, C_2)$$

Endif

$$D_1 = C_2 a_2 + a_1$$

$$D_2 = S_2 a_2$$

If $F1 = 0$ then q_1 arbitrary

Else

$$C_1 = (D_1 t_{14} + D_2 t_{24})/F1$$

$$S_1 = (D_1 t_{24} - D_2 t_{14})/F1$$

$$q_1 = Arctg2(S_1, C_1)$$

Endif

$$q_{123} = Arctg2(t_{21}, t_{11})$$

$$q_3 = q_{123} - q_1 - q_2$$

End.

3 Direct differential model

FIRST CASE

$$E_1 = a_2 S_2 dq_3$$

$$E_2 = a_1 dq_1 - a_2 C_2 dq_3$$

$$d\theta = dq_1 + dq_2 + dq_3$$

$$E_3 = E_1 - a_2 S_2 d\theta$$

$$E_4 = E_2 + a_2 C_2 d\theta$$

$$dp_{x_o} = C_1 E_3 - S_1 E_4$$

$$dp_{y_o} = S_1 E_3 + C_1 E_4$$

$$l = a_3 t_{11}$$

$$m = a_3 t_{21}$$

$$dX = dp_{x_o} - m d\theta$$

$$dY = dp_{y_o} + l d\theta$$

SECOND CASE

$$E_1 = a_2 S_2 dq_3$$

$$E_2 = a_1 dq_1 - a_2 C_2 dq_3$$

$$E_3 = dq_1 + dq_2 + dq_3$$

$$E_4 = E_1 - a_2 S_2 E_3$$

$$E_5 = E_2 + a_2 C_2 E_3$$

$$dp_{x_o} = C_1 E_4 - S_1 E_5$$

$$dp_{y_o} = S_1 E_4 + C_1 E_5$$

$$l = a_3 t_{11}$$

$$m = a_3 t_{21}$$

$$dX = dp_{x_o} - m E_3$$

$$dY = dp_{y_o} + l E_3$$

4 Inverse differential model

FIRST CASE:

$$l = a_3 t_{11}$$

$$m = a_3 t_{21}$$

$$dp_{x_0} = dX + m d\theta$$

$$dp_{y_0} = dY - l d\theta$$

$$E_3 = C_1 dp_{x_0} + S_1 dp_{y_0}$$

$$E_4 = -S_1 dp_{x_0} + C_1 dp_{Y_0}$$

$$E_1 = E_3 + a_2 S_2 d\theta$$

$$E_2 = E_4 - a_2 C_2 d\theta$$

If $S_2 = 0$ then dq_3 arbitrary and $E_1 = 0$

Else

$$dq_3 = E_1 / a_2 S_2$$

$$dq_1 = (E_2 + a_2 C_2 dq_3)/a_1$$

$$dq_2 = d\theta - dq_1 - dq_3$$

Endif

End.

SECOND CASE:

$$l = a_3 t_{11}$$

$$m = a_3 t_{21}$$

E_3 arbitrary

$$dp_{x_0} = dX + m E_3$$

$$dp_{y_0} = dY - l E_3$$

$$E_4 = C_1 dp_{x_0} + S_1 dp_{y_0}$$

$$E_5 = -S_1 dp_{x_0} + C_1 dp_{y_0}$$

$$E_1 = E_4 + a_2 S_2 E_3$$

$$E_2 = E_5 - a_2 C_2 E_3$$

If $S_2 = 0$ then dq_3 arbitrary and $E_1 = 0$

Else

$$dq_3 = E_1/a_2 S_2$$

$$dq_1 = (E_2 + a_2 C_2 dq_3)/a_1$$

$$dq_2 = E_3 - dq_1 - dq_3$$

Endif

End.

5 Dynamic model

5.1 Step 1

$$\omega_{22z} = \dot{q}_1 + \dot{q}_2$$

$$\omega_{33z} = \omega_{22z} + \dot{q}_3$$

5.2 Step 2

$$a_{22z} = \ddot{q}_1 + \ddot{q}_2$$

$$a_{33z} = a_{22z} + \ddot{q}_3$$

5.3 Step 3

$$b_{1xx} = -\dot{q}_1^2$$

$$b_{2xx} = -(\omega_{22z})^2$$

$$b_{3xx} = -(\omega_{33z})^2$$

5.4 Step 4

$$\alpha_{11z} = -C_1 g$$

$$\alpha_{11y} = S_1 g$$

$$r_{11z} = \alpha_{11z} + b_{1zz} a_1$$

$$r_{11y} = \alpha_{11y} + \bar{q}_1 a_1$$

$$\alpha_{22z} = C_2 r_{11z} + S_2 r_{11y}$$

$$\alpha_{22y} = -S_2 r_{11z} + C_2 r_{11y}$$

$$r_{22z} = \alpha_{22z} + b_{2zz} a_2$$

$$r_{22y} = \alpha_{22y} + a_{22z} a_2$$

$$\alpha_{33z} = C_3 r_{22z} + S_3 r_{22y}$$

$$\alpha_{33y} = -S_3 r_{22z} + C_3 r_{22y}$$

5.5 Step 5 (off-line)

$$x^1 = m^1 x_1$$

$$x^2 = m^2 x_2$$

$$x^3 = m^3 x_3$$

5.6 Step 6

$$F_{3x}^3 = F_x^4 + b_{3zz} x^3 + m^3 \alpha_{33z}$$

$$F_{3y}^3 = F_y^4 + a_{33z} x^3 + m^3 \alpha_{33y}$$

$$F_{2x}^3 = C_3 F_{3x}^3 - S_3 F_{3y}^3$$

$$F_{2y}^3 = S_3 F_{3x}^3 + C_3 F_{3y}^3$$

$$F_{2x}^2 = F_{2x}^3 + b_{2zz} x^2 + m^2 \alpha_{22z}$$

$$F_{2y}^2 = F_{2y}^3 + a_{22z} x^2 + m^2 \alpha_{22y}$$

$$F_{1y}^2 = S_2 F_{2x}^2 + C_2 F_{2y}^2$$

5.7 Step 7

$$C_{3z}^3 = C_z^4 + b^3 a_{33z} + x^3 \alpha_{33y} + a_3 F_y^4$$

$$C_{2z}^2 = C_{3z}^3 + b^2 a_{22z} + x^2 \alpha_{22y} + a_2 F_{2y}^3$$

$$C_{1z}^1 = C_{2z}^2 + b^1 \ddot{q}_1 + x^1 \alpha_{11y} + a_1 F_{1y}^2$$

5.8 Step 8

$$\Gamma^3 = C_{3z}^3 + F^{33} \dot{q}_3 + H^{33} sgn(\dot{q}_3)$$

$$\Gamma^2 = C_{2z}^2 + F^{22} \dot{q}_2 + H^{22} sgn(\dot{q}_2)$$

$$\Gamma^1 = C_{1z}^1 + F^{11} \dot{q}_1 + H^{11} sgn(\dot{q}_1)$$

6 Generalized workspace D_q and aspects A_i

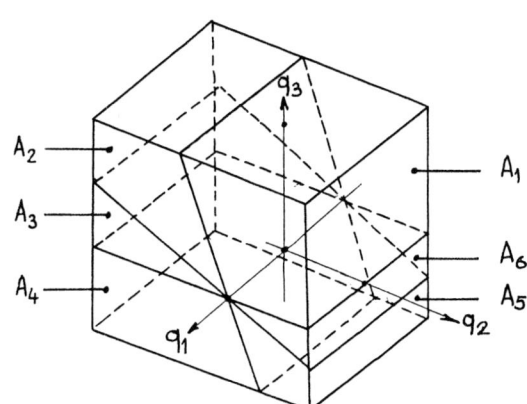

$-60° \leq q_1 \leq 60°, -90° \leq q_2 \leq 60°, -60° \leq q_3 \leq 90°$

a) Aspects A_i

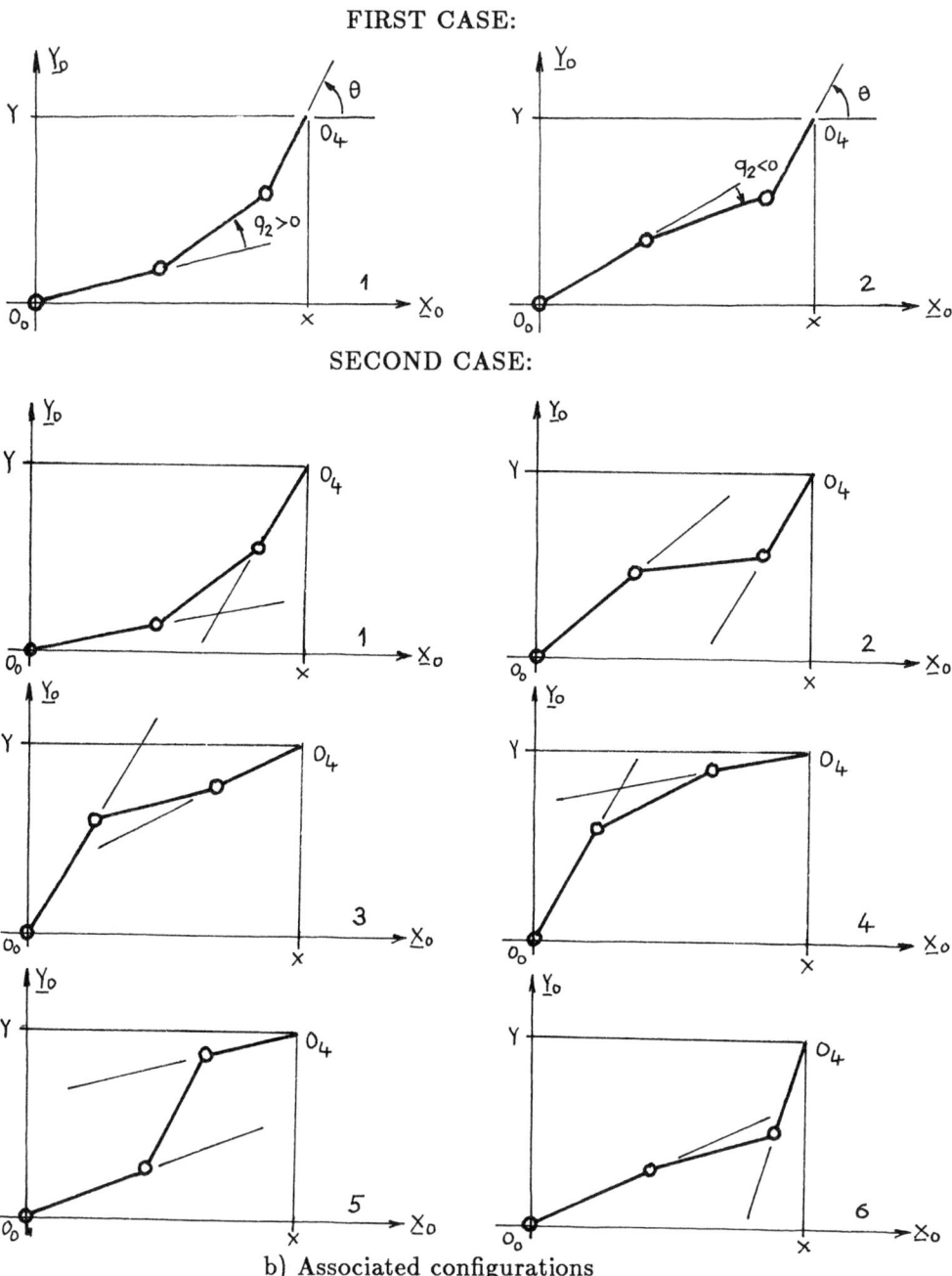

b) Associated configurations

Fig. A.3

7 Images of aspects $f_p(A_i)$ and operational workspace in position D_{x_p}

FIRST CASE:

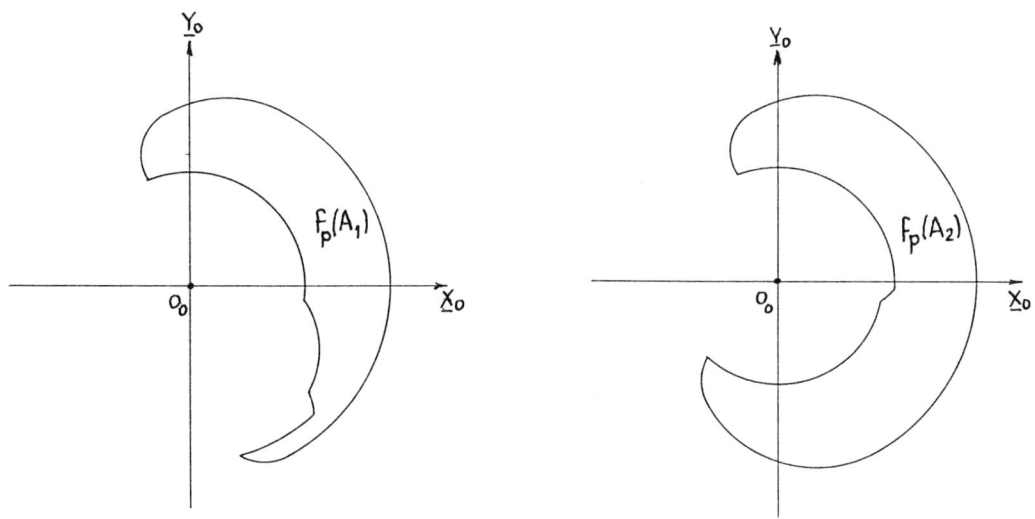

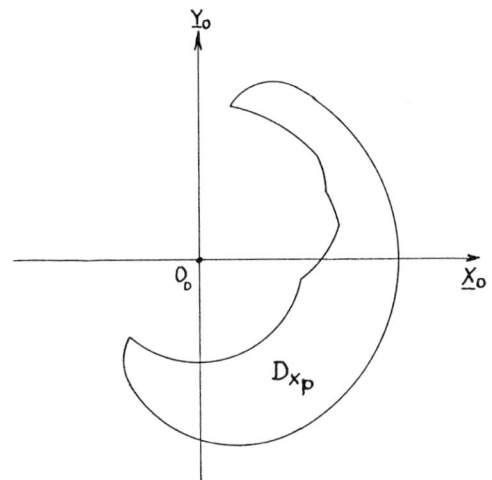

SECOND CASE:

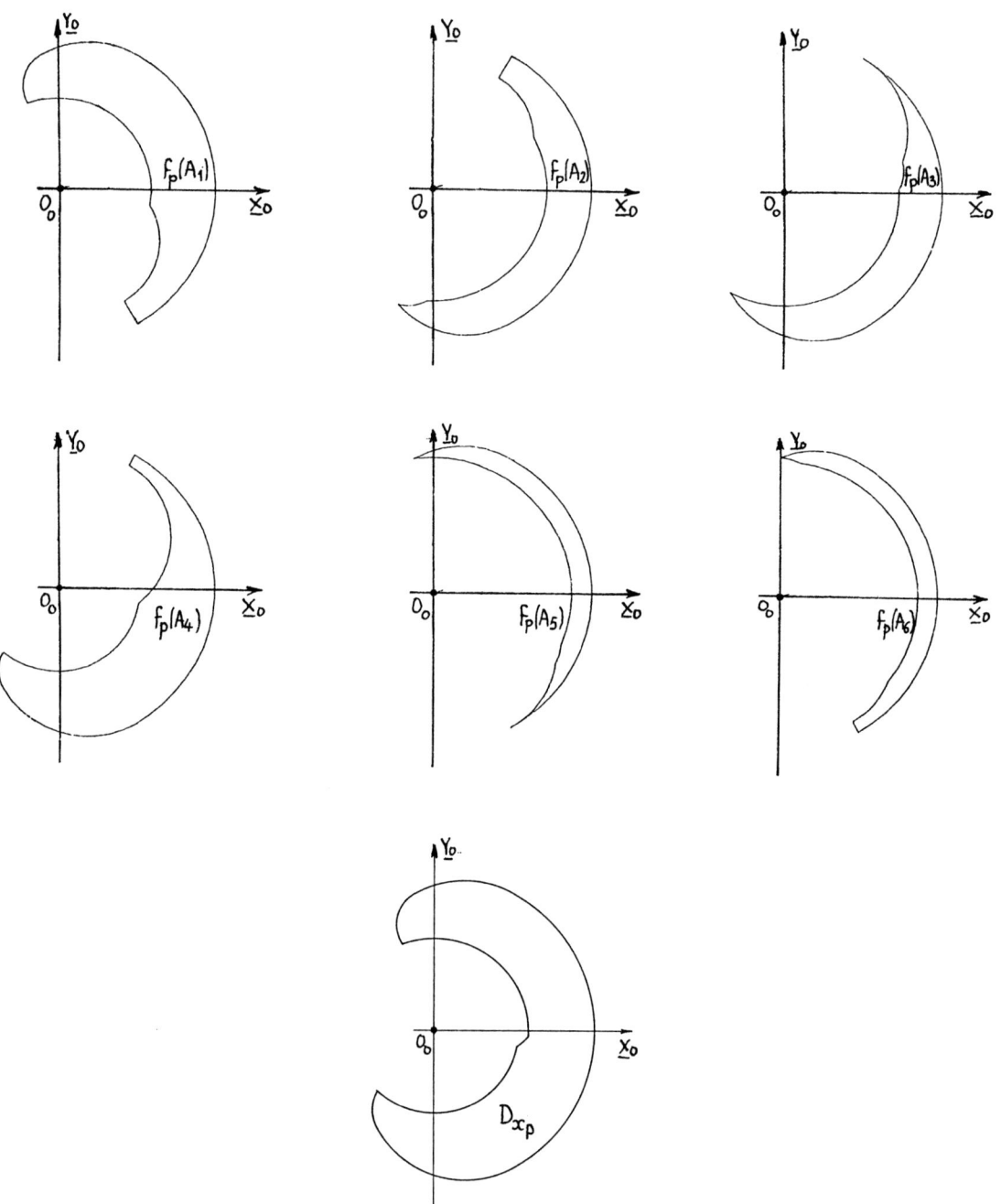

Fig. A.4 Images of aspects A_i and operational workspace in position D_{x_p}

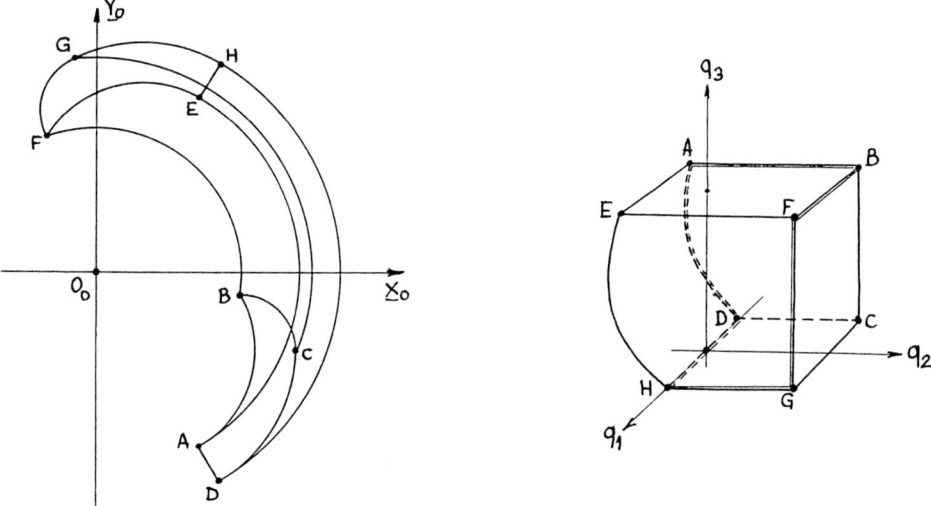

Fig. A.5 Particular case of aspect A_1 (second case)

Modelling Positioning Uncertainties

Isabelle Mazon

Groupe Robotique et
Intelligence Artificielle
LAAS-CNRS
7, Avenue du Colonel Roche
31077 Toulouse Cedex

Abstract

Robot programs generated by a human or by an automatic planner, must execute in a real environment which differs slightly from the model used at programming time. Thus we need to represent this uncertainty. This representation can be used at programming time to directly produce a valid program, or to verify the validity of a program afterwards.

In this paper, two methods were developped to represent uncertainties in robotics. Both methods have been implemented: one at LIFIA (Grenoble France), the other at LAAS (Toulouse France).

1 Introduction

1.1 Automatic Programming of Assembly Robots

Research aimed at developing automatic programming aids for robotic systems, has foccused interest on high level programming: "task level". Task level programming is characterized by a symbolic description of the task in terms of geometrical constraints to be satisfied. This symbolic description is associated with a world model describing the robot, its environment and the initial configuration. This model is essentially a geometrical one, and includes morphological and spatial descriptions of the objects, the sensors, and the manipulator(s) [Bro82], [AC86], [LB85], [Cho86], [Lau87], [Maz87a].

Such a system generally, transforms the symbolic description into a lower level program which is executable by the robot. It has to solve the following problems :

1. Trajectory Planning (collision avoidance, mechanism limitations)

2. Layout of the work space

3. Synthesis of fine motion strategies. Use of sensor-based strategies to cope with uncertainties on the shape and position of objects, and on robot motions (which could lead to the plan failing).

In fact, these problems are not independent, and cannot be solved separately. Since they are very complex, each is solved separately, and then a higher level procedure verifies the validity of the whole plan and backtracks if necessary [LB85].

One criteria of validity is the robustness of the plan with respect to uncertainties induced by the complicated physical system used (robot, sensors) and by the environment model, which is an imperfect representation of the real world.

We have to face two kinds of errors : positioning and shape. The second class are important for fine motions, and we can supposed that they are taken into account by the fine motions synthesis module. But positioning errors still remain to be explicitly considered.

In certain cases (for example to analyze the feasibility of a plan of actions [TP87], [Maz87b]), it is necessary to take into account the different sources of uncertainties which can play a role during the manipulation, in order to evaluate a given uncertainty. To do so, we need an explicit representation of uncertainties, and various operators to reason about them.

1.2 Mobile Robots

A mobile robot, in contrast to manipulator robots, manoeuvers in an unknown or partially known environment. It moves in the environment which it perceives to be free.

One problem consists of building an environment model, which the robot explores, by means of sensory data using vision (stereo, dynamic or simple)), ultra-sonic sensors, and knowledge about its moves (odometry)

All these data are uncertain (slippage make the odometry imprecise, sensory resolution). This uncertainty must be taken into account in the environment model and in the decision processes [CL85], [SC87].

In fact, if we neglect the perception uncertainty, it will be practically impossible to model the environment in an intelligent manner:

- The object representation are determined by one sensor will never correspond to the representation of the same object obtained by another sensor

- The environment model will therefore include more objects than there really are

Certain algorithms implicitly take into account these uncertainties (stereo-vision for example), but it is necessary to include of an explicit representation of uncertainties in the environment model.

1.3 Outline of the paper :

In the first section, we will describe the environment modelling we have chosen

In the second section, two different ways of modelling uncertainty modelizations will be detailed.

In the last section, we will explain the calculation methods used to reason with these uncertainties.

An application example concerning assembly robotic can be found in [TP87] [Maz89]

2 Environment Model

In manipulation robotics, as well as in mobile robotics, the cartesian positions of objects are described relative to a fixed reference or to other objects. In the general case, the positions are characterized by six independent parameters:

- three for the cartesian position (translation)
- three for the orientation

The spatial representation of the environment must allow us to determine all the positional relations between objects.

For example from :

- object-A position relative to table-1
- object-B position relative to table-2
- table-1 position relative to table-2

We must be able to deduce object-A position relative to object-B

To do so, we model the environment by a directed graph where nodes represent coordinate frames, and arcs represent the relative position of the child node (coordinate frame) relatively to the parent node.

This graph must represent the real state of the environment; the only physical existing relations will be represent by arcs. The fixed objects will be connected to an arbitrary chosen reference (generally the base of the manipulation robot, or the "home" of the mobile robot).

The evaluation of a new relative relation, from physically existing ones, can be done in two steps:

- Compute a path in the non-oriented graph (which generally admits only one connected component $\Rightarrow$ the path exists)
- Execute a composition (or inversion) of relations to obtain the new relation

All the relative relations are only imperfectly known, so we will refer to the arcs as uncertain relations or transformations.

3 Operators on the State Graph

We have chosen to represent the relation between two objects by a homogeneous matrix.

A homogeneous matrix P_{ij} defines the position (translation) T_{ij} and the orientation Rot_{ij} of the coordinate frame R_j in the coordinate frame R_i as follows :

$$P_{ij} = \begin{pmatrix} Rot_{ij} & T_{ij} \\ 0 & 1 \end{pmatrix}$$

This representation allows a simple composition of relative certain positions using matrix multiplication, while establishing new translation or orientation vectors.

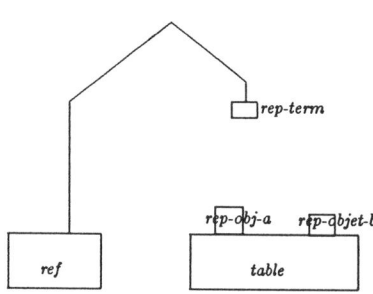

Figure 1: Environment Example

3.1 Use of the State Graph

A classical operation on the state graph consists of determining a new relative ralation from known relations.

From the graph defined in figure 2, which represents the environment of figure 1, we would like to know the relation between the robot gripper and object-A.

If the graph was composed of certain relations, we would have :

$$\text{t-robot-obj-a} = (\text{t} - \text{robot})^{-1} * \text{t-table} * \text{t-table-object-a}$$

with '*' representing the matrix product.

So we define two operators on uncertain relations :

- Inversion of an uncertain transformation
- Composition of uncertain transformations

3.2 Transformation by an Action

There are two classes for robots actions:

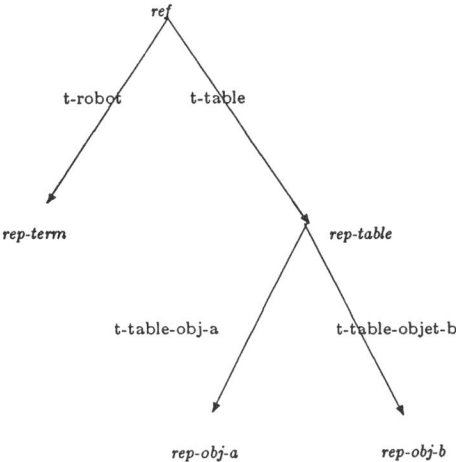

Figure 2: Graph representing previous environment

perception actions which do not physically modify the environment but permit us to improve our knowledge about it.

The sensors provide complementary or redundant information about the environment state. We have to define a new operator, that will allow us to insert these data into the state graph in a coherent manner. This operator is refered to as "multi-sensory fusion"

real actions which physically modify the environment.

An action is intended to modify the environment state while performing a given target goal. Therefore, we must modify the state graph to follow the evolution of the real world :

- Add or suppress a physical relation between two elements (eg pick or place)
- Modification of the values of certain relations (eg pushing an object on a table)

If we would like to analyze the feasibility of a plan of actions, or to properly characterize the importance of uncertainties, it is necessary to know the effects of actions on uncertainties.

Therefore, for each different type of action, we must know :

- Modifications to carry out on the state graph.

- Modifications to carry out on the positioning errors

For example the action " put an object currently held by the robot on the table" would require the followings changes:

- delete the arc between the robot gripper and the object.(the physical relation is broken when the gripper is opened)
- add the arc between the table and the object (the value is known, because it is a parameter of the action)
- reset the translational error of the relation table-object, along the normal to the plane of contact (This can be deduced from the definition of an action as a set of contacts to be established)

4 Modelling Uncertainties

An error represents the gap between the nominal position and the real position of the coordinate frame under consideration. This error can be considered as the sum of two kinds of errors:

- A systematic error
- A random error

In the following, we will suppose that systematic errors have been suppressed (by identification or calibration) and we will represent a real transformation as the product of the nominal transformation and a transformation representing the random error.

$T^{real} = T^{nom} * \mathcal{E}$ where: $\begin{cases} T^{reql} \text{ is the real transformation} \\ T^{nom} \text{ is the nominal transformation} \\ \mathcal{E} \text{ is a homogeneous matrix representing the error} \end{cases}$

The error matrix is characterized by six independent parameters, three for the position (translation) and three for the orientation. The uncertainty that will be associated with the real transformation T^{real} will be the characterization of the variations of these parameters. Therefore, it depends on the choice of the six parameters.

For the position, the parameters will be the three components of the translation vector.

For the orientation we can choose among :

- Euler angles associated with the rotation matrix
- The vector of rotation (whose direction is the axis of rotation and whose the value is the angle of rotation)

We describe here two kinds of characterization of the error :

- Set-oriented : we are interested in the minimun and maximum values which can be taken by the error parameters. The sets are defined in R^6. This representation is used at LIFIA.
- Probabilistic : we work on the probability distributions associated with the vector composed of the six parameters of $\mathcal{E}$. This is used at LAAS.

4.1 Set-oriented Representation

4.1.1 General Rule

In this framework, an uncertainty is considered as *the set of errors ε that can occur* .

A geometric transformation can be defined by two vectors and an angle of rotation. The first vector represents translation (belongs to R^3), and the second one the axis of rotation (belongs to the unit sphere : S(1)).

Therefore, the set of transformations is isomorphic to the cartesian product $R^3 \times S(1) \times [-\pi, +\pi]$. We will represent an uncertainty by a sub-set of this cartesian product. As this product space has few attractive algebraic properties (it cannot be described by a structure of euclidian vectorial space), it is difficult to characterize a set of any form. So we approximate a set by a product of the sub-set : $T \times U \times A$, where T is a particular sub-set of R^3, U a particular subset of $S(1)$ and A a subset of $[-\pi, +\pi]$. Thus, the problem consists of deducing the approximation of a subset E of $R^3 \times S(1) \times [-\pi, +\pi]$ in the new space. The problem is solved as follow :

1. Project E on each of the spaces R^3, $S(1)$, and $[-\pi, +\pi]$, that gives three subset Tr', U' and A' respectively of R^3, $S(1)$, and $[-\pi, +\pi]$.

2. Approximate Tr', U' and A' , by some sets Tr, U and A of particular forms (see below).

The approximation of E will be the product $Tr \times U \times A$.

It is important to note that at the time of each approximation, we choose a set *including* the real set, this means that we are increasing uncertainties. This is consistent with our definition of uncertainty : Any positioning error that can occur must be included in the set defining the uncertainty.

In practice, we will never completely specify the sub-set of $R^3 \times S(1) \times [-\pi, +\pi]$, but we will calculate its approximation. In a natural manner, we decompose the positional uncertainty of an object into two terms : translation and orientation. This operation leads directly to the set Tr, U, A.

4.1.2 Examples

solid on a plane:

Consider a cube lying on an "horizontal" plane, $z = constant$. Its position on the plane is perfectly known in the direction z, due to the contact between the plane and one face of the cube.

So the uncertainty in translation we will be represented by a disc whose axis is along z and radius is of length ε. The rotation will be represented by a unit vector along z (the only rotation physically possible), the angle of rotation A bounded by $[-\alpha, +\alpha]$.

solid in space :

If we now consider a free solid in space, the translation uncertainty has no more intrinsic direction, so it will be represented by a three dimensional subspace, for example a sphere. This will be the same for the rotation, A which could be represented by an interval $[-\alpha, +\alpha]$.

4.1.3 Particular Form of the Sets Tr, U, et A

The composition and inversion operators on transformation, presented in section 3 (and later in section 5.1), lead to operations on the sets Tr and U. The difficulty of calculations on these sets, has led us to choose particular forms for these sets, in order to be able to perform rapidly the calculations required.

For the set Tr, the type of calculations needed are :

1. image by a known rotation
2. increasing of the set (Minkowsky's sum of two sets) by another set
3. image by a set of rotations

For the set U, the typical operations are:

1. cross product by a known vector
2. image by a set of rotations

The complexity of the operations is large for any set of dimension 3. So we choose to approximate real sets by simpler ones, on which the previous operations can be done, not in an exact manner but at least, with a good approximation. The choice of the sets is dictated by the type of uncertainty relation met in the practice. In particular, the type of contacts between objects, constrains the form of sets Tr and U.

So, the selection of approximation sets for Tr are:

- line segments (monodimensional translation uncertainty)
- discs (bidimensional translation uncertainty, plane on surface contact)
- spheres and cylinders

The selection of approximation sets for U are the complete unit sphere $(S(1))$, or a simple unit vector.

4.2 Probabilistic Representation

We suppose that each parameter chosen to be characterized in $\mathcal{E}$ is a random variable whose density function is a priori known.

One possible representation consists of choosing the Euler angles in order to characterize the rotation part of the error matrix. The Euler angles we use are defined by :
$Rot_{ij} = Rot(x, \delta_x) * Rot(y, \delta_y) * Rot(z, \delta_z)$.

The six parameters form a random vector $\bar{\epsilon}$:

$$\bar{\epsilon} = \begin{pmatrix} dx \\ dy \\ dz \\ \delta x \\ \delta y \\ \delta z \end{pmatrix} = \begin{pmatrix} \overline{dx} \\ \overline{\delta} \end{pmatrix}$$

then we have the following error matrix :

$$\mathcal{E} = \begin{pmatrix} cos(\delta y) * cos(\delta z) & -cos(\delta y) * sin(\delta z) & sin(\delta y) & dx \\ sin(\delta x) * sin(\delta y) * cos(\delta z) + cos(\delta x) * sin(\delta z) & -sin(\delta x) * sin(\delta y) * sin(\delta z) + cos(\delta x) * cos(\delta z) & -sin(\delta x) * cos(\delta y) & dy \\ -cos(\delta x) * sin(\delta y) * cos(\delta z) + sin(\delta x) * sin(\delta z) & cos(\delta x) * sin(\delta y) * sin(\delta z) + sin(\delta x) * cos(\delta z) & cos(\delta x) * cos(\delta y) & dz \\ 0 & 0 & 0 & 1 \end{pmatrix}$$

The uncertainty will then be characterized by the variance-covariance matrix associated with the random vector $\bar{\epsilon}$ as follows : $\Delta_\epsilon = E\left(\epsilon \epsilon^t\right)$ where E is the expectation operator.

As the translation vector and rotation vector do not play the same role in the subsequent calculations, we decompose this matrix as:

$$\Delta_\epsilon = \begin{pmatrix} v_{tt_\epsilon} & v_{t\delta_\epsilon} \\ v_{\delta t_\epsilon} & v_{\delta\delta_\epsilon} \end{pmatrix}$$

For simplicity, we suppose in the rest of this paper that the probability distribution of ϵ is gaussian in the six parameters. The identification of the parameters of the function is assumed to be done from the calibration data, or from a priori knowledge (eg tolerance on the size of objects, or sensors characteristics).

5 Definition of the Operators

5.1 Composition Operator

The problem consists of determining the uncertainty of the "product" of two uncertain transformations.

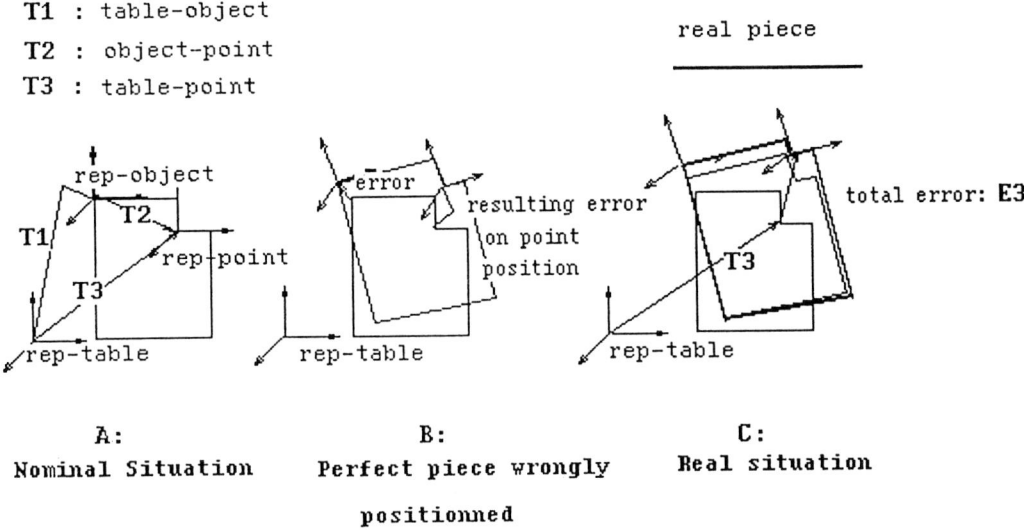

Figure 3: composition example

To illustrate this problem, consider the example of figure 3. We know the position of an object on the table, as well as a position of an interesting point of this object in an intrinsic coordinate frame linked to the object (CAD reference for example). We attach a coordinate frame to this point, and we would like to know the uncertainty of the relation table-point, starting from the knowledge of the uncertainties on the relations table-object (the object was put on the table by an imprecise robot), and object-point (tolerances).

Using the notation of the figure, we obtain :

$$T_3 = T_1 * T_2 = T_1^{nom} * \mathcal{E}_1 * T_2^{nom} * \mathcal{E}_2 = T_3^{nom} * \mathcal{E}_3$$

then :

$$\mathcal{E}_3 = \underbrace{\underbrace{(t_2^{nom})^{-1} * \mathcal{E}_1 * t_2^{nom}}_{\alpha} * \mathcal{E}_2}_{\beta}$$

The α calculation consists of computing the error in position of the point due uniquely to the imprecise positioning of the object on the table (see fig 3.b). The calculation β "adds" this error to the positioning error of the point respectively to the piece (see fig 3.c).

5.1.1 Calculation Details

We define : $T_2^{nom} = \begin{pmatrix} Rot_2 & t_2 \\ 0 & 1 \end{pmatrix}$ $\mathcal{E}_1 = \begin{pmatrix} Rot_{\mathcal{E}_1} & t_{\mathcal{E}_1} \\ 0 & 1 \end{pmatrix}$

The calculation gives : $T_2^{nom-1} = \begin{pmatrix} Rot_2^t & -Rot_2^t * t_2 \\ 0 & 1 \end{pmatrix}$

We find the α term :

$$\alpha = \begin{pmatrix} Rot_2^t * Rot_{\mathcal{E}_1} * Rot_2 & Rot_2^t * (Rot_{\mathcal{E}_1} * t_2 - t_2 + t_{\mathcal{E}_1}) \\ 0 & 1 \end{pmatrix} = \begin{pmatrix} Rot_\alpha & t_\alpha \\ 0 & 1 \end{pmatrix}$$

As we work on small errors, we can approximate to the first order :

$$Rot_{\mathcal{E}_1} = I + \mathcal{R}_{\mathcal{E}_1} \quad ; \quad \mathcal{R}_{\mathcal{E}_1} = \begin{pmatrix} 0 & -a & b \\ a & 0 & -c \\ -b & c & 0 \end{pmatrix}$$

Where $\mathcal{R}_{\mathcal{E}_1}$ is pre-cross-product matrix associated with vector $\overline{v} = \begin{pmatrix} a \\ b \\ c \end{pmatrix}$

We have the following equivalences, between the different possible representations:

	euler angles	rotation vector u, angle θ
a	δ_x	$u_x * \theta$
b	δ_y	$u_y * \theta$
c	δ_z	$u_z * \theta$

In the following, we let $\delta_{\mathcal{E}}$ represent the vector $\overline{v}$ associated with the rotation part of an uncertainty $\mathcal{E}$.

Therefore, we obtain the α parameters using properties of homogeneous transformations :

<u>rotation part</u> : $Rot_2^t * Rot_{\mathcal{E}_1} * Rot_2$ is a rotation of axis : $Rot_2 * \delta_{\mathcal{E}_1} = \delta_\alpha$
then, we have $Rot_\alpha = I + \mathcal{R}_{\delta_\alpha}$

<u>translation part</u> : $Rot_2^t * (Rot_{\mathcal{E}_1} * t_2 + t_{\mathcal{E}_1} - t_2)$ which can be approximated (to first order) as: $Rot_2 * (\mathcal{R}_{\mathcal{E}_1} * t_2 + t_{\mathcal{E}_1}) = Rot_2^t * (\delta_{\mathcal{E}_1} \wedge t_2 + t_{\mathcal{E}_1}) = t_\alpha$.

The second part of calculation consists of computing the product of two uncertainties :

$$\mathcal{E}_3 = \beta = \alpha * \mathcal{E}_2$$

Using the first order approximation, we find:

$$\mathcal{E}_3 = \begin{pmatrix} Rot_\alpha * Rot_{\mathcal{E}_2} & Rot_\alpha * t_{\mathcal{E}_2} + t_\alpha \\ 0 & 1 \end{pmatrix} = (I + \begin{pmatrix} \mathcal{R}_{\delta_\alpha} & t_\alpha \\ 0 & 0 \end{pmatrix}) * (I + \begin{pmatrix} \mathcal{R}_{\delta_{\mathcal{E}_2}} & t_{\mathcal{E}_2} \\ 0 & 0 \end{pmatrix})$$

$$\mathcal{E}_3 = (I + \begin{pmatrix} \mathcal{R}_{\delta_\alpha} + \mathcal{R}_{\delta_{\mathcal{E}_2}} & t_{\mathcal{E}_2} + t_\alpha \\ 0 & 0 \end{pmatrix}) = I + \begin{pmatrix} \mathcal{R}_{\delta_\alpha + \delta_{\mathcal{E}_2}} & t_{\mathcal{E}_2} + t_\alpha \\ 0 & 0 \end{pmatrix}$$

So, the $\mathcal{E}_3$ parameters are:

- translation part : $t_{\mathcal{E}_3} = t_{\mathcal{E}_2} + Rot_2^t * (\delta_{\mathcal{E}_2} \wedge t_2 + t_{\mathcal{E}_1})$
- rotation part : $\delta_{\mathcal{E}_3} = \delta_{\mathcal{E}_2} + Rot_2 * \delta_{\mathcal{E}_1}$

5.1.2 Set Implementation

Calculation of Term α :

The rotation part of T_α is the image of the rotational vector of T_{ϵ_1} by the rotation Rot_2, leaving the rotation angle of uncertainty not modified. The rotation part can be expressed as : $U_\alpha = Rot_2(U_1)$ and $A_\alpha = A_1$.

To calculate the set Tr_α representing the set of vectors: $T_\alpha = Rot_2^{-1} * (\delta_{e_1} \wedge t_2 + t_{\epsilon_1})$ we will proceed in three steps:

- Cross product of set U_1 by the vector t_2, and similarly by the half-amplitude of A_1
- Minkowski sum of the obtained set and Tr_1
- Transformation of result by the inverse of Rot_2

Calculation of Term β :

rotation part : as the expression of $\delta_{\mathcal{E}_3}$ is not a simple function of the two previous rotation vectors and angles, we perform the following simplifications :

- if $U_\alpha = U_2 = \overline{u_0}$ a single vector, then $U_3 = \overline{u_0}$; in all others cases $U_3 = S(1)$
- if $A_1 = [-\alpha_1, +\alpha_1]$ and $A_2 = [-\alpha_2, +\alpha_2]$, A_3 will be equal to $[-(\alpha_1+\alpha_2), (\alpha_1+\alpha_2)]$

the vector of translation $T_3 = Rot_\alpha t_{\mathcal{E}_2} + t_\alpha$ is computed in two steps:

- image Tr_2 by the rotations set Rot_α
- Minkowski sum of the resulting set and Tr_α

5.1.3 Probabilistic Implementation

We are looking for the variance-covariance matrix associated with $\mathcal{E}_3$. To do so, we assume that errors on T_1 and on T_2 are independent.

$$E(\epsilon_3 * \epsilon_3^t) = \begin{pmatrix} E(t_{\mathcal{E}_3} * t_{\mathcal{E}_3}^t) & E(t_{\mathcal{E}_3} * \delta_{\mathcal{E}_3}^t) \\ E(\delta_{\mathcal{E}_3} * t_{\mathcal{E}_3}^t) & E(\delta_{\mathcal{E}_3} * \delta_{\mathcal{E}_3}^t) \end{pmatrix}$$

In order to find the expressions of $E(\epsilon_3 * \epsilon_3^t)$, we express t_3 and δ_3 as a matrix product :

$t_{\mathcal{E}_3} = t_{\mathcal{E}_2} + Rot_2^t * (P_2 * \delta_{\mathcal{E}_1} + t_{\mathcal{E}_1})$, where P_2 is the pre-cross-product

$\delta_{\mathcal{E}_3} = \delta_{\mathcal{E}_2} + Rot_2 * \delta_{\mathcal{E}_1}$ matrix associated with t_2

After all the calculations, we obtain:

$$\begin{aligned}
E(t_{\mathcal{E}_3} t_{\mathcal{E}_3}^t) &= Rot_2^t * P_2 * v_{\delta\delta_{\mathcal{E}_1}} * P_2^t * Rot_2 + Rot_2^t * v_{tt_{\mathcal{E}_1}} * Rot_2 + \\
& \quad Rot_2^t * v_{t\delta_{\mathcal{E}_1}} * P_2^t * Rot_2 + Rot_2^t * P_2 * v_{\delta t_{\mathcal{E}_1}} * Rot_2 + v_{tt_{\mathcal{E}_2}} \\
E(t_{\mathcal{E}_3} * \delta_{\mathcal{E}_3}^t) &= Rot_2^t * v_{t\delta_{\mathcal{E}_1}} * Rot_2 + Rot_2^t * P_2 * v_{\delta\delta_{\mathcal{E}_1}} * Rot_2 + v_{t\delta_{\mathcal{E}_2}} \\
E(\delta_{\mathcal{E}_3} * t_{\mathcal{E}_3}^t) &= E(t_{\mathcal{E}_3} * \delta_{\mathcal{E}_3}^t)^t \\
E(\delta_{\mathcal{E}_3} * \delta_{\mathcal{E}_3}^t) &= Rot_2^t * v_{\delta\delta_{\mathcal{E}_1}} * Rot_2 + v_{\delta\delta_{\mathcal{E}_2}}
\end{aligned}$$

5.2 Inversion of an Uncertain Transformation

This operation is necessary when the search for a new relative relation in the state graph meets an arc to be traversed in the reverse direction. The uncertainty of the direct transformation is known, so we need to calculate the uncertainty of the inverse transformation.

We search $\mathcal{E}'$ such that:

$$(T * \mathcal{E})^{-1} = T^{-1} * \mathcal{E}'$$

We find :
$$\mathcal{E}' = T * \mathcal{E}^{-1} * T^{-1}$$

This relation is a composition of uncertainty : we compose the relation T^{-1} with the identity certain relation.

Therefore, we only need to characterize $\mathcal{E}^{-1}$ to solve the inversion problem. We find the parameters of $\mathcal{E}^{-1}$ by assuming that :

$$\mathcal{E} = \begin{pmatrix} Rot_\varepsilon & t_\varepsilon \\ 0 & 1 \end{pmatrix} \quad \mathcal{E}^{-1} = \begin{pmatrix} Rot_\varepsilon^{-1} & -Rot_\varepsilon^{-1} t_\varepsilon \\ 0 & 1 \end{pmatrix}$$

5.2.1 Set Implementation

If we let Tr', U', A', be the characterization set of the uncertainty of $\mathcal{E}^{-1}$, the results of previous developments leads to :

- $U' = U$ and $A' = A$ (because A is symmetrical)
- Tr' is the image of Tr by the set of rotations defined by U' and A'

5.2.2 Probabilistic Implementation

If we assume, a first order approximation, we have :

translation part $t_{\mathcal{E}^{-1}} = -t_\mathcal{E}$

rotation part $\delta_{\mathcal{E}^{-1}} = -\delta_\mathcal{E}$

As we have chosen to represent the parameter density functions as gaussians which are therefore symmetrical with respect to their means ($= 0$), the variance-covariance matrix is unchanged.

5.3 Multisensory Fusion

The problem consists of determining from different uncertain observations of the same state, a new approximation of this state, by synthesizing these observations so as in order to reduce the resulting uncertainty.

We treat the case of two observations, the generalization to many observations being easy.

Multi-sensory fusion, can be decomposed into two sub-problems:

- combine two measurements of the same type, where both sensors provide the same datum(or data)
- combine two different kinds of data and improve the known information about the relation between two objects

5.3.1 Set Implementation

The method consists of intersecting the sets defining the uncertainty of each measurement. Then we approximate the resulting uncertainty sets by the chosen sets (disc, sphere , unit vector ...).

This operation can only reduce the uncertainty of the data common to both measurements.

5.3.2 Probabilistic Implementation

Here we will treat only the first case where we are able to express the measurements in the same coordinate frame, without any infinite uncertainty on one of the parameters.

The second class of problem is very complex, and will not be treated here.

Same kind data fusion

We assume that the measurement permits us to determine the six parameters of the relation (This limitation does not appear in the set implementation).

We want to estimate the vector of six parameters of the relation from the two parameter vector of the measurements. The new vector is denoted θ .

There are two probabilistic methods to solve this estimation problem:

Bayesian Theory : which minimizes a loss function associated with the error in the estimation of the parameter (here the vector θ).

Kalman's filter : which minimizes the standard deviation of the committed error on the parameter estimation. This method is a particular case of the Bayesian Theory.

Principle of Bayesian Theorem

The method consists of minimizing the value of a loss function associated with the error on the estimation a of the parameter θ

This loss function, $L(\theta, a)$ can take the following form:

α) $L(\theta, a) = (\theta - a)^2$, in monodimensional case

$L(\theta, a) = (\theta - a)^t * Q * (\theta - a)$ in the multidimensional case.

β) $L(\theta, a) = |\theta - a|$

γ) $L(\theta, a) = \begin{cases} K1 * |\theta - a| & \text{if } \theta \leq a \\ K2 * |\theta - a| & \text{if } \theta \geq a \end{cases}$

For example, the loss due to an underestimation of the distance between the robot and an object is ($K1$ / $K2$) more important than an overestimation of this distance.

δ) The maximun likelihood principle.

This method needs a priori knowledge about the parameter to estimate

- its density function

- a density function with includes all the possible values of the parameter.

This knowledge (referred to as the a priori density law) is necessary because the Bayesian principle uses conditional probability to produce an a posteriori law of the parameter. Thus, the function changes according to the measurements.

In every case, it is necessary to know the sensor laws. These laws represent the probability of measuring the real value θ of the parameter to measure, and they are referred to as $f(x|\theta)$

Implementation for Gaussian probability laws (cf KALMAN)

We assume that we know:

- the a priori density function of the parameter to estimate, designated by $\Pi(\theta)$
- the conditional probability function of observation x knowing the true value θ of the parameter, noted $f(x|\theta)$

The aim of this method is to determine the a posteriori law of the value θ having observed x, this in order to be able to calculate the loss function.

The a posteriori law is :

$$\Pi(\theta|x) = \frac{\Pi(\theta) * f(x|\theta)}{\int_{-\infty}^{\infty} \Pi(\theta) * f(x|\theta) \, d\theta}$$

If we choose a gaussian model of the sensory data (sensor measurements are assumed to be independent), and for the a priori density function, it can be demonstrated that for a loss function of the kind (α or β or γ), the a posteriori distribution is also gaussian.

Application to the Fusion Problem

Let x_1 and x_2 are two observations; their known variances are respectively Δ_1 and Δ_2. The best estimate will be x whose variance is Δ :

$$\Delta = (\Delta_1^{-1} + \Delta_2^{-1})^{-1}$$

To determine x we have to extract the six parameters of each observation.

- $\overline{P_1}$ parameters of x_1
- $\overline{P_2}$ parameters of x_2

We have :

$$\overline{P} = \Delta * (\Delta_1^{-1} * \overline{P_1} + \Delta_2^{-1} * \overline{P_2})$$

From $\overline{P}$, we can deduce x.

Remarks

If one of the { Δ_1 , Δ_2 } matrix cannot be inverted, this means that the sensor does not provide all information about the relations, so the assumptions are violated.

When some information is not provided by one of the observations, but the determination of x is possible (eg the missing informations is complementary), we can solve the fusion problem in the following manner :

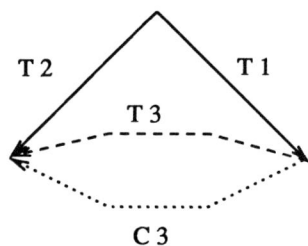

Figure 4: Example of fusion

- fuse the common data
- leave unchanged the variance associated with a parameter of the transformation, which is missing. This one is deduced from only one of the observations

5.3.3 Problem Induced by Multisensory Fusion

The fusion algorithm induces the problem of maintaining the coherence of state model. This problem is due to the fact that the composition operator, is not conservative : the resulting uncertainty of a composition will depend on the path chosen in the graph. This means that cycle in the state graph are unwished : **The graph has to be a tree.**

This restriction cause no problem, at the time of building the initial state tree, for the following reasons :

- For a mobile robot, the initial environment is unknown or is the result from a previous exploration which is assumed to be consistent ($\rightarrow$ we have a tree)
- For a manipulation robot, if the initial state graph is not a tree, the operator can choose an arc to eliminate among the arcs belongings to cycle(s). (Generally, these arcs are due to unnecessary measurements or calibrations)

Consistency Problem

Let us consider the example of figure 4 with T_1, T_2 transformations. The fusion algorithm permits us to improve the relation T_3 (a fictitious but computable relation) by the measurement C_3. But if we now insert an arc (fusion of C_3 and T_3) T'_3, we create a cycle and so destroy the consistency of the graph as follows:

$$T_3 = (\text{composition (inversion } T_1) \ T_2 \ )) \neq T'_3$$

An uncertainty computation on this graph would produce a value which depends on the path chosen, when physically there is only one.

In order to maintain model consistency, two solutions have been proposed in the literature:

- Modifying the uncertainties associated with T_1 et T_2, in a manner inversely proportionally to their uncertainties [Dur86]

- Destroy one of the sources of conflict (either a new observation or the current value of the transformation) : "intelligent forgetting " [Fa86]

Set Implementation

The solution we have chosen is close to the one developed by Durrant-Whyte's one. The problem consists to choose a solution among an infinity of solutions.

> for each arc of the cycle ARC_{kl}
> we fuse this arc with,
> the path $k - l$ which includes the arc associated with the observation and not the arc ARC_{kl}.
> the result of the fusion is associated with the arc ARC_{kl}.

The selection of this solution is based on the analogy of distortion propagation in a network of elastic links.

Consider a set of flexible links, forming a cycle. If we stiffen one link, the possible moves of the other links will be modified, as well as their rigidity, as follows :

- If all links are rigid except one, it is this one that will become more rigid. The rigidity of others links will not be modified

- If there are two or more "elastic" links, we cannot predict their behavior

Probabilistic Implementation

We have used the method of "intelligent forgetting", which permits us to simply maintain the consistency of the graph and is very well adapted to manipulation tasks.

Indeed, cycles created by perception actions, have only short existences : they are created to carry out an action which itself modifies the environment and generally destroys one or more relations appearing in the cycle; in addition, the perception information is generally not used after the action has succeeded.

Thus it is possible to locally modify the graph when an unintended cycle is created, by destroying one of the cycle arcs.

The drawback of this method appears, when the cycle has a long life, as in mobile robotics, The suppression of the arc would be done according to local considerations, which could be found "wrong" later on.

6 Conclusion

Comparison of the two methods

calculations : In the set method, we have simple operations on sets (Minkowski sum, union, intersection ..). In the probabilistic method the calculations rapidly become extensive (particularly in case of dependent observations).

statistical error : By its nature, the calculations of the probabilistic method compensates for the statistical error in observations. On the other hand, the overestimation of uncertainties is less important in the probabilistic approach.

action : Generally, the uncertainty modifications due to an action are modelled by a projection on the degrees of freedom staying free. Operations very simple in both cases.

It might be necessary to know the maximun value of an uncertainty (or one of its components), for the analysis of assembly motion. This value can be obtained from the probabilistic method, only as a percentage.

Applications

The operators we have defined can be used for an applications involving robot manipulations robotics field where the environment is known, simplifying the fusion operation (for the probabilistic method).

One such application concerns the problem of automatic verification of the feasibility (with respect to uncertainties) of a plan of actions expressed in a high level language (Assemble object A and object B), can be found in [TP87], [Maz89].

As for mobile robotics, it is necessary before developing an application to completely solve the multi-sensor fusion problem (non-independent sensor fusion, fusion of data of any kind) and to improve the system to maintain consistency.

Acknowledgements The work presented in this paper is a concatenation of work done by P. Puget at LIFIA (Grenoble, France) and by I. Mazon at LAAS (Toulouse, France). I would like to thank P. Puget for instructive discussions and his collaboration on the draft version of this paper.

References

[AC86]   R. Alami and H. Chochon. Programming of flexible assembly cells: task modeling and system integration (reprinted from ieee ra - st-louis 1985). In *ROBOTICS and Industrial Engineering - Selected Readings - Volume II*, 1986.

[Bro82]   R.A Brooks. Symbolic error analysis and robot planning. *International Journal of Robotics Research*, 1(4), December 1982.

[Cho86]   H. Chochon. *Programmation de tâches d'assemblage robotisées: modélisation et processus décisionnels*. Thèse de l'Université Paul Sabatier, Toulouse (France), Laboratoire d'Automatique et d'Analyse des Systèmes (C.N.R.S.), November 1986.

[CL85]   R. Chatila and J. P. Laumond. Position referencing and consistent world modeling for mobile robots. In *IEEE, International Conference on Robotics and Automation, St Louis (USA)*, April 1985.

[Dur86]   H.F. Durrant-Whyte. Consistent integration and propagation of disparate sensor observations. In *IEEE, International Conference on Robotics and Automation, San Francisco (USA)*, April 1986.

[Fa86]   O. D. Faugeras and al. Building visual maps by combining noisy stereo measurements. In *IEEE, International Conference on Robotics and Automation, San Francisco (USA)*, April 1986.

[Lau87]   C. Laugier. *Raisonnement géométrique et méthodes de décision en robotique. Application à la programmation automatique des robots*. Thèse d'état, Institut National Polytechnique de Grenoble, December 1987.

[LB85]   T. Lozano-Perez and R.A. Brooks. *An approach to automatic robot programming*. A.I Memo 842, Artificial Intelligence Laboratory, MIT, April 1985.

[Maz87a]   E. Mazer. *HANDEY : un modèle de planificateur pour la programmation automatique des robots*. Thèse d'état, Institut National Polytechnique de Grenoble, December 1987.

[Maz87b]   I. Mazon. *Modélisation des incertitudes de positionnement. Application à l'aide à la programmation de tâches de manipulation robotisées*. Rapport interne 87-395, Laboratoire d'Automatique et d'Analyse des Systèmes (C.N.R.S.), December 1987.

[Maz89]   I. Mazon and R. Alami. Representation and propagation of positioning uncertainties through manipulation robot programs. integration into a task-level programming system. *Submitted in IEEE, International Conference on Robotics and Automation*, 1989.

[SC87]   R.C. Smith and P. Cheeseman. On representation and estimation of spatial uncertainty. *International Journal of Robotics Research*, 5(4), Winter 1987.

[TP87]   J. Troccaz and P. Puget. Dealing with uncertainty in robot planning using program proving techniques. *Robotics Research: The Fourth International Symposium, Santa Cruz (USA)*, September 1987.

Contact manipulation and geometric reasoning

A. GIRAUD, D. SIDOBRE
Groupe Robotique et Intelligence Artificielle
LAAS-CNRS
7 avenue du Colonel Roche, 31077 Toulouse Cedex, France

Abstract

In assembly tasks, part-mating involves contact between parts with size tolerances and for some assembly strategies, jamming results. A way to avoid such problems is to explicitly use the contact during motion with an active compliant procedure. This paper deals with contact manipulation.

The first part models the assembly task taking into acount the contact relationship, and provides a solution to program active compliance. An assembly strategy is a set of motions constrained by the contact. The second part models the kinematics of contact relationships between polyhedral parts. The motion is made in space where the degrees of freedom are limited.

An assembly strategy is then a sequence of contact relationships. The papers describes a method to generate such strategies via the concept of a contact graph describing the changes in contact relationship when a new constraint is obtained along a trajectory. Thus motion occurs in a space with one degree of freedom. When a zero degree of freedom contact is obtained, a new connexe contact relationship with one degree of freedom is chosen.

1 Introduction

Work on assembly robotics has often focused on methods of position control and programming for manipulators designed to accomplish assembly tasks. However, the poor performance of these methods, particularly in terms of accuracy, dimensions and positions, has led to the concept of "fine motion synthesis" characterizing the need for a better understanding of the complexity of the physical phenomena involved in the contact between manipulated objects.

The physical reality shows that contact force and position are closely related in assembly robotics (objects are not in contact, partially in contact, or attached) and that contact forces of given signs but of approximate value very often suffice to obtain a successful "blind" manipulation.

To efficiently perform assembly, force generating manipulators should be preferred to their position generating counterparts. But the current technology is more amenable to the realization of position generating manipulators in spite of the fact that the latter are not sufficiently accurate for the assembly of adjusted objects with size tolerances .

On-going research issues at LAAS in assembly robotics are the control of a robot manipulator equipped with a force sensor, assembly task programming, and the synthesis of automatic assembly strategy through methods of geometric reasoning .

This paper reviews these areas of robotic research and presents on-going developments. Also, emphasis is placed on current trends, the theoretical solutions obtained, and particular practical solutions.

The section on contact manipulation defines the notion of "primitive" for bringing manipulated objects into contact, provides a particular solution for programming in the task space, and defines the link with the manipulator control. The concept of contact is an efficient means of defining the relative position of the manipulated objects. The section on "kinematics of a contact link" details the method allowing definition of the number of degrees of freedom (DOF) of an object in contact with another object, and the method for generating the motion that allows maintenance of contacts. These motions (translation or rotation) ensure transition from a qualitative contact position to another situation.

The "assembly strategy" section reviews this concept and poses the problem of the generation of assembly strategies from a 3-D description of objects to be assembled. The practical approach is close to the one given in [1], but the work is extended to the 3D space. It is based on a theorem given by [2] and [3]. The simplest form of this theorem states that if two objects in contact can be moved to another configuration in which they are in contact, then there is a way to move them from the first configuration to the second configuration such that the objects remain in contact throughout the motion (if translation and rotation of objects are allowed). The decomposition and retractation approaches in [4] are used. When two objects constrained by a contact are in a zero degree of freedom configuration, a vertex of a contact graph is defined. All the edges connected to this vertex define a new contact with one degree of freedom. A trajectory including translation and rotation can be computed for each edge and a motion along this trajectory leads to an another vertex of the contact graph.

2 Contact manipulation

Contact manipulation occurs in assembly whenever the absolute positions of the manipulated objects and of the working environment (grippers, devices, etc...) are not sufficiently accurate relative to the dimensional tolerances of the object to be assembled. Contact then entails forces which usually distort the manipulator and/or the compliant structure that may be attached to it. Because of the complex nature of the force- distortion mechanisms (due to manipulator motion; entailing geometry, surface

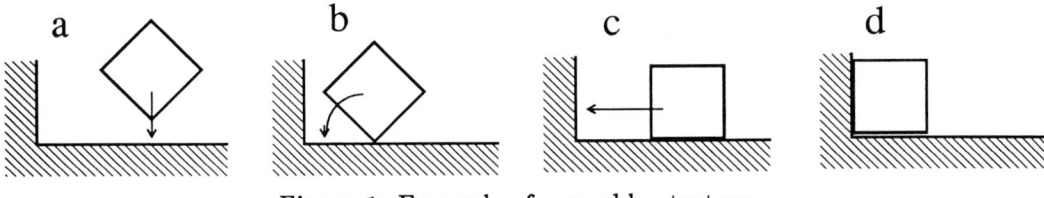

Figure 1: Example of assembly strategy

condition, object material) these forces may lead to the readjustment of the relative position, thereby favoring insertion or, conversely, jamming the relative motion of these objects.

The execution of assembly tasks with a reasonable rate of success requires applying one or several of the following approaches according to the context:

- refining the object dimensional accuracy (notably that of the manipulator and its gripper) and that of the dimensional data required for programming. Application of calibration and/or adjustment procedures.

- use of a passive compliant structure favoring the auto centering of the objects during insertion and avoiding jamming as much as possible. Use of insertion aid devices.

- slaving of the manipulator motions to the contact force, estimated through measurement from a force sensor (force feedback control: active compliance).

2.1 Principal objective of the expanded active compliance

Insertion primarily entails the accuracy of the relative position of the manipulated objects. This is done by bringing these objects into contact and by sliding the latter on their contact surfaces. The objective of active compliance is to arrive at a characteristic contact relationship through handling of the contact forces via the manipulator motion: point on plane, plane on plane. A succession of such contacts defines an insertion strategy (see Fig. 1).

Objects are brought into contact by the motion of the manipulator endpoint M (see Fig. 2) evolving according to six DOF as it is supposed attached to the 6 DOF compliant structure. The manipulator motion is defined by its position pM, velocity vM and acceleration gM.

Because force is taken into account during contact manipulation, modeling is carried out in a dynamic fashion: motions are characterized by the accelerations (for example, gA/B for the relative motion of A with respect to B) in the coordinate system (not necessarily orthonormed) chosen to describe contact.

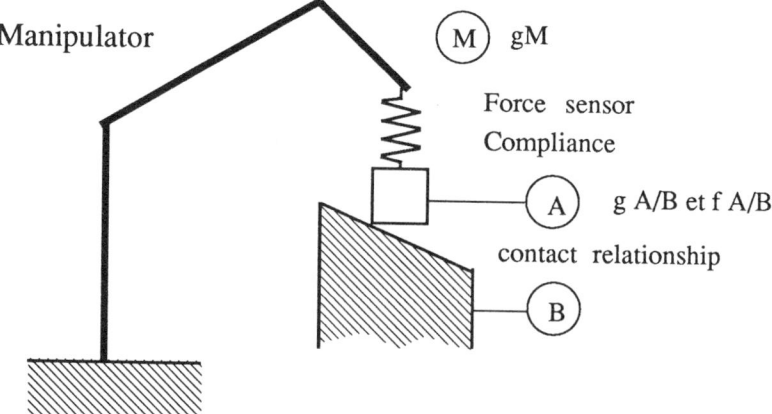

Figure 2: Mechanical structure of the contact manipulation

2.2 Modeling of the contact relationship

The stability of the manipulator's motion slaving to the measured force relies on the mechanical process that monitors motions, contacts and distortions. In this process, the so-called kinematic model of the contact relationship plays a preeminent part and can be characterized by the following items:

- list of vertices, edges, and contact surfaces

- degrees of freedom of the mechanical link defined by the maintenance of the preceding contacts. Object A will then be subject to acceleration gA/B according to these degrees of freedom.

- forces fA/B (generally unilateral) supported by the degrees of freedom which have been blocked in order to maintain contact.

- forces in the link space of the degrees of freedom, induced by dry or wet friction phenomena: these are not modeled yet.

2.3 Modeling of the manipulator motion relationship-contact relationship

Contact force fA/B is induced by the deformability of the compliant structure whose model is a nonsingular 6×6 matrix. Since contact forces are linearly a function of the distortion of the compliant structure, it is convenient to consider in this modeling the second derivative $f"A/B$ of contact forces which, through certain aproximations (omission of the successive derivatives with respect to time gM and of gA/B), is proportional to gM.

An approximated dynamic model can then be established as:

$$\begin{pmatrix} f"A/B \\ gA/B \end{pmatrix} = MAT \times gM + Cte \qquad (1)$$

where gA/B characterizes the motion of A relative to B, $f"A/B$ an approximation of the second derivative relative to the force time of the contact and Cte a term characterizing the dynamics of the pendular system composed of part A and the compliant structure.

2.4 Manipulator control for contact manipulation

gM must be defined such that part A takes the desired position relative to B while exerting the required contact force. Upon establishment of the state of the contact relationship derived from the interpretation of both the manipulator's position and velocity measures and the force measures, we apply a proportional-derivative control in the task space to arrive at the position and velocity goals by generating gA/B and $f"A/B$. The model (1) allows definition of the gM goal for manipulator control. The latter can be completed by the force fM generated by the task on the manipulator if the latter is force-controlled or provided with dynamic control.

2.5 Manipulator control to achieve a contact relationship

Modeling of the contact relationship is realized for the desired contact, irrespective of the current contact relationship. The preceding control is then applied. Note however, that as manipulation entails structural variations in the contact relationhip, stability and convergence problems may be encountered. The latter, or at least the simplest cases, may be addressed by applying an assembly strategy as defined above.

With respect to principles, stability is obtained through use of on-line identification of the contact relationship, followed by real time modeling of the motion relationship between the manipulator motion and the contact motion/force. In practice, it is clear that slow contact manipulation based on small gains ensures stability and that a higher level of complexity of real time control is not desirable.

2.6 Practical approach for contact manipulation

The objective of this practical approach is primarily to provide the user of an assembly manipulator with easy to use primitives for bringing objects into contact; these primitives also greatly facilitate handling of the force sensor. A second objective is to reduce the computation load required for slaving in order to allow for real time implementation of the position, force slavings (for clarity, velocity and, subsequently, proximetry slavings have been intentionally omitted as they can easily be included in the proposed approach). The salient features of the implementation are described as follows:

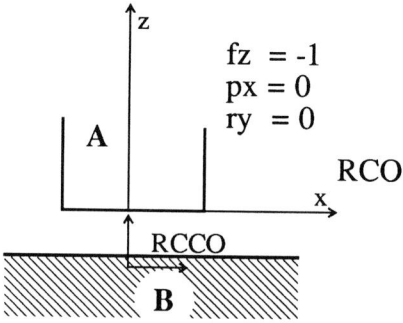

Figure 3: Control to obtain contact primitive

2.6.1 Characterization of the contact relationship

It is basically achieved through two orthonormed frames RCO and $RCCO$ located on the mobile object A and on the fixed object B, respectively. The latter frame allows definition of the relative position of two parts A and B (see Fig. 3). The force measure obtained from the force sensor located between the manipulator M and A is translated into contact force following correction for the forces due to gravity; inertial forces are discarded.

A plane on plane contact is characterized by a slaving of the type: position px equal to zero for x-axis, force fz of one Newton exerted in z-axis, torque fmy equal to zero for y-axis.

This type of description covers many contacts that are useful in assembly robotics. Note, however that for the Fig. 5 manipulation, a finer characterization is required.

2.6.2 Position and force slaving for each DOF of RCO

For the user, the proposed slavings are of the position and force type in order to improve the manipulated object's motion control.

Real time control will lead to the selection position or force slaving according to the current state of the position and force.

Position slaving leads to the definition, for each RCO axis (for example z-axis) of one acceleration of A relative to B provided by a proportional-derivative control:

$$gpz = a \times epz + b \times vmz \qquad (2)$$

where vmz is the velocity of A relative to B projected onto the z-axis of RCO, epz a position error varying according to the position goal (see Fig. 4), the current position, two thresholds actually defining an area of uncertainty where the position goal is considered reached.

Force slaving is similarly defined as:

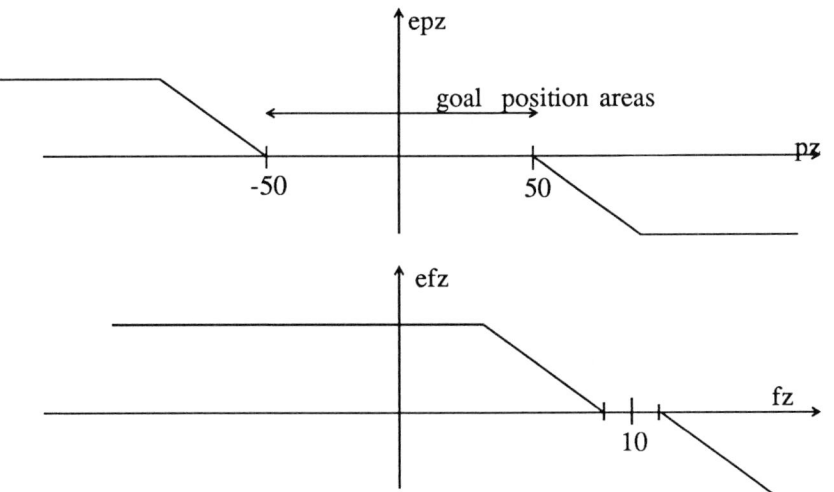

Figure 4: Slaving: position and force error along the z-axis

$$gfz = a \times ke \times efz + b \times vmz \qquad (3)$$

where efz is a current error, ke an elasticity constant such that a displacement of $ke \times efz$ along the z-axis of RCO causes a distortion of the compliant structure proportional to that caused by a force equal to efz. Here, notice an approximation concerning the linear model $MECO$ of the compliant structure evaluated in the RCO frame (ke is a diagonal term of $MECO$ matrix).

2.6.3 Position or force slaving for each DOF of RCO

Selecting the appropriate slaving is achieved in real time with the help of a simple decisional process which is described in the following. It may be noted that gpz and gfz reflect two actions that must be exerted on motion A. The action with priority is that which primarily affects a virtual actuator displacing object A along the z-axis. The following algorithm is used to select acceleration gz (component of gA/B) needed from the manipulator to displace object A:

- $gz = max(gpz, gfz)$ if gpz and gfz are positive or zero,
- $gz = min(gpz, gfz)$ if gpz and gfz are negative or zero,
- $gz = gpz + gfz$ if gpz and gfz are of opposite signs. In this case, position and force slavings are contradictory. The selection of gz leads the manipulator toward a situation of equilibrium that favors safety, the position and force goals being generally not satisfied.

Only discussed in this paper are the simple additions to the decisional process which allow for stability in spite of on-line switching of slaving laws that are assumed to be stable.

Interestingly from a practical point of view, notice that the decisional process implemented in real time allows the user to modify the manipulator's motion since it suffices to exert a suffiently disrupting effort in the desired displacement direction. Object A will than maintain its new position within the thresholds of uncertainty defined for position slaving along the RCO axes.

2.6.4 Determining manipulator motion gM

The motion of part A is defined by acceleration gA/B, each component of which is defined by the preceding process. The link between the point M of the manipulator and part A is assumed to be rigid and the manipulator motion gM is obtained through simple change in gA/B frame. The model of compliant structure which plays a key role in the stability of the active compliance is briefly taken into account by gains ke.

2.6.5 Decisional process between RCO axes

The aim of this process is to facilitate accomplishment of insertion tasks. It consists of switching in real time the position and force slaving parameters along one of RCO axes, for example z (as in Fig. 4) when the position and force goals are reached on another RCO axis, say the x-axis. Thus in Fig. 1, seeking a left vertical contact can only be achieved if the horizontal plane on plane contact is established.

2.7 Contact manipulation programming

Given the current contact relationship between the two objects to be manipulated, it is necessary to select the contact relationship toward which it is possible to converge with the aid of position and force slaving. Programming then requires to choose those orthonormed frames RCO and $RCCO$ which best express the contact to be made, the position and force goal values for each of the generalized coordinates implicitly introduced by RCO and the uncertainty areas associated with these reference inputs. The manipulator control defined with the help of these data then allows for a new contact relationship.

This programming based on contact primitives relies on a 3-D description of objects from which, given the relative positions of objects, it is possible to deduce the numerical values required. It seems possible to envisage, from a 3-D database, an efficient aid to programming and in the long run, automatic generation of assembly strategies .

3 Kinematic modeling of a contact link

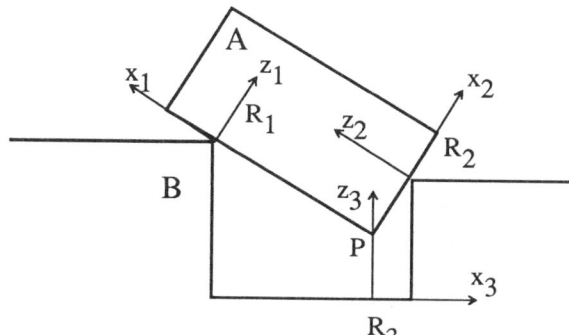

Figure 5: Points contacts

3.1 Principle

3.1.1 Contact of point on surface

The contact between objects A and B is initially analyzed as a set of contact points (sometimes infinite) on planes tangent to the contact surfaces.

At each point of contact, we define a fixed frame and a mobile frame relying on this point, which are superposed at the instant considered but linked to either a solid mobile or a fixed solid (R_{1A} and R_{1B} in Figure 5). One of the preferred frame directions is the normal to the contact often associated with direction z.

Allowing the motion to maintain contact requires that the velocity between the two frames in the direction of the normal to the plane tangent to contact, be zero

This may be written as: $C \times V^t = O$ where V is the velocity vector of the mobile frame relative to the fixed frame and all components of C are zero except for the one corresponding to the direction of the jammed translation (generally the z-axis).

This condition is a linear constraint on the translation velocity. A linear system is obtained for one contact and has to be reduced .

This linear relationship may be expressed irrespective of the couple of frames which are superposed at the instant considered and linked to either a mobile solid or a fixed solid.

Example: Contact constraint associated with frames R_{1A} and R_{1B} in Figure 5.

$$\begin{pmatrix} 0 & 0 & 1 & 0 & 0 & 0 \end{pmatrix} \times \begin{pmatrix} v_x \\ v_y \\ v_z \\ vr_x \\ vr_y \\ vr_z \end{pmatrix} = 0$$

If this relationship is verified, the motion maintains contact in O_1.

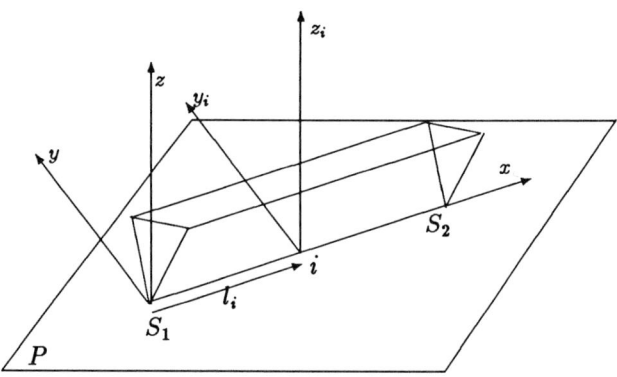

Figure 6: Line/plane contact

3.1.2 Equivalent systems of constraints

Two linear constraints $C_j \times V_j^t = O$ and $C_i \times V_i^t = O$ expressed in a single frame are equivalent if they are linearly linked.

By substituting frames it is possible to derive from the translation constraints, those rotation constraints which simplify the expression and the system of constraints.

Example of a line in contact with the plane:

In the case of a segment $S_1 S_2$ in contact with plane P (see Fig. 6) contact frames R_i can be defined at each points i and the corresponding constraints C_i.

In the local frame, a constraint is expressed as:
$C_{i.Repf_i} \times V_i = \begin{pmatrix} 0 & 0 & 1 & 0 & 0 & 0 \end{pmatrix} \times V_i = 0$

In frame RS_1, this constraint is expressed as: $\begin{pmatrix} 0 & 0 & 1 & 0 & l_i & 0 \end{pmatrix} \times V_{S1} = 0$

The contact of the line with the plane can therefore be modeled by:
$\begin{pmatrix} 0 & 0 & 1 & 0 & 0 & 0 \\ 0 & 0 & 0 & 0 & 1 & 0 \end{pmatrix} \times V_{S1} = 0$

We obtain a system consisting of more than six constraints. Generally, (see concluding remarks), the complement to six yields the number of DOF of the contact link.

3.1.3 Representation of the object position: Generalized coordinates

An object position is generally defined by the relative position of two orthonormed frames. One expression is a vector translation (tx, ty, tz) and a rotation vector (rx, ry, rz)

parallel to the rotation axis, and whose norm is the rotation value. A generalized coordinate is one component of one vector for any frame couple, and a position is defined by six independent generalized coordinates. Such an approach is convenient to express the position of parts in an assembly task.

For example, in Fig. 5, the position of P along coordinate z_3 of R_3 defines the position of A relative to B in the motion that maintains the contacts by $z1 = 0$, and $z2 = 0$.

3.1.4 Choice of generalized coordinates of motion

In the case of Figure 5, the generalized coordinate z_3 defines the state of the insertion and completes the modeling of the assembly to be performed. One wishes to obtain $z_3 = 0$ by maintaining contacts 1 and 2.

In the general case it is necessary to choose the generalized coordinates that complete the contact system to obtain six independent linear relationships.

3.2 Kinematic model: $C \times V^t = O$

A linear contact relationship was established by means of a contact point and a goal to be reached. It is now necessary to reduce this six equations system to obtain a Jacobian matrix such that:

$V_{A/B} = JAC \times CG$ where CG represents the generalized coordinates of the contact and goal.

Hence there are two operations:

- simplification of the system of constraints. A constraint by means of blocked DOF is obtained.
- verification of the compatibility of the goal generalized coordinates with the contraints.

Computing matrix JAC entails the linear computation with, in particular, a $3O \times 3O$ matrix inversion.

3.3 Motion generation: $V_{A/B} = JAC \times CG$

The relationship $V_{A/B} = JAC \times CG$ directly yields the velocity of the contact motion by taking as value for CG:

- 0 for the generalized coordinate which express a contact constraint.

- q (in general $q \neq 0$) for the goal generalized coordinates.

Integrating this velocity permits computation of the trajectory of a motion maintaining contact. If the contact has one DOF, the trajectory is unique.

4 Expression of an assembly strategy

4.1 Succession of contacts

An assembly strategy consists of a succession of motions constrained by contacts. A motion always results in a new contact. The latter may constitute a limit situation, for example in the case of an edge contact on side, the encounter of an edge vertex with the edge side. In all cases, motion results in a contact with one DOF less.

4.2 Contact graph

A particular case of interest is that of the survey of contacts with one or zero DOF. A trajectory corresponding to a one-DOF contact is limited at its two endpoints by a zero DOF contact.

At each zero- DOF contact, it is possible to associate one- DOF subcontacts. For those subcontacts which define a possible motion, a new trajectory is obtained which possesses a zero-DOF contact as second endpoint.

If zero- DOF contacts are associated with the graph vertex and one- DOF contacts are associated with a graph edge, the contact graph is obtained.

The latter represents the boundary of the possible space configurations of the positions of objects A and B.

The study of contact parts with one or zero DOF reduces to that of a graph whose arcs consist of one-DOF motions. See in Figure 7 the start of a graph. Vertices are shown with squares.

4.3 Seeking assembly strategies

Seeking assembly strategies is therefore equivalent to searching a path on the contact graph between a vertex where objects are assembled and a vertex where objects are not assembled, or more precisely, a vertex that can definitely be obtained.

It is possible to pass from a disassembled position (no contact) to a position defined by a zero- DOF contact by introducing fictitious contacts.

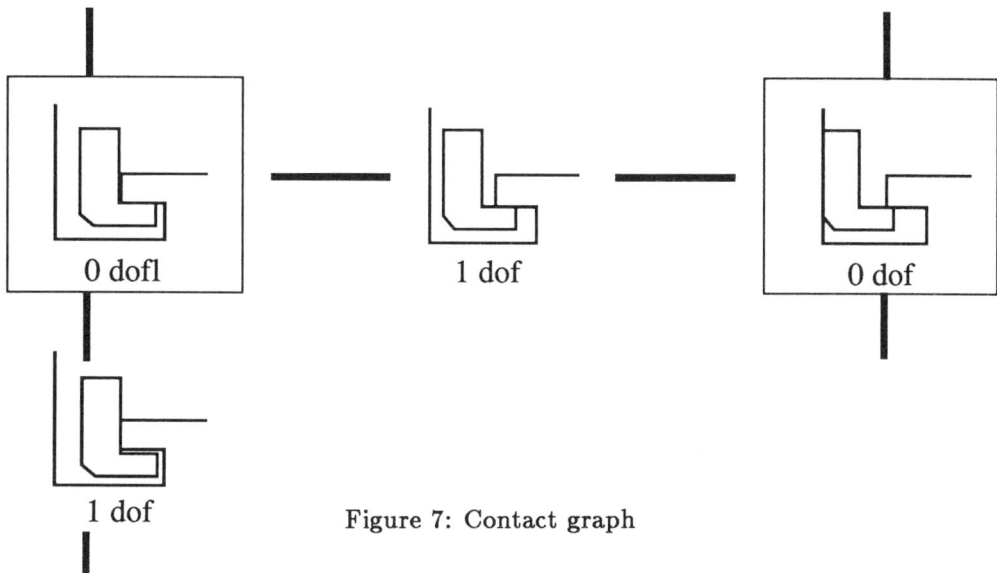

Figure 7: Contact graph

Figure 8 shows an assembly strategy in the graph. In this example, there are two vertices which define the assembly position. The disassembled vertex can easily be reached from the disassembled vertex without contact by maintaining the two fictitious contacts on the edges in dotted lines.

5 Concluding remarks

5.1 Advantages of the notion of contact

Rotations may be treated implicitly, thereby avoiding discretization of space configurations.

The notion of contact is directly linked to contact forces and hence to the programming of force for manipulators. The linear model can directly take the form of a reference control for a position, velocity and force controlled manipulator.

- a contact constraint by means of a force goal.
- a displacement along a generalized coordinate by a velocity goal.
- a virtual contact constraint by a position goal.

The notion of assembly strategies is easily defined and the concept of contact allows for their constructions.

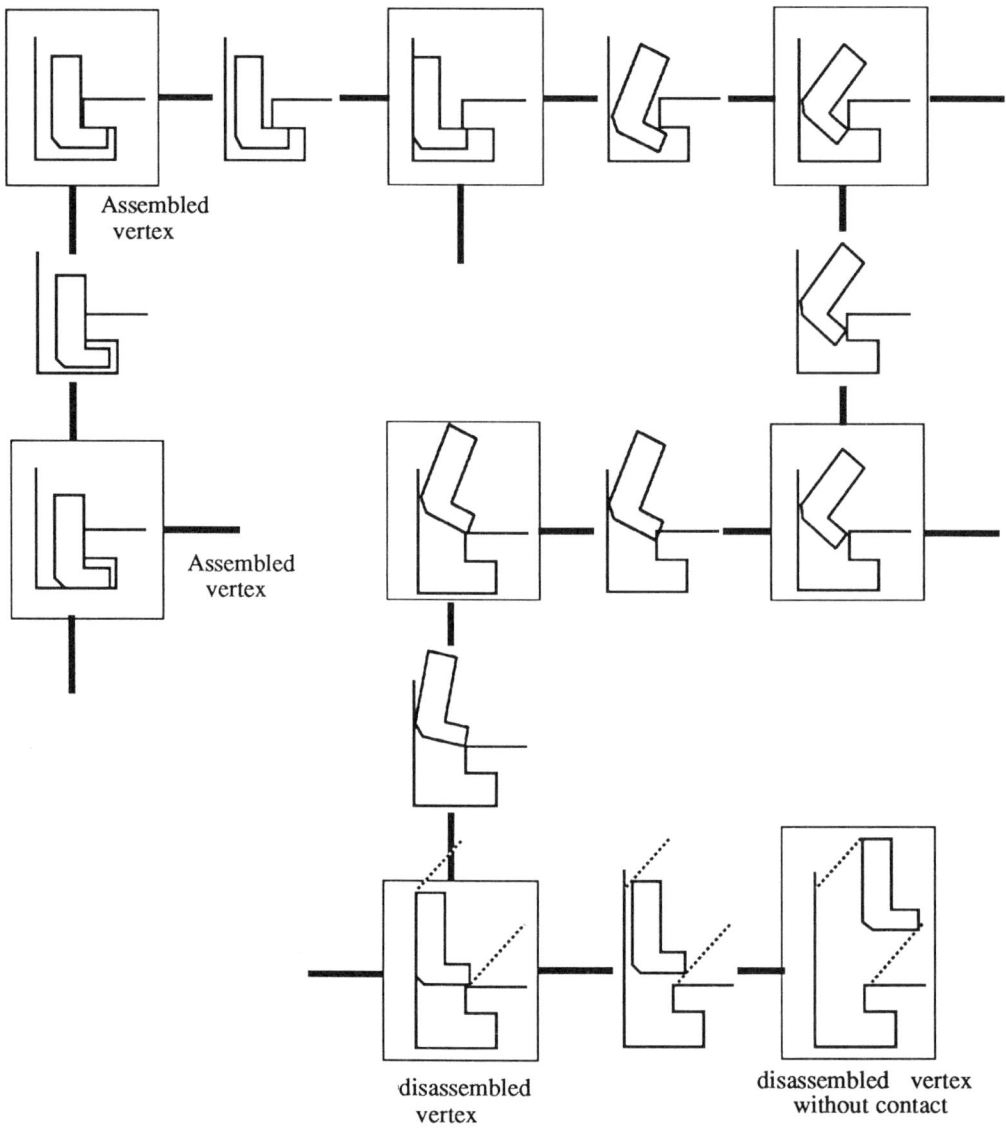

Figure 8: Assembly Strategy

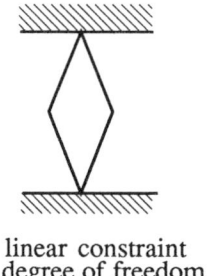
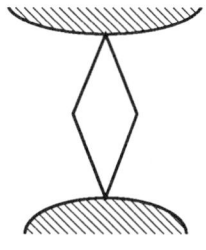

1 linear constraint
1 degree of freedom

1 linear constraint
0 degree of freedom

Figure 9: Limit of the linear model

6 Points requiring further study

Limits of the kinematic model:

In Figure 9, the linear model associated with both cases is the same and reduces to a single linear contact relationship.

On the other hand, the number of DOF is not he same in both cases.

To correctly model these particular cases, it is necessary to introduce a second order model taking into account the radius of curvature of the surfaces.

Calculation of the number of DOF in these particular cases is of limited interest for the study of the contact graph. These limiting cases may be surveyed by introducing a small displacement and checking whether it results in a loss of contact.

Heuristics to compute an assembly trajectory:

The size of the contact graph does not currently allow its complete calculation like it is possible in the 2D space [1]. Heuristics must be developed to compute only locally the contact graph and to search a path.

Existence of solutions in the contact graph:

If a solution exists for the assembly, a theorem in [2] states that a solution exists where the parts remain in contact. But how to prove the existence of a solution in relation to the contact graph connectivity remains an open question.

Calculation of contact graph:

Transition from a 3D model of the objects to the contact graph is not straight-froward and necessitates the introduction of algorithms for generating a graph and for searching it. A 3D C.A.D. data base and an efficient detection of two intersecting objects are needed (this computation is carried out at each increment of displacemnt along a graph edge). Heuristics are mandatory because of the elevated number of vertices. Also, notice that it is difficult to define the criteria that a good strategy would have to satisfy.

Note that in spite of the theoretical aspects and apparent volume of computations, the results are usually simple. In the case of Fig. 5, the first calculated arc yields an assembly strategy and in the Fig. 8 example, the strategy obtained does correspond to the trajectory that would be arrived at manually. In addition, a position and force controlled robot can immediately carry out this strategy.

References

[1] F. Avnaim and J-D. Boissonnat. A practical exact motion planning algorithm for polygonal objects admist polygonal obstacles. In *IEEE International Conference on Robotics and Automation, Philadelphia (USA)*, 1988.

[2] Hopcroft and Wilfong. On the motion of object in contact. In *Robotics Research*, MIT Press, 1984.

[3] J.E Hopcroft and G.Wilfong . Reducing multiple object motion planning to graph searching. *SIAM Journal Comput.*, 15(3), Août 1986.

[4] Chee-Keng Yap. *Algorithmic and Geometric Aspects of Robotics.* Volume 1, Lawrence Erlbaum Associates, London, 1987.

Geometric reasoning in motion planning*

Christian LAUGIER[†]
LIFIA/IMAG
46 Avenue Felix Viallet
38031 Grenoble Cedex, FRANCE.

Abstract

Automating the programming of assembly robots necessitates to develop methods for planning robot motions. In this paper we describe the geometric models and the reasoning techniques we have implemented as part of the SHARP system (SHARP is an automatic robot programming system currently under development at the LIFIA laboratory). We first present which modelling facilities are required for constructing a suitable representation of the robot world. Then, we show how this representation has been used for implementing two classes of reasoning functions: functions aimed at computing collision free trajectories for the robot and its payload, and functions allowing to automatically generate contact based motions under uncertainty constraints (i.e motions involved in grasping and in part-mating operations). Our method for solving the first motion planning problem operates in the configuration space. It is based on two types of techniques aimed at computing the valid ranges of values associated with some selected motion directions, and at constructing and searching a graph representation of the free space. Solving the second planning problem makes it necessary to construct an explicit representation of the involved contacts along with their associated moving constraints. It leads to reason on the morphological properties of the manipulated objects.

Keywords: automatic robot programming, motion planning, computational geometry, spatial reasoning, morphological reasoning, sensory interaction.

1 Introduction:

Programming a robot for a specific assembly task requires to determine which robot actions have to be executed, which sensing operations are necessary, and the way actions and sensing are to be combined. The main problem to solve for automating this programming process consists in determining all the parameters of the involved robot motions (trajectory, velocity, accuracy, involved forces and torques ...). A general formulation of

---

*This paper has also been presented at a NATO workshop on CAD based programming for sensory robots
[†]Senior Researcher at INRIA

this problem can be stated as follows: *given a description of the initial and of the goal situations along with a complete model of the robot world, find a robot motion allowing to reach the goal situation without generating any collision between the robot (the arm and the payload) and the objects belonging to the robot workspace; moreover, the generated solution must verify various constraints (contacts, accuracy, velocity, robustness ...) depending on the context of the motion to be executed.*

The major difficulty that have to be faced when solving this problem, relies in the high algorithmic complexity which is associated to the find path problem. On a theoritical point of view, Schwartz and Sharir [49] have shown that there exists polynomial time algorithms for solving the trajectory planning problem for any type of manipulator. However, the execution time associated with these hypothetical algorithms have terms which make them impractical, even in relatively simple cases. Fortunately, a more "practical" work done in the context of automatic robot programming [34] [33] [40] [30] [41] has shown that it is possible to consider three main instances of the problem which can be separately solved using more specific approaches:

(1) *Transfer motions*. They represent large motions which are executed in a moderately occluded environment. Planning such motions requires to take into consideration the whole robot arm, in order to generate "safe trajectories" for the robot and its payload (i.e trajectories which can be executed with a high velocity without generating any collision). This means that the computed solutions are located far enough from the obstacles in order to avoid unexpected collisions caused by the control errors. Consequently, it is possible to approximate objects by more simple shapes for reducing the amount of geometric computations.

Two classes of approaches have been developed for solving this motion planning problem: the local approaches and the global approaches. The *local approaches* lead to progressively construct a safe trajectory by reasoning at each step on the local characteristics of the encountered situations. In this case, the local decisions are made using either a "generate and test" scheme [47] [1] or a function allowing to associate some repulsive fields to the obstacles [24] [15]. On the other hand, the *global approaches* operate on an explicit model of the position constraints imposed by the physical environment. Most of them are based on the "configuration space" scheme allowing to represent a safe trajectory for the robot arm by a connected set of free configuration values expressed in a n-dimensionnal space (where n is the number of degrees of freedom of the robot) [36]. Several instances of the problem have been solved using such an approach: moving a bidimensionnal object in the plane [35] [9] [5], moving a polyhedral object in the three dimensionnal space [36] [6] [10], dealing with rotating joints [56] [18] [14] [29] [39]... A more detailed analysis of these methods can be found in [4] and in [32].

(2) *Grasping operations*. These operations involve small motions which are executed in a very constrained environment located in the vicinity of the robot gripper. These movements allow to both reach the chosen grasping position in the initial environment, and to remove the gripper from the final environment once the planned manipulation have been achieved. The related trajectories are often very simple, but they are constrained by several factors like the type of the involved contacts, the stability of the gripped object and the accessibility of the selected features.

Four different aspects of the automatic grasping problem have been studied in the literature [54]: (1) *The determination of the grasping features* is done by symbolically reasoning on the geometry of the object to be grasped; this reasoning is either based on sensing (visual or tactile) informations [19] [2] [22] [55] [52], or on a complete geometric model [34] [59] [25] [28] [27] [60]. (2) *The stability* of the object in the gripper have been studied using either mathematical models of static and friction [60] [16] [7] [44], or heuristics leading to apply some simple geometric computations [25] [54] [20]. (3) *The accessibility analysis* is aimed at determining if a given solution is reachable in the initial and in the final environments. The applied methods generally operate in the configuration space of the gripper [34] [27] [55]. (4) The problem of the *compatibility* of the selected grasp with the planned manipulation have currently received very few attention [20] [42].

(3) *Fine motions*. They represent sequences of small movements which are executed in a very constrained environment located in the vicinity of the manipulated object. Such motions have to be guided by the sensory data, since the position uncertainty may make the robot fail. The applied technique consists in considering the local environment as a "geometric guide" for the robot. The related trajectories are very simple, but they are constrained by the involved contacts and by the limited accuracy of the robot command and of the sensing operations.

Several types of techniques have been developed for dealing with uncertainty in robot programming. Some of these techniques are aimed at executing "compliant motions" involving both force and position parameters in the command [58] [43] [48] [17]. The other techniques were developped for the purpose of constructing complete fine motion strategies. A first approach for solving this problem consists in generating a solution by instanciating some predifined "procedure skeletons" using error bounds computations [50] [34] [40], or by assembling a set of partial strategies using learning techniques [11] [12]. A more general approach consists in constructing the fine motion strategies by reasoning on the geometry of the task [37] [13] [31]. As we will see further, our fine motion planner is based on this approach.

In this paper we describe the geometric models and the reasoning techniques we have implemented as part of the SHARP system (SHARP is an automatic robot programming system currently under developpement at LIFIA). We first present in section 2 which modelling facilities are required for constructing a suitable representation of the robot world. Then, we show in section 3 how this representation has been used for implementing two classes of reasoning functions: functions aimed at computing collision free trajectories for the robot and its payload, and functions allowing to automatically generate contact based motions under uncertainty constraints (i.e motions involved in grasping and in part-mating operations). Our method for solving the first motion planning problem is described in section 4. It is based on two types of techniques aimed at computing the valid ranges of values associated with some selected motion directions, and at constructing and searching a graph representation of the free space. Solving the second planning problem makes it necessary to construct an explicit representation of the involved contacts along with their associated moving constraints. It leads to reason on the morphological properties of the manipulated objects. This point is developed in section 5. Finally, section 6 describes the motion planners which have been implemented in the SHARP system.

2   World modelling:

Several aspects of the world model have to be constructed and maintained by the system in order to make possible the geometric reasoning involved in motion planning (see [32]): the geometry and the physical properties of objects, the evolution of objects relationships, the robot motions, and the robot states (including sensory informations) associated to each world state. Most of these informations are not explicitly represented in the initial CAD models. Consequently, one of the first task of the system is to construct a more suited model of the robot world combining three main representations: a geometric model including topological informations on objects, a structured representation of world states, and a model of robot motions. All these informations are used at the planning time for constructing a problem oriented representation of the robot world. We will call "planning space" such a representation (see section 3.1).

2.1   The geometric models:

The geometric models required for motion planning are basically the same that those previously developped in the context of off-line robot programming [27] [53]: boundary representations for solid objects made of polyedra, parallelepipeds, cylinders, cones and spheres; numerical data specifying geometric parameters (radius, dimensions, coordinates ...) and geometric transforms.

Since motion planning requires to apply a lot of geometric computations involving volumic and topological properties, these models have been completed by two types of constructions aimed at reducing the amount of computations [54] [32]: a hierarchy of elementary surrounding volumes (parallelepipedic boxes), and an explicit representation of the topology of objects (spatial hierarchy, local structures of the type "winged edge", and matter distribution in the vicinity of the elementary elements like faces or edges). Such constructions allow the system to apply fast interference checking algorithms, and to easily extract local informations on object shapes. For example, the grasp planner can check for the possibility of establishing a contact between a jaw of the gripper and a particular feature of the object, by applying very simple algebraic computations involving a small number of geometric entities (for example: a face and its normal external vector, and the counterclock oriented edges located in the vicinity of the analyzed object feature).

2.2   The world states:

Since the robot modify its environnement when operating, the world model has to be changed according to the executed actions. If we make the assumption that the robot operates in a "closed world" (i.e any world modification is the consequence of a robot action), it is possible to associate each world change to a particular robot sensing or manipulating operation. Since the purpose of motion planning is to find sequences of actions allowing the robot to progressively reach a goal state from an initial state, it is possible to only represent the world states obtained after each elementary robot action.

In our system, each world state is represented by a directed graph where nodes denote cartesian frames associated to objects, and arcs represent objects relationships. Position informations on objects are expressed in terms of nominal geometric transforms and of associated uncertainties (see [32]). The basic structure of the graph is a hierarchy where the root is the reference frame of the robot world. The other arcs represents physical constraints existing between couples of objects (joints, contacts ...). In the current implementation of the system, contact relations are automatically determined and updated using a set of "demons" (see [51]).

2.3 Modelling the robot motions:

2.3.1 The motion parameters:

In order to guarantee that the planned motions will achieve their expected goals, the system must reason on a model including physical informations like forces and frictions, control and sensing errors, characteristics of the applied command, and termination predicates. In our system, we have implemented very simple models for representing these parameters [32]: friction cones based on a rough approximation of the static coefficients, error bounds for control and sensing, generalized spring type of command, and termination predicates represented by cartesian products of the type $P_a \times F_a$, where P_a is a subset of position in $\Re^3 \times SO^3$ and F_a is a set of reaction forces. Such a model is sufficient for solving the assembly problem we are concerned with, if we make the assumption that the involved contacts can be unambiguously identified using position and force data. More complex models may be required for applications having possible sensing interpretation ambiguities (see [13] and [8]).

2.3.2 The configuration space:

Geometric aspects of motions have also to be modelled in order to make motion planning possible. The developped representation is based on the *configuration space* scheme, first introduced in [36].

Definition 2.1 *Let A be a mobile system composed of l ($l \geq 1$) rigid elements moving in a cartesian space $\Re^k$ ($k = 1, 2$ ou 3). A configuration $\dot{c}$ of A is a minimal set of parameters, allowing to unambiguously specify the position and/or the orientation in $\Re^k$ of each rigid componant of A.*

A configuration c of A is represented by a vector in a n-dimensionnal space. We will say that A has n degrees of freedom (d.o.f), and that $A(c)$ represents the "position" of the whole system A in $\Re^k$ when A is in the configuration c.

Definition 2.2 *One call* **configuration space** C_A *of a mobile A, the set of the possible values of the configuration vector c. One call* **free space** EL_A *the set of configurations c such as $A(c)$ do not generate collision between the components of A and the objects belonging to the environment of A.*

EL_A is a subset of C_A, and C_A is a n-dimensionnal set defined by the cartesian product $I_1 \times I_2 \cdots I_n$, where I_i is the set of possible values for the parameter p_i of c. If A is a solid object in $\Re^3$, C_A may be seen as a subset of $\Re^3 \times SO^3$ (SO^3 is the group of the orthogonal rotations); if A is six d.o.f revolute arm, C_A is a subset of a 6-dimensionnal space often called "6-tore" because of its particular topological structure. Using the notations introduced in [36], the image $CO_A(B_i)$ of an obstacle B_i in C_A (called C-obstacle) and the free-space EL_A may be characterized as follow:

$$CO_A(B_i) = \{c \in C_A : A(c) \cap B_i \neq \emptyset\}$$
$$EL_A = C_A - \cup_{i=1}^{m} CO_A(B_i)$$

The techniques which have been developed in SHARP for computing the C-obstacles and the free-space associated with an articulated robot operating in $\Re^3$ are described in section 4.

3 Outline of our motion planning approach:

3.1 The concept of planning space:

The models described in the sections 2.1 and 2.2 cannot be directly used by the motion planners, because they do not explicitly contain all the needed informations. Consequently, the system has to construct a more suited representation of the robot world before starting to plan. This representation is problem oriented, and it is called the *planning space* associated with the motion planning problem to solve. It characterizes *the sets of robot states that can be realized using the commands provided by the system, and that can be identified using the available sensor devices*. Only position and force sensory informations are considered in our system. Consequently, we chose to represent a *robot state* E_p associated with a commanded position p ($P \in \Re^3 \times SO^3$), by the set of sensory couples (p^*, f^*) which can be read on the position and the force sensors, once the command has been executed by the robot. If $\xi(\Re^3 \times SO^3)$ represents an Euclidian approximation of $\Re^3 \times SO^3$ [13], a robot state E_p can be defined as follow [32]:

$$E_p = \{(p^*, f^*) : p^* \in B(p, \varepsilon_p + \varepsilon_m) \wedge f^* \in cone(n, \phi + \vartheta_f)\}$$

where n is the external normal vector in p to the contact surface, and ϕ is the angle associate with the friction cone. ε_p, ε_m and ϑ_f are respectively the position control error bound, the position sensing error bound, and the force sensing error bound. Then, the planning space E can be defined as set of robot configurations p in C_{robot} which verify the following properties:

- $E_p \cap E_{p'} = \emptyset \quad \forall p, p' \in E$

- $C_{robot} = Closure (\cup_{p \in E} P_p)$

Two major difficulties have to be faced when constructing such a space [8]: the set of robot states is not enumerable, and the dimension of the space do not allow to exactly represent all the constraints drawn from the object surfaces. This is the reason why we will construct an approximate model, by grouping together the robot states which can be considered as "equivalent" relatively to the type of motion to be planned. In order to be exploited by the planners, this model is structured as a *state graph*, where each node n_i defines a set E_i of "equivalent states" and each arc a_{ij} represents the motions allowing the robot to move from any state e_i in E_i to any state e_j in E_j. Using this representation, it becomes possible to consider the motion planning problem as an instance of the graph search problem.

Since motions involving contacts necessitate to take into consideration both position and force criteria whereas free-space motions (i.e motions executed at a distance to obstacles greater than $\varepsilon_p + \varepsilon_m$) only rely on position criteria, two types of representations have to be considered when planning. In our system, these representations are constructed using the following criteria [32]:

- A class of "equivalent states" for free-space motions is defined as a *convex set of robot configurations which generate no collision with the environment*. An important property of this approach is to guarantee that any trajectory corresponding to a straight line (in the C-space) between two configurations of a given class is collision free. As we will see in section 4.5, the classes constructed in SHARP are hyperparallelepipeds obtained by discretizing the joint space.

- A class of "equivalent states" for motions involving contacts is defined as a *connex set of robot configurations which generate similar reaction forces* [8]. Using this definition, a face, an edge or a vertex of an object may be considered as a potential basis for defining a set of equivalent states (the condition to verify in this case is that each sensory couple (p^*, f^*) can be unambiguously associated to a single set of contact points). Several levels of details may be required depending on the characteristics of the motion to be planned. In our system, we choose to operate on a minimal set of classes, by only considering the different contacts that can be realized between the mobile and the concerned fixed objects (see section 5).

The motion planners that we have developed on the basis of these models are *consistant* since the generated solutions are guaranteed to work despite sensory and control errors. Conversely, they are not *complete* since our representation eliminates two types of solutions: those which are ambiguous because of the interpretation mechanisms, and those which are missed because of the applied approximations. But the completeness property is not of prime importance, as long as the missed solutions do not make the system fail in practical cases. Fortunately, the processed mechanical assemblies generate a large set of possible solutions that can be found by the system, provided that the applied approximations are accurate enough. A refining process may be applied in case of failure.

3.2 The geometric reasoning involved in motion planning:

The problem to solve is to automatically generate the various robot motions which are involved in the basic manipulation operations (transfer, grasping and part-mating). As mentioned above, two types of motions have to be considered at planning time: free space motions and contact space motions. The reasoning techniques required for planning these different types of motions are called "spatial reasoning" and "morphological reasoning".

The spatial reasoning techniques are aimed at computing safe trajectories for the robot (i.e trajectories which are both collision free and robust relatively to sensing and control errors). A first class of computational tools have been developed for solving this problem in the context of grasping and part-mating operations. These tools are aimed at computing the "valid ranges of positions" which can be achieved by the gripper and the manipulated object without colliding with the other objects. They operate on an "exact" representation (relatively to the used world model), since involved contacts do not allow to apply too large approximations. In this case, the computed solutions are composed of simple trajectories leading to create or to destruct a set of contacts. The other class of computational tools have been developed in the context of transfert motions. These tools operate on a complete model of the robot configuration space. But in this case the algorithmic complexity inherent in the general trajectory planning problem is reduced by applying various approximations.

The morphological reasoning techniques are aimed at computing the robot states involved in grasping and part-mating operations. The computational tools developed for that purpose allow to compute three main properties of contacts [32]: local accessibility, mechanical constraints and motion constraints. They are based on simple geometric and topological computations performed using local informations on object shapes.

The planning of free space motions can be done by applying pure spatial reasoning techniques, whereas the planning of motions involving contacts requires to combine morphological and spatial reasoning techniques in order to master the algorithmic complexity. The basic idea consists in progressively guiding the search choices, by successively analysing more and more detailed constraints drawn from the geometry of objects. This method has been implemented in SHARP by applying an ordered set of *simple geometric filters*. It leads to separate the computation of potentially reachable positions and valid movements, from the determination of those which are really executable by the robot. The two related reasoning phases are the following:

1. In a first step, the system computes a set of *potential solutions* by analysing the local properties of the implicated contacts. For exemple, it will generate a set of "potential grasps" by studying potentials contacts between the gripper and the object to be grasped. It also will determine "potential moving directions" associated with a set of contacts, by analysing the related topological constraints. This computation is useful when generating a sequence of compliant motions aimed at mating two parts. All the computational techniques applied during this phase are called morphological reasoning techniques, since they lead to reason on local morphological properties of objects.

2. The second phase leads to evaluate the *global accessibility* of the suggested solutions. It analyses the global constraints drawn from the task, in order to reject or to validate the previous choices. The retained solutions are then refined and completed according to the results of the applied computations. For example, the obstacles which constrain the movements of the jaws of the gripper, will lead to

prune the set of possible grasping positions. The computation of the valid ranges of position associated with a potential moving direction, will also lead to specify the missing motion parameters. All the computational tools applied during this phase are derived from the spatial reasoning techniques, since they lead to reason on the spatial constraints drawn from the robot environment.

4 Spatial reasoning techniques:

4.1 Presentation and notations:

The purpose of this section is to describe the basic techniques which have been developed in SHARP for computing the images $CO_A(B)$ of obstacles B in the configuration space C_A of A, and for constructing an explicit representation of the free-space EL_A. We will make use within the presentation of the basic notations and definitions developed in [36].

The main difficulty associated with the computation of C-obstacles $CO_A(B)$ comes from the fact that these sets represent hypervolumes in a n-dimensionnal space which is generally non isomorphic to the Euclidian space $\Re^n$. In general, an exact determination of these sets requires to make use of mathematical tools, which can not be really applied because of their associated algorithmic complexity. This is the reason why we often reason on subspaces of C_A, in order to successively explore different subsets of the d.o.f of A. Such an approach allows to iteratively construct approximations of the sets $CO_A(B)$, when exact representations cannot be computed. We will represent by C_A^{xyz} the subspace of C_A associated with the three translations along axes x, y and z, and by CO_A^{xyz} the related C-obstacles. In the same way, we will represent by $C_{A_q, A_1(v_1) \cdots A_{q-1}(v_{q-1})}^q$ the subspace associated with the joint q of an articulated structure A, when the anterior joint variables are fixed to the values $v_1, v_2, \ldots v_{q-1}$, and the posterior links are not considered. In order to simplify the notations, we will represent by C_A^q this subspace, and by $CO_A^q(B)$ the related C-obstacles.

4.2 Computing valid ranges of positions:

The problem to solve consists in determining all the ranges of position of a mobile object A, that can be reached in a given direction without colliding with other objects B_i. A more formal definition may be stated as follows:

Definition 4.1 *One call valid ranges of position $V_A(q)$ associated with the mobile A moving along a direction q, the set of values q such as: $A(q) \cap B_i = \emptyset$.*

Definition 4.2 *Let c_j be a set of joint values of the type $(v_1, v_2, \ldots v_{j-1}, v_{j+1}, \ldots v_n)$. One call valid ranges of position $V_{A[c_j]}(q_j)$ associated with an articulated mobile A moving along a direction q_j of C_A, the set of values q_j such as: $A(v_1, v_2 \ldots v_{j-1}, q_j, v_{j+1} \ldots v_n) \cap B_i = \emptyset$*

The determination of the sets $V_A(q)$ requires to apply interference and collision checking algorithms, derived from those developed in [3]. These algorithms allow to determine the possible contacts between two polyhedral objects A and B, by analysing the intersections between the geometric entities of B, and the curves and the surfaces described

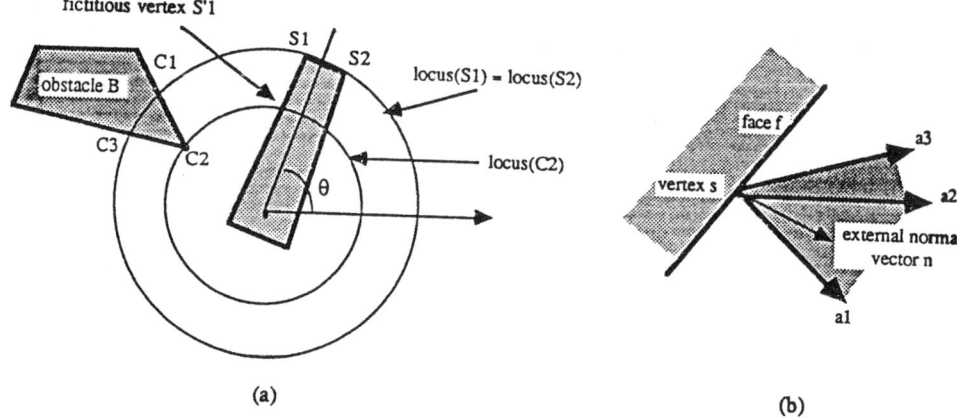

Figure 1: Computation of the contacts associated to a rotating object. (a) Potential contacts in a 2D exemple: (s_1, c_1), (s_2, c_2), (s'_1, c_2). (b) Contact (s, f) locally valid: $a_1 \cdot n \geq 0$ and $a_2 \cdot n \geq 0$ and $a_3 \cdot n \geq 0$.

by the vertices and the edges of A when A is translating or rotating. Each computed intersection is first considered as a potential contact between A and B (see figure 1). Then, very simple computations based on the vectorial calculus are applied in order to eliminate obviously wrong solutions (see figure 4.2).

This approach does not guarantee the faisability of the whole set of computed contacts when objects are not convex. In this case, the remaining ambiguities are processed by the last step of the algorithm which leads to construct the sets $CO_A^q(B)$ representing 1-dimensionnal C-obstacles in C_A^q along with their associated valid ranges of values $V_A(q)$ for $q \in D_q$ [32]:

$$V_A(q) = D_q - \cup_{i=1}^n CO_A^q(B_i)$$
$$CO_s^q(Bi) = \{q : s(q) \cap Bi \neq \emptyset\}$$
$$CO_s^q(Bi) = Closure\{\cup_{s \in A} CO_s^q(Bi)\}$$

where $CO_A^q(B_i)$ is a closed interval of $\Re$ representing the forbidden configurations of A defined by two consecutive contacts, and $s(q)$ is either a straight line (for translation) or a circle (for rotation). The figure 2 illustrates this computation. The algorithmic complexity in both the 2-dimensionnal and the 3-dimensionnal cases, may be represented by a term $\mathcal{O}(n^2)$, where n is the medium number of edges of A and B (see [32]).

4.3 Computing the C-obstacles associated with a rigid mobile:

Lozano-Perez shown in [38] that the C-obstacles of the type $CO_A^{xyz}(B)$ can be computed in a polyedral world in time $O(n^2 \log n)$, if n is the number of vertices of A and B. The applied method basically consists in constructing the convex hull of the set of vertices obtained by positionning the mirror image $\ominus A$ of A on the vertices of B (see figure 3):

$$CO_A^{xyz}(B) = conv(\{s \in \ominus A(s_i), \forall s_i \in B\})$$

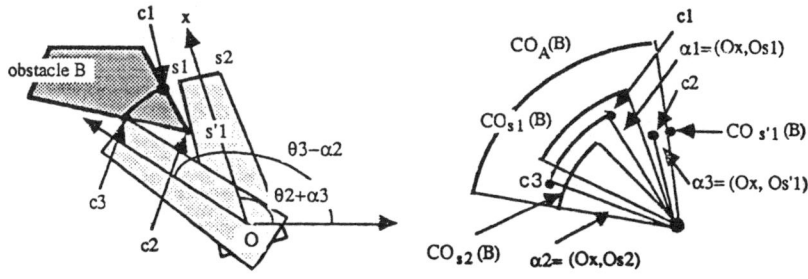

Figure 2: Computing $CO_A(B)$ associated with a polygonal rotating objects A.

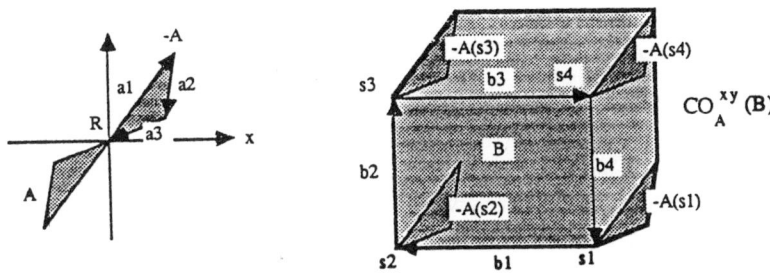

Figure 3: Construction of a C-obstacle of the type $CO_A^{xy}(B)$.

But this method fails when A can also rotate, because the related C-surfaces are topologically very different from the initial real surfaces. In order to adapt his method to rotating objects, Lozano-Perez [38] makes use of several *slices of orientation* for computing approximations of the C-obstacles. But his approach is not really applicable when the moving object is an articulated revolute arm.

An other approach initially developped in [29] and in [39], consists in computing the sets of *valid orientations* of a translating and rotating object A, after having discretized the translation domain. This approach leads to construct 1-dimensionnal slices in C_A (i.e slices having one d.o.f in rotation), by applying the following operator [32]:

$$C_A^q = Reduction(C_A^{D_q})$$

where D is a domain in $\Re^2$ or $\Re^3$, and q is an angular sector. This operator leads to compute the ranges of orientation of A which generate no collision with the obstacles B, when the reference point of A moves in D. It can be expressed using the following functions:

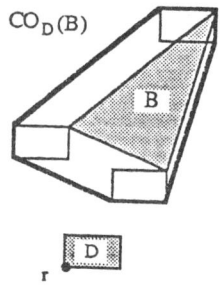 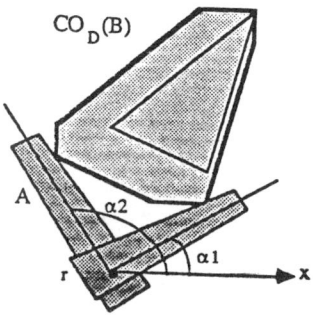

Figure 4: Computing the valid ranges of orientation associated to a translating and rotating polygonal object. (a) Determination of the set $CO_D(B)$. (b) Determination of the valid ranges of orientation of A relatively to $CO_D(B)$: $[0\ \alpha_1] \cup [\alpha_2\ 2\pi]$.

$$\forall p \in D : \quad F_{A,B}(p) = \{\alpha \in [a\ b] : A(p,\alpha) \cap B = \emptyset\}$$
$$G_{A,B}(D) = \bigcap_{p \in D} F_{A,B}(p)$$

where (p, α) is a configuration of A in $C_A^{D^q}$. The function $F_{A,B}(p)$ can be evaluated using the algorithms described in the previous section, but the function $G_{A,B}(D)$ cannot be directly computed since it represents the intersection of an infinite number of sets. A practical method for computing an approximation of $G_{A,B}(D)$, consists in reducing D to a point while expanding obstacles inversely to the shape of D:

$$G_{A,B}(D) = F_{A,B'}(r)$$

with:
$B' = CO_D^{xyz}(B)$.
r = reference point of D.

This method is illustrated by the figure 4. Its algorithmic complexity is in time $O(p+q)$, if p and q represent the terms associated to the growing transform and to the computation of the valid ranges of values.

4.4 Dealing with articulated mechanisms:

4.4.1 A method for constructing C-obstacles:

Our approach for dealing with an articulated mechanism A of the type "open kinematic chain", consists in successively analysing the constraints imposed by the obstacles on the different components A_i of the mechanism. For that purpose, the joints of A are ordered from the fixed extremity of A towards the free one. Then it becomes possible to apply a recursive algorithm, in which each step leads to study the behavior of a component A_i ($i = 1, 2 \cdots n$), after having fixed the positions of the anterior d.o.f q_j ($j = 1, 2 \cdots i-1$) to a set of chosen values $(v_1, v_2, \cdots v_{i-1})$. At this step, the posterior components A_k

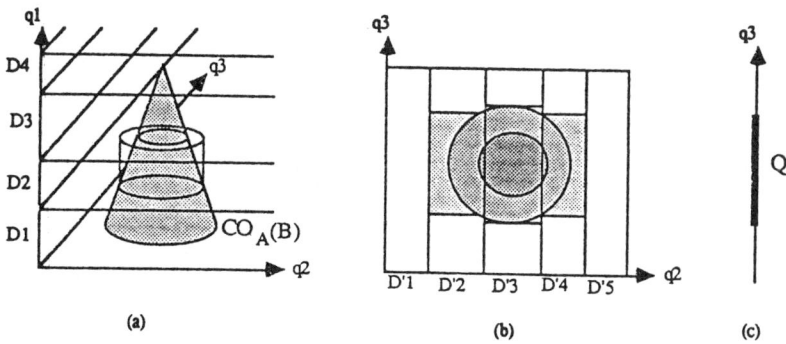

Figure 5: Computing an approximation of $CO_A(B)$ for a three joints robot A. (a) Constructing the slices associated to q_1. (b) Representing the approximation of $CO_A^{D_2}(B)$ associated with the slice D_2. (c) Representing the approximation of $CO_A^{D_2 \times D'_4}(B)$ associated with the slices D_2 and D'_4.

$(k = i+1 \cdots n)$ are ignored by the system. This computation is applied for each small interval of values obtained by sampling the domains D_j ($j = 1, 2 \cdots i-1$) associated with the joint variable q_j.

The trajectory planning methods described in [18], [14], [29] and [39] are based on this approach. Even if the applied computations and the constructed free-space representations are different, all these methods lead (1) to consider the C-space as a set of hyperparallelepipeds of the type $dq_1 \times dq_2 \cdots dq_n$, and (2) to compute an approximation of C-obstacles in this set using the recursive algorithm illustrated in the figure 5. In this example, $CO_A^{D_2}(B)$ represents the conjonction of the constraints imposed by B on the three joints of A, for all $q_1 \in D_2$. This set may be seen as the projection of the part of $CO_A(B)$ located between the planes P_1 and P_2 of C_A (a slice). Finally, the interval Q represents the values of Q_3 which may generate a collision with B, when $q_1 \in D_2$ and $q_2 \in D'_4$. Using this computation, $CO_A(B)$ can be approximate by the union of all the sets of the type: $D_i \times D'_j \times Q_k$.

In our system, the recursive algorithm which compute the C-obstacles is based on two techniques allowing to respectively *compute the position constraints* generated by the B_i, and to *propagate these constraints* along the robot arm.

4.4.2 Computing the position constraints:

This computation is done using the techniques described above. For that purpose, the local behavior of each link A_j ($j = 2, 3 \cdots n$) is analyzed after having fixed the position of the anterior links $A_1, A_2 \cdots A_{j-1}$, and after having associated a small variational domain dq_{j-1} to the joint variable q_{j-1}. Let R_j be the reference point of A_j (located on the joint axis for symplifying the computations), and D_j the locus of the positions of R_j when q_{j-1} takes all the values in dq_{j-1}. D_j is either an arc of circle or a straight line, depending on the type of the joint A_{j-1}.

These hypotheses lead to locally associate two combined motions to the link A_j: the translation along D_j, and the movement (translation or rotation) associated to the joint A_j. Consequently, it becomes possible to "locally" approximate the involved subset of the C-space by a less dimensionnal space of the type $C_{A_j}^{D_j \times q}$, where q is the variational domain associated to q_j. Then, this representation can be transformed in a more useful representation of the type $C_{A_j}^q$, by applying the "reduction operator" leading to expand obstacles B_i inversely to the shape of D_j (see section 3.2.3). Finally, the computation of the valid ranges of values for q_j can be executed using the G function [32]:

$G_{A,B}(D_j) = F_{A,B'}(r_j)$
with:
$B' = CO_{D_j}^{xyz}(B)$
r_j = reference point of D_j

This approach gives good results as long as the dq_i are small enough. That means that too large computed domains have to be split into sets of small intervals, before beeing considered for the next joint of the arm. But in practice, we reduce the amount of geometric computations by applying two different steps (see next section):

1. Each link A_j ($j = 1, 2 \cdots n-1$) is processed using a simplified growing transformation and an interference checking function, both applied on a set small intervals dq_i (there exists one set for each considered slice of the type $dq_1 \times dq_2 \cdots dq_{j-1}$).

2. Link A_n is processed using the complete algorithm which leads to compute all the valid ranges of values associated to dq_n (i.e for each slice of the type $dq_1 \times dq_2 \cdots dq_n$).

4.4.3 Propagating the position constraints:

Since joints interact to each other, each set of position constraints associated to a joint A_j have to be propagated towards the next joints $A_{j+1}, A_{j+2} \cdots A_n$. That means that one must take into consideration all the previous growing operations, when computing the grown obstacles associated to the joint A_j. In order to simplify this computation, our system applies a recursive algorithm for expanding obstacles [32] (see figure 6):

$B^0 = B$
For $k = 1, \cdots j - 1$:   Compute $B^k = CO_{D_j^k}^{xyz}(B^{k-1})$

where D_j^k is the locus of R_j for $q_k \in dq_k$.

Each step of the algorithm leads to apply very simple growing transformations, since the related domain D_j^k is either a small straight line or a small arc of circle (which can be approximated by a straight line or a polygonal line).

Theoritically, the algorithm should be applied for each link A_j and for each slice of the type $dq_1 \times dq_2 \cdots dq_{j-1}$. But in practice, we reduce the amount of computation by determining an "upper bound" of the domains D_j, and by applying only one simple growing transformation for each slice of the type $dq_1 \times dq_2 \cdots dq_{n-1}$:

$B' = Gros_r(B)$
$r = MAX_{i=1\ldots n} \{length(locus\ R_i)\}$

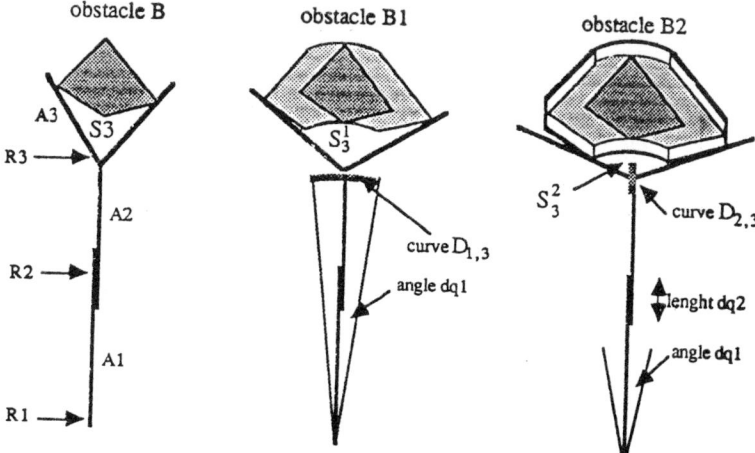

Figure 6: Propagating the position constraints along the arm.

where $Gros_r(B)$ represents an uniform expansion of B by the distance r. This approach may be extended to the whole space, when the splitting algorithm generates slices having an uniform size. Lozano-Perez [39] makes use of a similar approach by expanding the robot links with a distance equal to the maximum displacement computed using the jacobian matrix:

$$\delta = MAX_{Q \in I_1 \times I_2 \cdots I_n} \|J(Q) \cdot dQ\|$$

The complexity of the algorithm is in $O(N^{p-1})$, where N is the medium number of slices associated to a joint, and p is the number of d.o.f of the robot.

4.5 Constructing a free-space representation:

The free-space EL_A is composed of "cells" of the type $dq_1 \times dq_2 \cdots dq_n$. These cells are constructed using the following recursive function:

$FREE\ (j,\ D_{i_1} \times D_{i_2} \cdots D_{i_j})$
$\iff$
$FREE(j-1,\ D_{i_1} \times D_{i_2} \cdots D_{i_{j-1}})$
and
$A_j(q_i) \cap B = \emptyset,\ \forall q_i \in D_{i_j}$

with:
$FREE(1,\ D_{i_1}) \iff A_1(q_1) \cap B = \emptyset,\ \forall q_1 \in D_{i_1}$

In this algorithm, the domains D_{i_j} associated to the joint variables q_j $(j = 1, 2 \cdots n-1)$ are obtained by splitting the variational domains I_j into small intervals; the sets D_{i_n} are computed using the G function. The next step consists in constructing a graph structure allowing to connect the "adjacent" free cells:

$$ADJACENT\ (j,\ D_{i_1} \times D_{i_2} \cdots D_{i_j},\ D'_{i_1} \times D'_{i_2} \cdots D'_{i_j})$$
$$\iff$$
$$ADJACENT(j-1,\ D_{i_1} \times D_{i_2} \cdots D_{i_{j-1}},\ D'_{i_1} \times D'_{i_2} \cdots D'_{i_{j-1}})$$
$$et$$
$$D_{i_j} \cap D'_{i_j} \neq \emptyset$$

with:
$$ADJACENT(1,\ D_{i_1},\ D'_{i_1}) \iff D_{i_1} \cap D'_{i_1} \neq \emptyset$$

4.6 Searching for a safe trajectory:

Each path in the free-space graph defines *one class* of safe trajectories for the robot and its payload. This class is represented by a sequence $(C_1, C_2 \cdots C_m)$ of connected free cells. Then, searching for a safe trajectory requires to apply two computational steps respectively aimed at finding an "optimal" path in the graph, and at choosing a "good" trajectory among those represented by the generated path.

The graph search step is executed using a A^* algorithm [46]. Several types of solutions may be obtained when modifying the characteristics of the cost function (see [32]). In SHARP, this cost function is based on an Euclidian distance allowing to minimize the end effector displacements. Let c_{init} be the intial configuration in the cell C_1, c_{end} the goal configuration in the cell C_n, c_i the configuration corresponding to the middle of the cell C_i, and $(C_1, C_2 \cdots C_j)$ the path computed at the step j. The costs associated to this intermediate solution are the followings:

length of the path: $d(c_{init}, c_2) + d(c_2, c_3) + \cdots d(c_{j-1}, c_j)$
distance to the goal: $d(c_j, C_{end})$

The determination of a trajectory in a path $(C_1, C_2 \cdots C_n)$ consists in choosing a "good" set of configurations (i.e as "short" and as "smooth" as possible), allowing to connect c_{init} to c_{end} within the subset of C_A represented by $(C_1, C_2 \cdots C_n)$. Since any cell C_i is an hyperparallelepiped in C_A, any couple of points located on the boundary of C_i may be connected by a *straight line in* C_A. Consequently, the problem to solve may be converted into a more simple problem, consisting in determining a set of suitable configurations on the boundaries shared by the consecutive couples of cells of the computed path. A straightforward solution is to choose at each step j the configuration located at the middle of the shared boundary $C_{j-1} \cap C_j$. However, this approach leads to generate inappropriate movements, especially when crossing through large cells. In order to partly avoid this problem, we have developped a method allowing to determine at each step j the configuration of $C_{j-1} \cap C_j$ which is the closest to the last computed configuration [32].

5 Morphological reasoning techniques:

5.1 Principle of the reasoning:

The purpose of this section is to describe the basic techniques which have been developed in SHARP, for reasoning on the contacts involved in the grasping and the part-mating operations. Three main characteristics of these contacts have been considered:

- *The local accessibility* of objects features. The problem to solve is to verify that the analyzed contacts are locally feasible according to some matter distribution criteria. The related computations evaluate a set of simple topological and geometrical properties.

- *The mechanical constraints* associated with contacts. The problem to solve is to verify that the analyzed solutions are valid according to a set of stability criteria. The related computations are mainly based on heuristics allowing to qualitatively evaluate some static properties.

- *The motion constraints* associated with contacts. The problem to solve is to determine which motions are locally possible in order to achieve (or to leave) the considered contact situation. The involved computations make use of an analytical representation of the potential motions that can be locally executed by the moving object.

The accessibility and the mechanical properties are mainly used for determining the potential grasping configurations, whereas the computation of the motion constraints is mainly executed in the context of fine motion planning.

5.2 Representing contacts:

The contacts between two objects O_1 and O_2 are represented as follows:

$$CONTACT(O_1, O_2) = \{((e_1, e'_1)(e_2, e'_2) \cdots (e_m, e'_m)), T, P\}$$

where each set (e_i, e'_i) is a couple of intersecting geometric entities, T is the type of the contact (point, line or surface), and P represents the geometric parameters which characterize the contact.

The geometric entities (GE) are those belonging to the B.R component of the model: vertices, straight and curved edges, planar and curved faces (cylindrical, conical and spherical surfaces). The retained combinations for these entities are only those representing robust physical situations, i.e situations which can be potentially achieved despite the control and sensing errors. Since the couples of non colinear edges are not relevant for the fine motion strategies, we will only take into consideration the contact relations of the type *face-X*, where X = face, edge or vertex. These couples are called *contact elements*. The other combinations of entities are considered as unsteady situations, corresponding to degenerated contacts called *adjacent contacts* (see figure 7). This type of contact is not

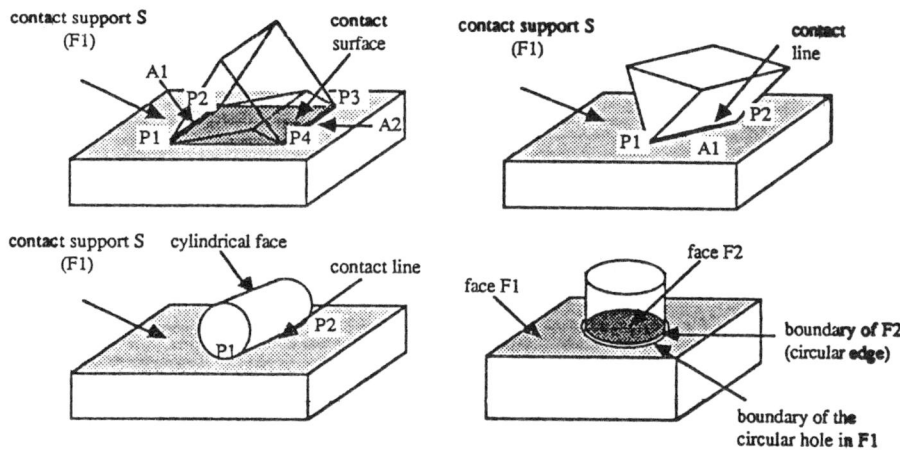

Figure 7: Example of some contacts: (a) Planar contact generated by a planar face and two edges. (b) Linear contact generated by a planar face and an edge. (c) Linear contact generated by a planar face and a cylindrical face. (d) Adjacent contact generated by two circular edges.

physically feasible because of the control and sensing errors, but its symbolic determination is necessary for finding the discontinuities generated by the geometry of the objects.

The type of the contact is defined by the topology of the set of contact points (point, curve or surface). The characteristics of this set depend on the type, the number and the relative positions of the involved GE. For example, a planar contact may be obtained by combining two contact elements generated by two non linear straight edges and a planar face (see figure 7). This multiple contact may be assimilated to a planar contact, because it has similar static properties. In particular, the convex hull of its contact points determines a "sustentation polygon", which have the same stability properties as a contact having a similar convex polygonal surface [23]. Since there exits a small number of different cases, it is not necessary to compute the convex hull of the contact points for determining the type and the geometric characteristics of the contact. It is sufficient in our context to analyse the sets $e_i \cap e'_i$ along with their possible combinations (see [26] and [21]).

The geometric parameters characterize the contact by a couple $(S, \delta S)$, where S is the surface defined by the set $E_1 \cap E_2$ of contact points, and δS is the boundary of the set $conv(E_1 \cap E_2)$. S is called the *contact support*, and δS is called the *contact surface*. Intuitively, we will say that the mobile object can slide on S, without "breaking" the contact, i.e without modifying the topology of δS. This property is very useful for planning fine motions.

In some cases, S cannot be unambiguously defined by the set $E_1 \cap E_2$ (for example when the contact is generated by a planar face and a cylindrical face). In this case, we chose to consider that S is defined by the geometric entities of the fixed object, i.e the object which is not manipulated by the robot. As an example, let consider the contact (c) of the figure 7. If the cylindrical object is manipulated by the robot, then the contact will be characterized as follows:

(CONTACT ((FACE F1 FACE FC2))
        (TYPE droite)
        (SUPPORT plan (0 0 1))
        (BOUNDARY (P1 P2)))

5.3 Reasoning on accessibility and mechanical properties:

The accessibility and the mechanical properties are evaluated in order to determine the combinations of contacts which are *locally accessible, stable and robust*. These combinations of contacts are used for constructing several sets of potential solutions for performing the grasp operations (see [25], [28], [54] and [32]). Such sets are called *potential grasps*.

Local accessibility property: One says that a combination of contacts is locally accessible, if the involved contacts are simultaneously feasible according to the local morphology of the two objects. Evaluating this property leads to verify that some simple topological and geometric constraints are verified in the vicinity of the features in contact: local convexity and P-convexity, parallelism and angular constraints, distance and dimensionnal constraints. All these properties are characterized by very simple algebraic formulas. For example, the local convexity of an edge A_1 belonging to two faces F_1 and F_2 is characterized by the formula $V_{A_1} \cdot (N_1 \wedge N_2) \leq 0$, where V_{A_1} is the edge A_1 oriented counterclockwise on the face F_1, and N_1 and N_2 are the external normal vectors of F_1 and F_2. In the same way, we will say that the edge A_1 is locally P-convex (i.e it is possible to place a plane P onto A_1, such as P do not intersect the object in the vicinity of A_1) if A_1 verify the following property:

$$V_{A_1} \cdot (N_1 \wedge N_2) \leq 0 \quad and \quad \exists a, b \in \Re : \ a < 0, \ b < 0, \ N_p = aN1 + bN2$$

where N_p is the external normal vector to P (P represents in this case the internal surface of one jaw of the gripper).

Stability property: This property allows to verify that the created physical situation is stable, i.e that the resulting forces and torques associated to the involved contacts are equal to zero. Instead of evaluating this property using an explicit representation of the forces and the torques, we chose to apply very simple geometric computations based on a qualitative knowledge of the static. The basic idea consists in verifying that the resulting reaction force of each contact "goes through" the contact surface δS of an other contact of the studied set. In the case of a set of contacts created by a two parallel jaws gripper, this condition may be evaluated by projecting the contact surfaces on a plane parallel to the jaws. If the obtained domains intersect, then the combination of "frictionless" contacts is considered as stable (see figure 8). This property is called the *mutual visibility property*, since it assumes that the contacts are located "in front of each other".

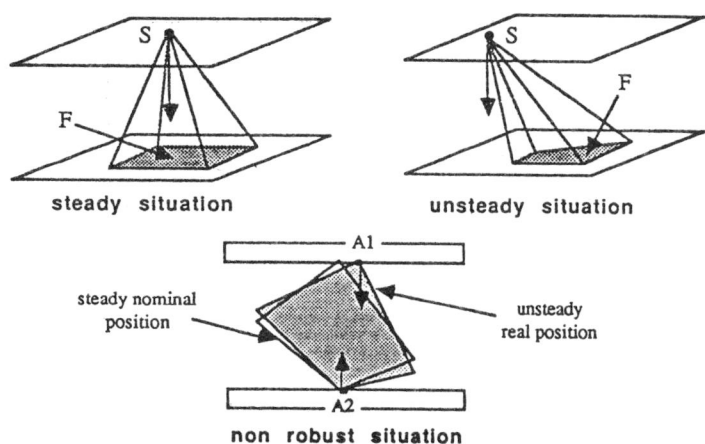

Figure 8: Stability and robustness properties. (a) Stable couple (S, F): $proj(S) \in F$. (b) Unstable couple (S, F): $proj(S) \notin F$. (c) Non robust couple (A_1, A_2): $proj(A_1) = proj(A_2) = straightline$.

Robustness property: This property allows to verify that the combination of contacts remains stable, when slightly moving the object. Our approach for planning robust grasps (i.e grasping points which avoid the "twisting" of the object inside the jaws), consists in choosing a combination of contacts which generates no d.o.f along the directions having some possible associated position errors. Consequently, we will eliminate the solutions which keep free a rotation around an axis not normal to the internal faces of the jaws, or not parallel to the revolute axis of some objects (see figure 8).

All the computations involved in the determination of the local morphological properties described above, have been implemented in SHARP using an ordered set of *geometric filters* (see [54] and [32]).

5.4 Reasoning on motion constraints:

5.4.1 The concept of potential motion:

Each contact reduces the number of d.o.f of the moving object (i.e the gripper or the manipulated object). In order to determine the next motion to execute from a given contact situation, the system must reason on the directions of movement which are constrained by the contacts. Consequently, it is necessary to explicitly represent the valid movements which can be *locally* associated to a contact situation. In order to simplify the reasoning, we will consider that these movements are either pure translations or pure rotations which are defined independently of their possible amplitudes. If A is a mobile object in contact with B, we will define a potential motion for A as *"a motion having an amplitude ds greater than the maximum control error, and which generates no collision between the features in contact"*. In practice, this definition has led us to develop an analytic support allowing to explicitly represent the forbidden motions, those which preserve the contacts and those which break them. This formalism is described below.

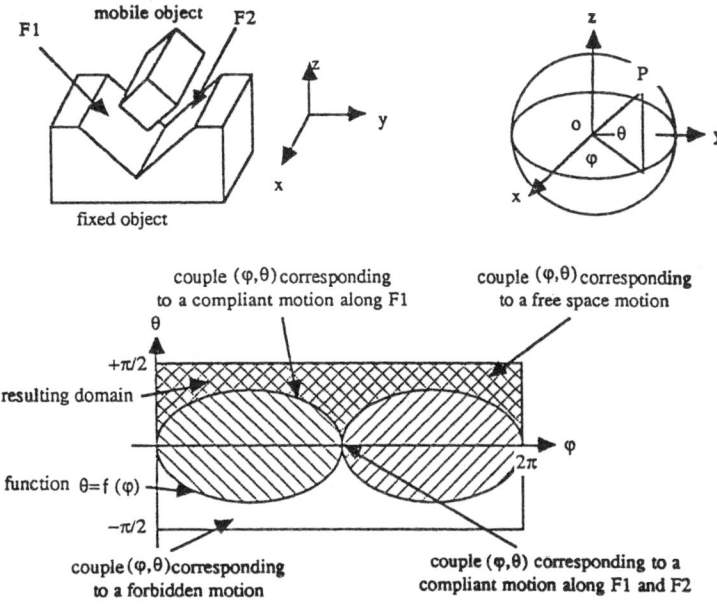

Figure 9: Representing the potential motions associated to a couple of planar contacts.

5.4.2 Representing potential translations:

A translation can be represented by a couple (v, a), where v is an unitary vector of $\Re^3$ and a is a real number. Then, a set of translations can be represented by a subset of $S(1) \times \Re$, where $S(1)$ is the unitary sphere. Since our purpose is to reason on sets of translation directions, we will only represent the related domains on $S(1)$. Then, a point on the sphere defines a particular translation direction, and a spherical domain characterizes a set of possible translating motions. For example, a planar contact generates an half-sphere domain, and a couple of contacts determines a domain obtained by intersecting those associated to each contact (see figure 9). Such a computation may be executed by projecting the constructed domains on a plane (φ, ϑ), where φ and ϑ are the spherical coordinates.

Contacts having a planar support:
When contacts are of the type "planar face - X" (X represents any type of GE), the computation of the resulting domains on $S(1)$ may be executed using the following function [51]:

$$\vartheta = \arctan((N_x \cdot \cos\varphi + N_y \cdot \sin\varphi)/ - N_z) \qquad (1)$$

where (N_x, N_y, N_z) is the normal external vector to the contact plane (see figure 9).

Contact having a curved support:
In this case, the applied computations depend on the type of the involved surfaces. For example, a cylindrical contact will only generate two possible translating motions along the revolute axis of the contact surface. On the other hand, a conical contact will generate a more complicated domain characterized by the function:

$$\cos\vartheta(a\cos\varphi + b\sin\varphi) + c\cos\vartheta - \cos\alpha = 0 \qquad (2)$$

where (a, b, c) and α are the cone axis and the cone top angle.

Other types of contacts may also be generated by non planar entities. But it is possible to apply the same type of computation, by using a first order approximation of the potential movements. This approximation is obtained by considering the fictitious planar contacts defined by the tangent planes associated to the contact surfaces [32]. For example, a contact generated by two partial cylindrical surfaces defines a domain of the type "$D(P_1) \cap D(P_2) \cap D(P_3) \cup D(cylinder)$", where $D(P_1)$, $D(P_2)$ and $D(P_3)$ are the domains associated to the three tangent planes located on the boundaries of the contact surface, and $D(cylinder)$ represents the two directions associated to the cylinder revolute axis.

5.4.3 Properties of the representation:

Let $S = \{C1, C2, \ldots Cn\}$ be a contact situation, and C_i the related contacts. The potential motions associated to S are represented by a domain D. This domain is defined as the intersection of the domains $D_1, D_2 \cdots D_n$ associated to the contacts $C_1, C_2 \cdots C_n$. Each D_i is computed using the function (1) or (2). Let D_i^* be its inside part, and δD_i its boundary ($D_i = D_i^* \cup \delta D_i$).

An important property of this representation is to clearly differenciate the different types of motions which can be associated to S (see figure 9):

- Each couple (φ, ϑ) outside of D represents a forbidden motion.

- Each couple (φ, ϑ) in D represents a motion which leads to break all the contacts.

- Each couple (φ, ϑ) on δD represents a compliant motion which leads to slide on some contact surfaces: if $(\varphi, \vartheta) \in \delta D_{i_1} \cap \delta D_{i_2} \cap \cdots \delta D_{i_j}$ and $(\varphi, \vartheta) \notin \delta D_k$ ($k \neq i_1, i_2 \cdots i_j$), then (φ, ϑ) defines a motion which maintains the contacts $C_{i_1}, C_{i_2} \ldots C_{i_j}$ (the other contacts are broken).

An other property of this representation is to be *complete* for contacts having a planar support S. Then, any existing solution is contained in D, but it is impossible to guarantee that this solution will be found in a finite time. Our approach for dealing with this problem is described in the next section.

The completeness property is not preserved when the support of the contact is not a planar surface. But the applied first order approximation is reasonnable, since it allows to locally represent the potential compliant motions as a set of tangential motions. This approach is consistant with the fact that such motions are always executed tangentially to the surfaces which support the contacts.

5.4.4 Dealing with rotations:

Dealing with rotations is more complicated, since the content of the potential rotating domains depends on both the orientations and the positions of the rotation axes. For example, a rotation axis belonging to the contact plane but located outside of the contact surface δS, will generate motions leading to break the contact; the same axis located tangentially to the boundary of δS will generate motions allowing to modify the type of the contact.

This characteristic of the rotations makes the previous approach impractical for the potential rotating motions, since several representations may be associated to a single contact. On the other hand, it suggests to develop an other approach consisting in grouping together the rotation axes having an "homogeneous behavior" relatively to the contacts. For example, we will group in a single set all the rotation axes which generate motions allowing to maintain the analysed contact. In the processed contacts, these sets are easilly derived from the type of the contact surfaces. For example, the rotation axes which maintain a planar contact are those which are normal to the contact plane; those which maintain a cylindrical or a conical contact are located on the revolute axis of the surface.

This approach allows to easily deal with a large set of practical cases. However, it needs to be developped in a more complete way in the future.

5.5 Constructing the solution space:

5.5.1 The set of potential grasps:

A *potential grasp* represents a set of grasping positions based on the same combination of contacts. In order to guide the selecting process, a partial order based on an heuristic evaluation of the "quality" of the involved combination of contacts is established among the potential grasps. This approach allows to first analyse the potential solutions which are more likely to generate a feasible grasping position. But it does not allow to generate a sequence of different grasping positions based on several combination of contacts, when a single solution cannot be found by the system. Such a method (called regrasping) have been developped in [41].

Representing potential grasps:

Let A be the object to be grasped and P the robot gripper. A potential grasp $\Pi(A/P)$ is represented as follows [28] [54] (see figure 10):

$$\Pi(A/P) = (\pi, P_c, P_m, h)$$

where $\pi = (C_1, C_2 \cdots C_m)$ is the set of the involved contacts, P_c represents the "preshaping" parameters of P (i.e the configuration of P before grasping and the jaws action to execute), P_m represents the orientation and the moving constraints associated to P, and h is an heuristic assessment of the *quality* of the solution (stability and robustness) and of its *degree of accessibility*. At this step, all the geometric parameters of the contacts C_i are not specified, since the position of P have not been already computed by the system.

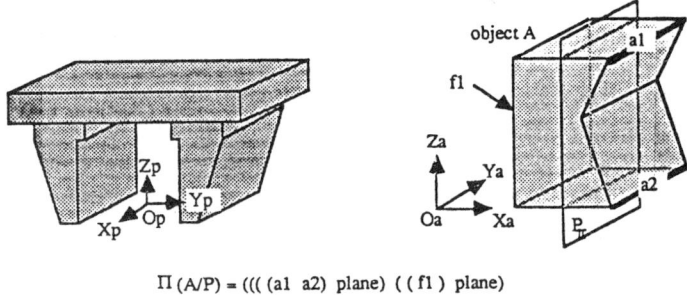

Π(A/P) = ((((a1 a2) plane) ((f1) plane)
(close 1.5)
((0.75 0.0 0.0) (1.0 0.0 0.0)) h)

Figure 10: **Example of a potential grasp.**

As we make the assumption that the grasping operations are executed using a two parallel jaws gripper, P_m may be characterized by a plane called the *gripping plane* P_π. This plane is parallel to the two planar contact surfaces (i.e the internal faces of the jaws), and it is located at an equal distance to this two surfaces. Then, the constraints P_m can be characterized by the expression "$O_P \in P_\pi$ and $X_P O_P Z_P \parallel P_\pi$", where $(O_P\ X_P\ Y_P\ Z_P)$ is the reference frame associated to P as shown in the figure 10.

Searching for a solution in $\Pi(A/P)$:

Searching for a solution in $\Pi(A/P)$ consists in determining a particular configuration of P which verify the constraints represented by π and P_π, and which can be realized in the initial and the final environments of the object to be grasped. Let $P(X, \Delta)$ be the configuration of P corresponding to $O_p = X$ and $O_p Z_p = \Delta$, W the set of GE belonging to A and to the obstacles located in the initial and the final environment of A, and E the set of GE involved in π. The configuration $P(X, \Delta)$ is unambiguously defined by the couple (X, Δ), since $X_p O_p Z_p$ is parallel to P_π.

Property 5.1 *Let p be a point in P_π and Δ a vector in P_π. A couple (p, Δ) will be considered as a particular solution in $\Pi(A/P)$, iff:*

- $P(p, \Delta) \cap e \neq \emptyset, \quad \forall e \in E$
- $P(x, \Delta) \cap \{W - E\} = \emptyset, \quad \forall x \in half-line(p, \Delta).$

The first expression verifies that all the contacts of π are simultaneously realized in the configuration $P(p, \Delta)$. Such a configuration we be considered as *valid*. The other expression verifies that the involved object features can be reached along the direction Δ. Such a configuration will be considered as *safe*.

Then, searching for a particular solution in $\Pi(A/P)$ can be done using three main steps: (1) choose a direction Δ in P_π, (2) construct the set S_π of the points $p \in P_\pi$ such as (p, Δ) verify the property 5.1, and (3) choose a point p in S_π such as $P(p, \Delta)$ minimizes the risks of sliding. This method is detailed in section 6.2.

5.5.2 The state graph associated to the fine motions:

The sets of contacts and of their associated potential motions are combined in order to construct the *state graph associated to the fine motions*. This graph represents all the combinations of motions and robot states which have been selected as *potential elements of solution* for the problem to be solved. This graph does not explicitly contain all the possible solutions, but we will see further that it may be locally refined in case of failure. Such a failure occurs during the search phase, and it means that no feasible fine motion strategy is represented in the current state graph.

Representing the state graph:

Let A and B be the two objects to assemble. The state graph $G(A/B)$ associated to the fine motions allowing to assemble A on B, is represented by a directed graph $(\mathcal{S}, \mathcal{A})$. Each node in $\mathcal{S}$ represents a robot state E_p (see section 3.1), and each arc in $\mathcal{A}$ defines a robot motion allowing to move from a state to an other one.

A node s_i in $\mathcal{S}$ is characterized as follows:

$$s_i = (P, C, D, I, Q)$$

where P is a set of positions for the robot, C is a set of contacts, D is a set of potential motions, I is a qualitative information on the associated physical situation (for example: a cylindrical insertion), and Q is an heuristic assessment of the "quality" of this situation (robustness relatively to uncertainties and mechanical faults). P and C characterize the state E_p. D defines the motions which can be theoritically applied from E_p (this information is used when refining the graph). I and Q are used for guiding the search function.

An arc a_{ij} in $\mathcal{A}$ is characterized as follows:

$$a_{ij} = (T, C, A, Q)$$

where T is a geometric transform, C and A are two sets of faces on which the mobile object must respectively slide and stick, and Q is an heuristic weight evaluating the "quality" of the motion (collision and sticking risks when slightly modifying the environment, effects of the gravity ...). The parameters T, C and A characterize the motion allowing to move from E_{p_i} to E_{p_j}. The parameter Q is used for guiding the search function.

Constructing the state graph:

Our analytic representation of potential motions allows to characterize the whole set of movements which are potentially feasible from a given state E_i. But this representation cannot be directly used by the motion planner, since each constructed domain D represents an infinite set of possible solutions (and consequently an infinite set of possible arcs for each node in the state graph).

A classical technique dealing with this problem, consists in *discretizing the sets of potential solutions*. This technique leads to split each domain D into a finite set of small spherical domains of the type $\Delta\varphi \times \Delta\vartheta$. Each obtained domain ΔS represents a set of motions which will be "globally" analyzed by the system. This approach requires that all the motion directions in ΔS allows to theoretically achieve the same symbolic contact situation, when executing these motions from a given position P. If the objects are polyedra and the motions are pure translations, such a constraint may be evaluated using a "visibility" analysis technique [8].

But the high algorithmic complexity of this approach along with its inability to deal with rotations and curved surfaces, has led us to make use of an heuristic based approach. The basic idea consists in analyzing a subset of the possible solutions, by selecting in D the most promising motion directions. This approach is consistent with the fact that most of the required movements for mating two mechanical parts, are executed along some *privileged directions* defined by the contact surfaces. In particular, most of fine motions can be executed along a direction normal or parallel to the surfaces located in the vicinity of the involved contacts. This is the reason why the selected directions correspond to some characteristic points in D: points located at the intersections of several curves of the type 1 and 2, points periodically distributed on the boundaries of D.

The algorithm used for constructing the state graph is described in section 6.3.

Searching for a solution in $G(A/B)$:

Let IS be the set of nodes of $G(A/B)$ which have an empty set of contacts, and GS the state corresponding to the situation where A and B are assembled. Each node in IS is considered as a possible starting point for constructing a fine motion strategy, because the related physical situation can be directly achieved by the robot.

Any path in $G(A/B)$ starting from a node s in IS and ending at the node GS, may be considered as a fine motion strategy allowing to assemble A on B. Then searching for a solution in $G(A/B)$ can be done using the following algorithm:

1. Search for a "good path" SG starting from a node s in IS and ending at node GS.

2. Verify that each arc of SG generates a movement which can be executed by the robot (no collision, reachability of the goal). Refine $G(A/B)$ in case of failure, and goto (1).

3. Synthesize the fine motion program from SG.

This algorithm is detailed in section 6.3. In the general case, SG is a sub-graph which includes at each level all the states (nodes of $G(A/B)$) which can be reached after executing the selected motion. This situation comes from the fact that the control errors may sometimes make the robot stop in different contact situations. This point is discussed in [32]. It is not taken into consideration in the current search function, since the states and the motions selected when constructing the state graph avoid such situations.

6 Implementation of the motion planners:

6.1 The trajectory planner:

As explained in section 4, our method for computing *safe trajectories* operates in two phases allowing to successively construct a graph representing an approximation of the free-space, and to search this graph using a A^* algorithm. The main originality of the method lies in the fact that the computation of the free-space representation is done using a constraint propagation mechanism allowing (1) to recursively apply very simple growing transformations depending at each step on the type of the analyzed joint, and (2) to parametrize the algorithm according to several constraints drawn from both the environment and the mechanical structure of the arm (see [29]). In the current implementation, the parameters which evaluate the cluttering of the workspace have been arbitrarily fixed to constant values (i.e the discretization of each d.o.f is done using an uniform predifined value). As explained in section 4.4.3, this approach allows to reduce the amount of computation by applying a single growing transformation of the type $Gros_r(B)$ for each slice of the type $dq_1 \times dq_2 \cdots dq_{n-1}$. Then the missing domains dq_n associated to these slices are computed using the algorithms described in section 4.2 and 4.4.2. The search phase is executed as described in section 4.6.

This method can theoritically be applied for the whole arm. But in practice it is only used for planning the wrist trajectories, because of its exponential complexity in terms of the number of d.o.f. Then, the required rotations of the wrist are determined by applying two complementary planning steps based on heuristics [32]:

- Determination of the wrist movements (with a fixed orientation) in the vicinity of the initial and of the goal positions.

- Determination of the rotations of the wrist along the computed trajectory, by considering the volume swept by the gripper and the payload when rotating.

This approach works fairly well when the manipulated object is small relatively to the gripper. It necessitates to execute the reorientation operations in some predifined uncluttered areas, when the size of the swept volume makes the algorithm fail.

The trajectory planner have been implemented in LUCID-LISP on a SUN 260. Most of the experimentations have been executed in simulation (see figure 11). Some of them have give rise to real executions using a six d.o.f SCEMI robot. But the CPU time required for computing the free-space model in the current version of the system is not really significant, because most of the geometric computations are performed using some external functions belonging to the modelling system (these functions have not been optimized and compiled). For instance, 28 mn of CPU time was needed for processing the example shown in figure 11. This example has give rise to a graph having 480 free cells of the type $5 \times 10 \times 15$, where $5°$, $10°$ and $15°$ are the angular sectors which have been used for discretizing the joint space.

6.2 The grasp planner:

As explained in section 3.2, our method for computing *secure grasps* operates in two phases allowing to successively compute a set of potential solutions, and to determine the

Figure 11: A simple trajectory computed by the system.

dynamic parameters of the chosen solution.

The purpose of the first phase is to determine the contact sets which are locally accessible, stable and robust. This computation is executed using an ordered set of geometric filters implementing the accessibility and the mechanical properties described in section 5.3. This approach has been developed in order to reduce the size of the search tree, by first applying the more discriminant filters. The resulting potential solutions are represented as shown in section 5.5.1.

The second phase verifies that the generated potential solutions are reachable in both the initial and the final environments. It determines for that purpose the gripper configurations which are both *valid* and *safe*, according to the position constraints imposed by the environment (see section 5.5.1). This computation allows to reject the unfeasible solutions and to determine the missing parameters of the selected one (position and orientation of the gripper, approach and deproach directions).

The implemented algorithm is the following:

- **Phase 1**

    1. Determination of the potential contacts.
    2. Determination of the potential combination of contacts and construction of the related potential grasps of the type $\Pi(A/P)$.

- **Phase 2**
    1. Choice of a potential grasp $\Pi(A/P)$ compatible with the task constraints.
    2. Choice of a direction Δ in the gripping plane P_π.
       If all the directions have been analyzed, then goto (1).
    3. Determination of the valid configurations $P(X, \Delta)$ in $\Pi(A/P)$:
    $$V_\pi(\Delta) = \{p \in P_\pi : P(p, \Delta) \cap e \neq \emptyset, \forall e \in E\}$$
    If $V_\pi(\Delta) = \emptyset$, then goto (2).
    4. Determination of the safe configurations $P(X, \Delta)$ in $\Pi(A/P)$:
    $$S_\pi(\Delta) = \{p \in V_\pi(\Delta) : P(p, \Delta) \cap \{W - E\} = \emptyset\}$$
    If $S_\pi(\Delta) = \emptyset$, then goto (2).
    5. Choice of a point p in $S_\pi(\Delta)$, such as $P(p, \Delta)$ maximizes the surface of contact and minimizes the torque created by the gravity.

This algorithm is completed by a complementary step allowing to choose a gripping plane, when the grasping operation is executed on a revolute surface. The choice of the direction Δ is made using an heuristic which determines the free angular sectors associated to the center of gravity of the object to be grasped (see [27]). The computation of the sets $V_\pi(\Delta)$ and $S_\pi(\Delta)$ can be executed using the classical growing transformation CO^{xz} [55]:

$$V_\pi(\Delta) = \cap_{e_{ij} \in E} \, Proj_{P_\pi}(CO^{xz}_{F_i[\Delta]}(e_{ij}))$$

$$S_\pi(\Delta) = V_\pi(\Delta) - Proj_{P_\pi}(CO^{xz}_{P[\Delta]}(B))$$

where $Proj_{P_\pi}(X)$ is the orthogonal projection of X on P_π, $F_i[\Delta]$ is the internal face of the jaw i oriented according to Δ, $P[\Delta]$ is the gripper oriented according to Δ, and B represents the obstacles. In our implementation, this computation is executed in a two dimensionnal space, after having projected on the gripping plane several slices delimited by a set of parallel planes (see [27]). We have also developped a specific growing transformation CO^* allowing to reject the points in $S_\pi(\Delta)$ which cannot be reached by the gripper because of the robot arm. This new transformation consists in expanding the set $CO^{xz}_{P[\Delta]}(B)$ in the direction $-\Delta$, in order to include the "shadow" of the object (see [27]).

The grasp planner has been first implemented in MACLISP on a CII-HB 70 computer. Its integration in the SHARP system has been partly realized, but several experimentations have been done in simulation with simple objects having less than 20 faces. For instance, less than one minute of CPU time was needed for computing the solutions shown in figure 12, and only one second was consumed by the first phase for generating 12 potential grasps among 32! possibilities. But this execution time increases very fast when the system has to take into consideration the surrounding obstacles.

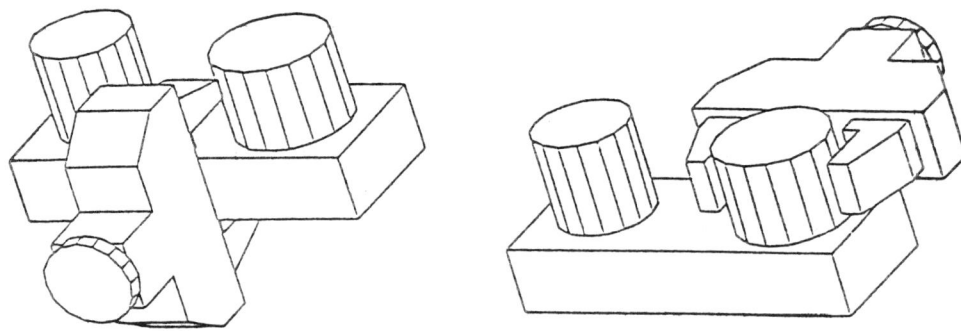

Figure 12: Two grasping positions computed by the system.

6.3 The fine motion planner:

As explained in section 3.2, our method for computing the *fine motion strategies* operates in two phases allowing to successively construct a state graph representing the set of potential solutions, and to search this graph in order to find a "good path" representing a feasible fine motion program.

The analysis phase allows to construct the state graph by reasoning on a fictitious dismantling of the assembly. For that purpose, the system determines at each step the different contact situations which can be reached from the current situation by applying a single motion. Only the moving constraints associated to the contacts are examined at this step (the other motion constraints are not considered). This computation is executed using the method described in sections 5.4 and 5.4.2. It leads to progressively decrease the number of contacts.

The search phase determines a "reverse path" in the graph (i.e a path starting from a node having an empty set of contact and ending at the node corresponding to the final assembly). This search phase is based on heuristics which attempt to optimize the selected solution in terms of efficiency (number of operations) and of reliability (robustness of the selected motions). In case of failure (for example one contact situation cannot be achieved because of the control errors), the graph is locally refined by introducing some new potential motions (see [31]).

The implemented algorithm is the following:

- **Analysis phase**

    1. Create the node GS and insert it in the list $OPEN$.
    2. If $OPEN = \emptyset$ then return $G(A/B)$, else process the node x located at the head of the list $OPEN$.
       Choose n directions $d_1, d_2 \cdots d_n$ in D_x.
       Create an arc a_{xi} for each chosen direction d_i.
    3. Create a node y_i for each arc a_{xi}.
       If the state E_i associated to y_i is already represented by a node z in $G(A/B)$, then **merge y_i and z**.

4. Insert the new nodes having an empty set of contacts in IS; insert the other nodes in the list $OPEN$.
   Goto (2).

- **Search phase**

   1. Search for a path SG starting from a node s in IS and ending at the node GS.
   2. Verify that each arc in SG represents a feasible motion (no collision, reachability of the goal).
      Refine $G(A/B)$ in case of failure, and goto (1).
   3. Synthesize the fine program represented by SG.

$OPEN$ represents the list of the next nodes to process, and D_x is the set of potential motions associated to the node x (see section 5.4). GS is the goal state (the parts A and B are assembled), and IS is the set of the possible starting states for the fine motion strategies (states having an empty set of contacts).

The moving directions are selected in D_x according to an heuristic function. For example, four directions will be initially generated by a couple of non-parallel planar contacts: $d_1 = N_1 \wedge N_2$, $d_2 = -d_1$, $d_3 = d_1 \wedge N_1$ and $d_4 = d_2 \wedge N_2$, where N_1 and N_2 are the external normal vectors to the contact faces F_1 and F_2. d_1 and d_2 define two compliant motions allowing to maintain the contacts; d_3 and d_4 define two motions leading to respectively break the contact associated to F_2 and to F_1. These moving directions are considered by the system only if they are included in D_x.

Each node in $G(A/B)$ is created after having applied a geometric computation allowing (1) to determine the involved contacts (parameter C), (2) to compute the associated set of potential motions (parameter D), and (3) to verify that the reached state E_i does not already exist in $G(A/B)$ and that it can be unambiguously distinguished from the other states represented in $G(A/B)$. Each arc in $G(A/B)$ is created after having applied a geometric computation allowing (1) to determine the valid ranges of position associated to the selected moving direction (parameter T), and (2) to determine all the contacts which are involved in the movement (parameters C and A).

The search phase is guided by a *cost function* which is combined with a set of *dynamic pieces of advices* implemented using production rules (see [31] and [51]). The cost function makes use of the heuristic weights associated to the nodes and to the arcs of $G(A/B)$. It leads to both minimize the number of operations and to maximize the reliability of the selected motions. The dynamic pieces of advices are activated when some particular situations are detected by the system. They mainly allow the system to deal with the physical situations which cannot be safely executed by the robot because of the position uncertainty (adjacent contact, closed obstacle ...). For example, if a contact situation cannot be directly left because of the presence of an obstacle in the vicinity of the moving part, then it is advised to first slide on the contact surface (and consequently to create new nodes and new arcs in the graph). In the same way, an adjacent contact in SG will lead the system to determine a set of intermediate contacts for guiding the robot. This approach allows to reduce the algorithmic complexity by only exploring "in detail" the branches of the graph which are really significant according to the selected solution. It

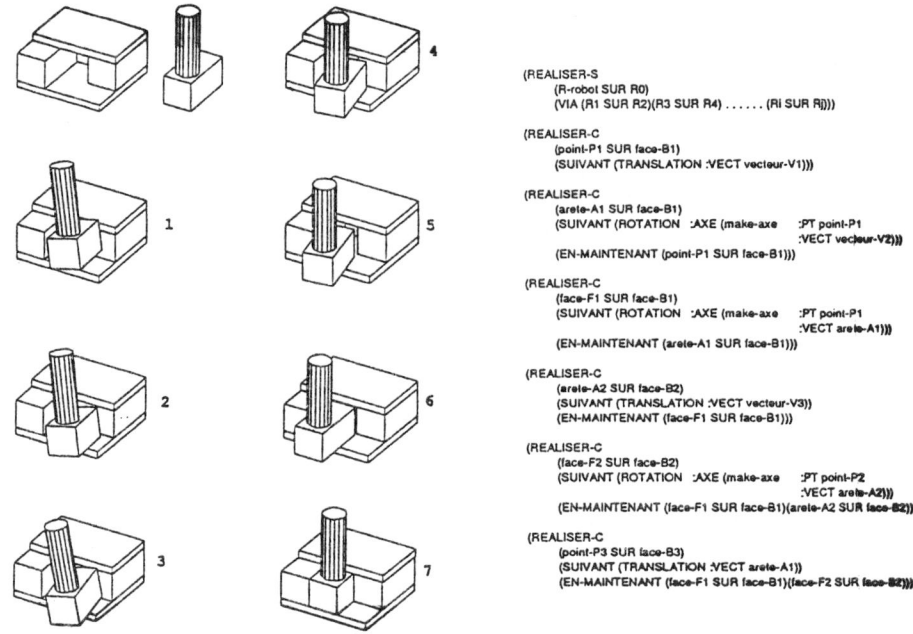

Figure 13: A fine motion strategy computed by the system.

also permits to integrate some well-known local strategies in the solution, when classical situations are recognized by the system (for instance: cylindrical insertion).

The collision and the reachability tests are executed using the following computations [32]:

$$Sweep(A, d) \cap Gros_\varepsilon(B) = \emptyset \quad \Rightarrow \text{ no collision}$$
$$A(p) \cap Gros_\varepsilon^{-1}(B) \neq \emptyset \quad \Rightarrow p \text{ is reachable}$$

where $\varepsilon = 2(\varepsilon_p + \varepsilon_i)$ and the terms ε_p and ε_i represent the control and the sensing error bounds; $Sweep(A, d)$ is the volume swept by A when moving along d, $Gros_\varepsilon(B)$ and $Gros_\varepsilon^{-1}(B)$ respectively represent the obstacles B grown and shrunk according to ε, and $A(p)$ is the object A in the configuration p.

The last step consists in computing the missing motion parameters of the selected movements, in order to synthesize a sequence of guarded compliant motions of the type:

MOVE <*objet-A*> ALONG <*T*> BY-MAINTAINING <*C*> UNTIL <*A*>

where T and C are the symbolic motion parameters recorded in the graph, and A represents the set of contacts which may stop the movement. The numerical values associated to these parameters at the execution time are computed using the geometric model and some predifined thresholds. For example, a face belonging to A will generate a condition of the type "$F_v > threshold$", where F_v is the projection of the reaction force on the moving direction v and v is assumed to be included in the friction cone (this condition is associated to the termination predicate). A similar computation is executed for the compliant parameters, but the needed thresholds are currently tuned by the operator.

The fine motion planner has been implemented in LUCID-LISP on a SUN 260. Most of the experimentations have been executed in simulation. Some of them have give rise to real executions using a six d.o.f SCEMI robot equipped with a force sensor. This is the case for the example shown in figure 13 which has been successfully executed by the robot. 9 mn of CPU time was needed for synthesizing the related fine motion program. The constructed state graph was made of 50 nodes and 80 arcs.

7 Conclusion:

In this paper we have shown that two types of geometric reasoning techniques have to be combined for planning the various robot motions involved in the basic assembling operations (transfer, grasping and part-mating operations). We first described the various computational tools that we have developed for implementing these two types of reasoning (called *spatial reasoning* and *morphological reasoning*). Then we explained how the proposed methods have been implemented and integrating in a single system (the SHARP system), aimed at automatically generating assembly robot programs. The obtained results and the current limitations of our approach have also been discussed. This approach has been partly validated by several experimentations executed in simulation, but very few results have been currently obtained in connection with the real robot.

Current work deals with three major points: the unsolved geometric problems mentioned in the paper (dealing with rotations for example), the algorithmic complexity (how to combine local and global reasoning techniques in order to reduce this complexity without increasing the risk of failure), and the connection between the geometric model of the task and the real robot (how to automatically compute the numerical values which are required for executing the planned motions).

Acknowledgements:

The work presented in this paper has been done by the members of the Robotics group at the LIFIA Laboratory. It was partly supported by the French National ARA project of the CNRS, the ADI agency, and the INRIA institute.

References

[1] N.Ahuja, R.T.Chien, N.Bridwell: *"Interference detection and collision avoidance among three dimensional objects"*, 1st American Association for Artificial Intelligence conference, Stanford, August 1980.

[2] J.D.Boissonnat: *"Stable matching between a hand structure and an object silhouette"*, IEEE Transactions on Pattern Analysis and Machine Intelligence, vol. PAMI-4, no 6, November 1982.

[3] J.W.Boyse: *"Interference detection among solids and surfaces"*, CACM, vol.22, nb.1, January 1979.

[4] J.M.Brady et al.: *"Robot Motion: Planning and Control"*, MIT Press, Cambridge, MA, 1982.

[5] R.A.Brooks: *"Solving the find-path problem by good representation of free space"*, 2nd AAAI conference, Carnegie Mellon University, August 1982.

[6] R.A.Brooks: *"Planning collision free motions for pick and place operations"*, International Journal of Robotics Research, vol 2, no 4, 1983.

[7] R.C.Brost: *"Planning robot grasping motions in the presence of uncertainty"*, Technical Report CMU-RI-TR-85-12, Carnegie Mellon University, The Robotics Institute, Pittsburgh, July 1985.

[8] S.J.Buckley: *"Planning and teaching compliant motion strategies"*, Ph.D Thesis, Artificial Intelligence Laboratory, MIT, January 1987.

[9] R.Chatila: *"Système de navigation pour un robot mobile autonome: modélisation et processus décisionnels"*, Thèse de Docteur Ingénieur, Toulouse, July 1981.

[10] B.R.Donald: *"Motion Planning with Six Degrees of Freedom"*, AI-TR-791. Cambridge, Mass.: Massachusetts Institute of Technology Artificial Intelligence Laboratory, 1984.

[11] B.Dufay: *"Apprentissage par induction enRobotique: Application à la synthèse de programmes de montage"*,Thèse de 3ème cycle, INPG, Grenoble, June 1983.

[12] B.Dufay, J.C.Latombe: *"An approach to automatic robot programming based on inductive learning"*, 1st International Symposium on Robotics Research, Bretton Woods, August 1983.

[13] M.Erdmann: *"On motion planning with uncertainty"*, AI-TR-810, Artificial Intelligence Laboratory, MIT,1984.

[14] B.Faverjon: *"Obstacle avoidance using anoctree in the configuration space of a manipulator"*, IEEE International Conference on Robotics and Automation, Atlanta, March 1984.

[15] B.Faverjon, P.Tournassoud: *"A local based method for path planning of manipulators with a high number of degrees of freedom"*, IEEE International Conference on Robotics and Automation, Raleigh, April 1987.

[16] R.S.Fearing: *"Simplified grasping and manipulation with dextrous robot hands"*, AI-Memo-809, Artificial Intelligence Lab., M.I.T., Cambridge, November 1984.

[17] C.Gandon: *"Introduction de la compliance dans la programmation des robots"*, Thèse de 3ème Cycle, INPG, Grenoble, October 1986.

[18] L.Gouzenes: *"Strategies for solving collision-free trajectories problems for mobile and manipulator robots"*, International Journal of Robotics Reserch, vol 3, no 4, 1984.

[19] H.Hanafusa, H.Asada: *"Stable prehension by a robot hand with elastic fingers"*, 7th International Symposium on Industrial Robots Tokyo, October 1977.

[20] H.Hanafusa, H.Asada: *"A robot hand with elastic fingers and its application to assembly process"*, IFAC Symposium on Information and Control Problems in Manufacturing Technology, Tokyo, 1977.

[21] A.Ijel: *"Inférence de l'équilibre d'un objet à partir d'une représentation des contacts"*, Convention IA 89, Paris, January 1989.

[22] K.Ikeuchi et al.: *"Determining grasp points using photometric stereo and the PRISM binocular stereo system"*, AI-Memo-772, Artificial Intelligence Lab., M.I.T., Cambridge, August 1984.

[23] M.Joyal, P.Provost: *"Statique"*, Masson & Cie, 1966.

[24] O.Khatib: *"Commande dynamique dans l'espace opérationnel des robots manipulateurs en présence d'obstacles"*, Thèse de Docteur-Ingénieur, Ecole Nationale Supérieure de l'Aéronautique et de l'Espace, Toulouse, December 1980.

[25] C.Laugier: *"A program for automatic grasping of objects with a robot arm"*, 11th International Symposium on Industrial Robots, Tokyo, October 1981.

[26] C.Laugier, B.Dufay: *"Geometrical reasoning in automatic grasping and contact analysis"*, PROLAMAT 82, Leningrad, May 1982.

[27] C.Laugier, J.Pertin: *"Automatic grasping: a case study in accessibility analysis"*, Published in "Advanced Software in Robotics", edited by A.Danthine and M.Géradin, North Holland, 1984.

[28] C.Laugier: *"Influence du raisonnement géométrique dans le choix d'une prise d'objet"*, Rapport de Recherche IMAG no. 414, December 1983.

[29] C.Laugier, F.Germain: *"An adaptative collision-free trajectory planner"*, '85 International Conference on Advanced Robotics, Tokyo, September 1985.

[30] C.Laugier, J.Troccaz: *"S.H.A.R.P.: A system for automatic programming of manipulation robots"*, 3rd International Symposium of Robotics Research, Gouvieux, October 1985.

[31] C.Laugier, P.Theveneau: *"Planning sensor-based motions for part-mating using geometric reasonning"*, ECAI'86, Brighton, July 1986.

[32] C.Laugier: *"Raisonnement géométrique et méthodes de décision en robotique. Application à la programmation automatique des robots"*, Thèse d'Etat, INPG, Grenoble, December 1987.

[33] L.I.Liebermann, M.A.Wesley: *"AUTOPASS: an automatic programming system for computer controlled mechanical assembly"*, IBM Journal of Research and Development, July 1977.

[34] T.Lozano-Perez: *"The design of a mechanical assembly system"*, AI-TR-397, M.I.T. Artificial Intelligence Laboratory, Cambridge, December 1976.

[35] T.Lozano-Perez, M.A.Wesley: *"An algorithm for planning collision- free paths among polyhedral obstacles"*, CACM, Vol.22, nb.1, October 1979.

[36] T.Lozano-Perez: *"Automatic planning of manipulator transfer movements"*, IEEE Transactions on System, Man and Cybernetics, SMC-11, no. 10, October 1981.

[37] T.Lozano-Perez, M.T.Mason, R.H.Taylor: *"Automatic synthesis of fine-motions strategies for robots"*, 1st International Symposium of Robotics Research, Bretton Woods, August 1983.

[38] T.Lozano-Perez: *"Spatial planning: a configuration space approach"*, IEEE Transactions on Computers, vol. C-32, no. 2, February 1983.

[39] T.Lozano-Perez: *"Motion planning for simple manipulator robots"*, 3rd International Symposium of Robotics Research Gouvieux, October 1985.

[40] T.Lozano-Perez, R.A.Brooks: *"An approach to automatic robot programming"*, AI Memo 842, Artificial Intelligence Laboratory, M.I.T., April 1985.

[41] T.Lozano-Perez et al.: *"Handey: a robot system that recognizes, plans and manipulates"*, IEEE International Conference on Robotics and Automation, Raleigh, April 1987.

[42] D.M.Lyons: *"A simple set of grasps for a dextrous hand"*, IEEE International Conference on Robotics and Automation, St Louis, March 1985.

[43] M.T.Mason: *"Compliance and force control for computer controlled manipulators"*, IEEE International Conference on Robotics and Automation, June 1981.

[44] M.T.Mason: *"Manipulator grasping and pushing operations"*, AI-TR-690, Artificial Intelligence Lab., M.I.T., Cambridge, June 1982.

[45] E.Mazer: *"HANDEY: Un modèle de planificateur pour la programmation automatique des robots"*, Thèse d'Etat, INPG, Grenoble, December 1987.

[46] N.J.Nilsson: *"Principles of Artificial Intelligence"*, Tioga, Palo Alto, 1980.

[47] E.G.Powell: *"An efficient collision warning algorithm for robot arms"*, 2nd American Association for Artificial Intelligence Conference, Carnegie-Mellon, August 1982.

[48] C.Reboulet, A.Robert: *"Hybrid control of a manipulator with an active compliant wrist"*, 3rd International Symposium on Robotics Research, Gouvieux, October 1985.

[49] J.T.Schwartz, M.Sharir: *"On the piano movers' problem II: General properties for computing topological properties of real algebraic manifolds"*, Dep. of Computer Science, Courant Institute of Mathematical Sciences, NYU, Report 41, February 1982.

[50] R.H.Taylor: *"Synthesis of manipulator control programs from task-level specifications"*, AIM 228, Stanford Artificial Intelligence Laboratory, July 1976.

[51] P.Theveneau: *Planification de mouvements fins de montage dans un système de programmation automatique de robots"*, Thèse de l'Institut National Polytechnique de Grenoble, 1988 (to appear).

[52] D.J.Todd: *"A method for grasping randomly oriented objects using touch sensing"*, Artificial Intelligence Lab., Queen Mary College, Univ. of London, June 1981.

[53] J.Troccaz: *"S.M.G.R.: a geometric and relational modeller for robotics"*, International Conference on Advanced Robotics, Tokyo, September 1985.

[54] J.Troccaz: *"Modélisation du raisonnement géométrique pour la programmation des robots"*, Thèse de l'Institut National Polytechnique de Grenoble, April 1986.

[55] J.Troccaz: *"On-line automatic robot programming: a case study in grasping"*, IEEE Conference on Robotics and Automation, Raleigh, April 1987.

[56] S.M.Udupa: *"Collision detection and avoidance in computer controlled manipulators"*, Proceedings IJCAI 77, Cambridge, August 1977.

[57] J.M.Valade: *"Raisonnement géométrique et synthèse de trajectoire d'assemblage"*, Thèse de Docteur-Ingénieur, Université Paul Sabatier, Laboratoire d'Automatique et d'Analyse des Systèmes, Toulouse, January 1985.

[58] D.E.Withney: *"Force feedback control of manipulator fine motions"*, Journal of Dynamic Systems Measurement and Control, June 1977.

[59] M.Wingham: *"Planning how to grasp objects in a cluttered environment"*, M. Phil. Thesis, University of Edinburgh, Scotland, 1977.

[60] J.D.Wolter, R.A.Volz, A.C.Woo: *"Automatic generation of gripping positions"*, RSD-TR-2-84, Center for Robotics and Integrated Manufacturing, Robot Systems Division, Ann Arbor (Michigan), February 1984.

JAN 1 1 1990